Fuzzy Systems Design Principles

IEEE Press
445 Hoes Lane, P.O. Box 1331
Piscataway, NJ 08855-1331

IEEE Press Editorial Board
Roger F. Hoyt, *Editor in Chief*

John B. Anderson	S. Furui	P. Laplante
P. M. Anderson	A. H. Haddad	R. S. Muller
M. Eden	R. Herrick	W. D. Reeve
M. E. El-Hawary	G. F. Hoffnagle	D. J. Wells
	S. Kartalopoulos	

Kenneth Moore, *Director of IEEE Press*
John Griffin, *Senior Acquisition Editor*
Lisa Dayne, *Assistant Editor*
Linda Matarazzo, *Assistant Editor*
Denise Phillip, *Associate Production Editor*

IEEE Neural Networks Council, *Sponsor*
NNC Liaison to IEEE Press, Stamatios Kartalopoulos

Technical Reviewers

Israel Alguindigue, *University of Tennessee, Chattanooga*
Evren Eryurek, *Fisher-Rosemount*
Rogelio Palomera-Garcia, *UPR-Mayaguez, Puerto Rico*
John Alexander, *Towson State University*

Also from IEEE Press . . .

Evolutionary Computation: Toward a New Philosophy of Machine Intelligence
David Fogel
1995   Hardcover   288 pp   IEEE Order No. PC3871   ISBN 0-7803-1038-1

Industrial Applications of Fuzzy Logic and Intelligent Systems
Edited by John Yen, Reza Langari, and Lotfi A. Zadeh
1995   Hardcover   376 pp   IEEE Order No. PC4002   ISBN 0-7803-1048-9

Intelligent Control Systems: Theory and Applications
Madan M. Gupta and Naresh K. Sinha
1996   Hardcover   856 pp   IEEE Order No. PC4176   ISBN 0-7803-1063-2

Understanding Neural Networks and Fuzzy Logic: Basic Concepts & Applications
Stamatios V. Kartalopoulos
1996   Paper   232 pp   IEEE Order No. PP5591   ISBN 0-7803-1128-0

Fuzzy Systems Design Principles
Building Fuzzy IF-THEN Rule Bases

Riza C. Berkan
MODiCO, Inc.

Sheldon L. Trubatch
Winston and Strawn

IEEE Neural Networks Council, *Sponsor*

The Institute of Electrical and Electronics Engineers, Inc., New York

This book may be purchased at a discount from the publisher
when ordered in bulk quantities. Contact:

IEEE Press Marketing
Attn: Special Sales
445 Hoes Lane, P. O. Box 1331
Piscataway, NJ 08855-1331
Fax: (908) 981-9334

For more information on the IEEE Press,
visit the IEEE home page: http://www.ieee.org/

© 1997 by the Institute of Electrical and Electronics Engineers, Inc.,
345 East 47th Street, New York, NY 10017-2394

All rights reserved. No part of this book may be reproduced in any form,
nor may it be stored in a retrieval system or transmitted in any form,
without written permission from the publisher.

Printed in the United States of America

10   9   8   7   6   5   4   3   2   1

ISBN 0-7803-1151-5

IEEE Order Number: PC5622

Library of Congress Cataloging-in-Publication Data

Berkan, Riza C.
  Fuzzy systems design principles : building fuzzy IF-THEN rule
bases / Riza C. Berkan, Sheldon L. Trubatch.
     p.   cm.
  Includes bibliographical references and index.
  ISBN 0-7803-1151-5 (cloth)
   1. System design. 2. Fuzzy systems. I. Trubatch, Sheldon L.
II. Title.
QA76.9.S88B47   1997
006.3—dc21   97-3658
                                                                   CIP

*To Vincent Besim
and
Elizabeth Victoria*

Contents

Foreword   xi

Preface   xiii

Chapter 1: Introduction   1

   1.1   Partial Truth and Fuzziness   1
   1.2   Foundation of Fuzzy Systems   4
   1.3   Fuzzy Systems at Work   7
   1.4   Fuzzy System Design   11
   1.5   How to Use This Book Effectively   17
   1.6   Terminology and Conventions   18
   References   21

Chapter 2: Theory   22

   2.1   Crisp Versus Fuzzy Sets   22
   2.2   From Fuzzy Sets to Fuzzy Events   26
   2.3   Fuzzy Logic and Linguistics   28
   2.4   Practical Fuzzy Measures   29
   2.5   Fuzzy Set Operations   34
   2.6   Properties of Fuzzy Sets   40
   2.7   Fuzzification Techniques   41
   2.8   Alpha Cuts   45
   2.9   Relational Inference   46
   2.10   Compositional Inference   52

2.11	Linguistic Variables and Logic Operators	56
2.12	Inference Using Fuzzy Variables	60
2.13	Fuzzy Implication	64
2.14	Fuzzy Systems and Algorithms	66
2.15	Defuzzification	70
2.16	Adaptive Fuzzy Systems and Algorithms	73
2.17	Expert Systems Versus Fuzzy Inference Engines	77
References		81

Chapter 3: The Basic Fuzzy Inference Algorithm — 83

3.1	Introduction	83
3.2	Overall Algorithm	85
3.3	Input Data Processing	87
3.4	Evaluating Antecedent Fuzzy Variables	92
3.5	Left-Hand-Side Computations	104
3.6	Right-Hand-Side Computations	109
3.7	Output Processing	126
Problems		130
References		131

Chapter 4: Conceptual Design — 132

4.1	Introduction	132
4.2	Fuzzy System Design and Its Elements	134
4.3	Design Options, Processes, and Background Requirements	139
4.4	Knowledge Acquisition	142
4.5	The First Principle of Fuzzy Inference Design	164
4.6	Linguistic Design Criteria	166
4.7	Application of the Design Criteria	175
4.8	Systems Ontology and Problem Types	175
4.9	Useful Tools Supporting Design	195
References		198
Recommended Books for Design		199

Chapter 5: Fuzzy Variable Design — 201

5.1	Introduction to Fuzzy Variable Design	201
5.2	Data-Driven Fuzzy Variable Design	206
5.3	Linguistic Fuzzy Variable Design	249
5.4	Practical Design Considerations	258

5.5	Summary	271
Problems		272
References		275

Chapter 6: Membership Function Shape Analysis — 276

6.1	Introduction to Shape Analysis	276
6.2	Membership Function Height	283
6.3	Membership Function Line Style	295
6.4	Overlapping	308
6.5	Summary	320
Problems		321

Chapter 7: Composing Fuzzy Rules — 323

7.1	Introduction	323
7.2	Basic Logic Operators	324
7.3	Logic Operator Design Issues	330
7.4	Rule Formation Per Inference Type	338
7.5	Rule Composition Strategies	357
7.6	Paradoxical Cases	382
7.7	Membership Function Shape Effects	391
7.8	Summary	400
References		401
Selected Bibliography		402

Chapter 8: Implication Design — 403

8.1	Introduction	403
8.2	Selecting Implication Operators	405
8.3	Behavioral Properties	425
8.4	Aggregation Design	437
8.5	Designing a Defuzzification/Decomposition Process	443
8.6	Interpreting Output Fuzzy Sets	452
8.7	Summary	455

Appendix: The Basic Fuzzy Inference Algorithm — 458

Index — 489

Foreword

In 1965, Lotfi A. Zadeh at University of California at Berkeley laid the foundations of fuzzy systems by generalizing the mathematical notion of a *set* leading to a *fuzzy set,* a set in which many degrees of membership are allowed. Fundamentally, sets are *categories.* Defining suitable categories and operations for manipulating them is a major task of modeling and computation. From image processing to measurement and control, the notion of category is essential in defining system variables, parameters, their ranges, and their interactions. Given modern day computational capabilities, the constraint of dual degree of membership to an all-or-nothing category is unduly restrictive and an obstacle to the development of systems that can be easily calibrated to the specifics of their environment or an emerging new behavior.

After two decades of relative indifference to fuzziness by the technical and scientific community, there is now a booming interest and a proliferation of fuzzy systems in all fields of science and engineering, as well as business and finance. *Fuzzy Systems Design Principles* successfully addresses the pressing need for technical literacy on fuzzy systems and is an important guide for the resolution of practical problems involved in the conception, design, and deployment of fuzzy systems.

Addressing design issues is a major theme of this book. Dr. Riza C. Berkan and Dr. Sheldon L. Trubatch ought to be commended for making a major contribution to fuzzy systems analysis and design from the practitioner's point of view. The book eloquently and convincingly illustrates the role of fuzziness as a fundamental feature of linguistic descriptions that abound in many modeling, control, decision-making, prediction, and simulation applications. The book's organization reflects the basic structure of fuzzy systems. The first three chapters introduce the reader to basic fuzzy theories. The next two chapters address important conceptual and fuzzy variable design issues with the final three chapters presenting a highly pragmatic exposition of various design issues enhanced by the inclusion of many insightful examples.

Fuzzy Systems Design Principles successfully illuminates the pros and cons of fuzzy system design and implementation. It is truly an important pedagogical book, addressing a variety of audiences, and written in a highly accessible manner. The authors certainly deserve our thanks and congratulations for producing such an outstanding piece of work.

Lefteri H. Tsoukalas,
Purdue University

Preface

Fuzzy logic is one of the fastest growing technologies in the world since the beginning of the computer era. Because of its simplicity and wide range of applicability, there is an increasing need for education about fuzzy logic, especially for novice practitioners. Among the many viable forms of educational material found in the literature, it is only recently that design-oriented books have started to emerge. This book is one of them, with a slightly different point of view than the existing trend. The focus in this book is on certain basic structures, and the reader's attention is narrowed down to simple design attributes.

Design books, in general, differ from theory or application books by content, style, language, outline, and most importantly by the general objective. In a design book, the topics to be covered must be well established in theory as well as in practice. In addition, the organization of the topics must help the reader navigate through a typical development process. In this process, the reader needs theoretical knowledge at a sufficient level, but not necessarily at the level of a theory book, and practical examples of reasonable complexity, but not as sophisticated and detailed as those found in application books. The book is intended to fill an important gap by focusing on the transition from theory to application while satisfying the typical requirements of a design book. Accordingly, the reader will encounter a balanced mixture of theory and application within a typical design framework.

The word *design* refers to almost any kind of creation in arts and sciences. In the artistic sense, design involves imagination, fiction, belief, and taste. In scientific terms, design applies theoretical principles in an appropriate and optimal manner, as in designing a car engine. Unfortunately, designing a fuzzy system has both scientific and artistic elements that make it hard to learn or teach the subject. The scientific element is based on mathematics such as fuzzy set theory. The artistic element comes from intuition, interpretation, insight, expectation, and so on, which we call *heuristics*. This book attempts to identify both artistic and scientific elements of design in their application to fuzzy systems, and to resolve problematic design issues when possible.

Fuzzy systems theory has progressed in two directions: 1) fuzzy inference, which is the main focus in this book; and 2) the application of fuzzy set theory to problems formulated strictly in mathematics. There is a fundamental difference of philosophy between these two directions, although they are often complementary. In the fuzzy inference approach, the solution of a problem comes from human interpretation (expertise, experience, data, etc.). The design challenge is to reproduce a specific solution by translating it from the original domain of knowledge (often natural language) to the calculus of fuzzy IF-THEN rules. In the second approach, the original solution of the problem at hand is expressed purely by crisp mathematics with very little, if any, involvement of heuristics. The design challenge is to apply fuzzy set theory to crisply defined concepts and formulas to improve their generality, robustness, and expressive power. In this book, the term *fuzzy systems* refers mostly to the first type of systems, which are governed by fuzzy IF-THEN rules.

The calculus of fuzzy IF-THEN rules is quite simple compared to other extensions of fuzzy set theory. There are well-established elements of the fuzzy IF-THEN structure, including fuzzy variables, membership functions, fuzzy rules, implication process, and decomposition. Putting these elements together forms a fuzzy inference algorithm. The manner in which this algorithmic flow is implemented is also well established in the literature and in industrial applications. This basic structure, which is called *the basic fuzzy inference algorithm* in this book, constitutes the backbone of design. However, the challenge starts when each element of the basic fuzzy inference algorithm is to be designed for a given global objective. This book was written with this view in mind. Thus, the basic fuzzy inference algorithm is presented before the detailed discussions on how to design each element. The reader is advised to follow this order while reading the book.

Despite the relatively rigid structure of the basic fuzzy inference algorithm, there is almost no structure or systematic method for designing its individual elements. This is because designing the elements, such as membership functions, involves a heuristic approach based on individualism, style, and expertise. Therefore, this book presents possible options to choose from or possible steps to be taken while designing different parts of the basic fuzzy inference algorithm.

The design principles suggested in this book are presented as "things to consider" instead of "things to do." Most of the design principles are stated to remind the reader what to expect when selecting certain design options. The information presented in the design principles is gathered from practical design experiences (i.e., academic research, consulting vendors, conference meetings, software development, classroom discussions, etc.) Most of the principles are simple to verify by commonsense reasoning without extensive mathematical proof.

Achieving a delicate balance between stating design principles and not restricting one's creativity is a challenge. The problem, which is seen in almost every field, is that creativity cannot be taught. It is very difficult, if not impossible, to learn how to dress elegantly from a book on fashion, or how to compose music from a book on composition. Thus, there is a limit in every design subject beyond which individual experimentation takes over. The point at which the arguments stop in this book is determined by studying the existing applications and realizing that there are, in fact, common assumptions made (i.e., all membership functions should have the same height) and common questions asked (i.e., what if I use trapezoidal membership functions instead of triangular ones?) by different designers. These assumptions and questions represent a transition stage from theory to application, a stage in which a designer is likely to stumble across many trivial and complex obstacles without knowing how to interpret them. *Fuzzy Systems Design Principles* was written to shed light on those design details that are often not explicitly or systematically reported in the literature. Beyond these matters, the designer will use his/her creativity.

It is important to point out that the calculus of fuzzy IF-THEN rules, which defines the basic fuzzy inference algorithm, is applicable with equal benefit to any decision-making problem one can imagine. There can be no distinction of quality if a pair of fuzzy rules estimates stock prices or navigates a robot through obstacles. The quality of performance depends on the embedded knowledge and its accurate representation in the rule base of a fuzzy system.

When topics like fuzzy control or fuzzy pattern recognition are encountered, it should not be assumed that a fuzzy inference algorithm must be designed according to the problem type. The properties of the calculus of the fuzzy IF-THEN structure are invariant in this respect. Similarly, the design of membership functions or any other element cannot be associated with the problem type. This book shows that there are linguistic criteria affecting design—criteria that are not susceptible to problem type, but are susceptible to the reasoning behind the embedded solution.

There are, in fact, drastic differences between fuzzy control, fuzzy diagnostics, fuzzy clustering, and so on, when these problems are formulated in ways other than fuzzy IF-THEN structure. Since every discipline can become a fuzzy discipline by applying fuzzy set theory to the original theory of the discipline, there is practically an infinite number of special-case fuzzy solutions. Due to the abundance of such methods and the difficulty of identifying common design attributes among them, this book is dedicated only to the design of fuzzy IF-THEN rule bases. Nevertheless, most of the commercial and industrial applications to date are based on the fuzzy IF-THEN structure because of its expressive power and simplicity.

This book was written with another important view in mind. Since its emergence a few decades ago, fuzzy logic has been a topic plagued with misconceptions. Particularly, two misconceptions are frequently encountered in practice. First, fuzzy logic is compared to other methods such as neural networks and genetic algorithms by the question: How does a fuzzy logic solution compare to the solutions obtained from other methods? This is an inappropriate question because fuzzy logic does not invent a solution, it implements a given one. Neural networks, on the other hand, invent a solution from a data set by means of learning, as do genetic algorithms by the principle of evolution. Thus, comparisons of this nature are inappropriate. The fact that fuzzy IF-THEN systems imitate a given solution is emphasized in the book in several discussions. The second misconception in practice is to assume that fuzzy logic is only useful in avoiding pointless precision to reduce cost. Although the majority of the existing applications serve this purpose, fuzzy logic theory is very broad and is perhaps the only theory that easily propagates into multiple disciplines such as sociology, biology, psychology, artificial intelligence, linguistics, engineering sciences, medicine, law, and so forth. This is due to the capability of computing with words in a manner no other method can cope with. In addition, the stigma that "fuzzy logic is fuzzy control" was deliberately rendered false by including unusual examples in the book to push the reader's imagination forward.

Another important emphasis in this book is the value of heuristics. One of my graduate students at the University of Tennessee came up with a brilliant idea one day. He decided to design a controller for a robotic-arm manipulator with three flexible joints simply by examining his own arm movements and transferring that knowledge into simple IF-THEN fuzzy rules. The fuzzy controller had an incredible success—superior to the solutions of known mathematical techniques—to manage redundancy as it was comparatively illustrated by simulation. The success was partly due to the fuzzy logic approach and partly because of the power of commonsense reasoning. Heuristic solutions, although they are individualistic and not easily reproducible, are undisputedly powerful and unavoidable in fuzzy system design. Accordingly, a conclusion might one day be reached that some aspects of scientific problems are best solved by treating them as social problems and by implementing heuristic solutions rather than by searching for piecemeal mathematical solutions expressed in pure symbolism.

The book consists of three major parts. The first part includes the introductory concepts in Chapter 1 and a review of fuzzy logic theory in Chapter 2. All design elements are included from the theory point of view. The second part includes Chapter 3, "The Basic Fuzzy Inference Algorithm." The details of the algorithm are included in the Appendix for the practitioners of computer programming. The basic fuzzy inference algorithm in the form presented in Chapter 3 can be directly implemented in practice. The third part of the book includes Chapters 4 through 8, in which design issues are discussed in detail. Reading the book from start to finish, the reader will encounter a topic three times: first the topic's definition in theory, then its place in the basic fuzzy inference algorithm, and finally its design.

The coauthor and I would like to thank the director of IEEE publishing, Dudley R. Kay, former acquisition editor, Russ Hall, editor John Griffin, assistant editor Lisa Dayne, and associate production editor, Denise Phillip for turning this project into reality. In the making of this book, we have received encouragement, assistance, and criticism from many experts, industrialists, academicians, artists, designers, businesspeople, and students who are themselves the promoters of fuzzy logic technology. We would like to express our gratitude to Prof. Lefteri H. Tsoukalas, Purdue University; Prof. Robert E. Uhrig, The University of Tennessee; Prof. Belle Upadhyaya, The University of Tennessee; Dr. Israel E. Alguindigue, University of Tennessee; Dr. Evren Eryurek, Fisher-Rosemount; Dr. Tanju Sofu, Argonne National Laboratory; Prof. Lamartine Guimaraes, University of Braz Cubas, Brazil; Prof. John Alexander, Towson State University; Dr. Daniele Ugolini, Power Reactor Nuclear Fuel Corporation, Japan; Dr. Omid Omidvar, National Institute of Standards; Rudolfo Bonilla, Bechtel Corp.; John Greenhill, Department of Energy Headquarters; Dr. Francois Pin, Oak Ridge National Laboratory; Mike Whitney, Westinghouse Savannah River Company; Roger Kisner, Oak Ridge National Laboratory; Darryl Horne, Horne Engineering and Environmental Sciences, Inc.; Dr. Cengiz Onuk, City University of New York; and Dr. Ann Hansen, H & R Technical Associates. Special thanks to Dr. Ahmet Unseren, Thomas Curtis, Dr. Srdan Simanovic, Paulo Brasko, Aucyone Da-Silva, Ahmet Yavas, Ali Riza Hangul, David L. M. Skov, Naz Oke, Selcuk Togul, and to our families for their patience and support.

Readers may contact the authors at www.modico.com

Riza C. Berkan

Chapter 1

Introduction

> This chapter begins with an overall view of the fundamental concepts in fuzzy systems theory. Next is a brief description of how fuzzy systems work in practice and the extent of their industrial applications. These introductory arguments lead to the topic of fuzzy system design as exercised today in its simplest form. Design elements, design process, and the required expertise are outlined. The conceptual alignment provided in this chapter will clarify the design path on which practical solutions find utmost utilization, sometimes without a complete theoretical proof.

The essence of science is to discover identity in difference.
F. S. Marvin

1.1 PARTIAL TRUTH AND FUZZINESS

A bat's hearing is better than a dog's. An owl can see better than a mouse. A man in front of a radar screen can predict an approaching hurricane better than a man watching the sky. Examples, which are countless, indicate that measurement capability (or quality) is relative. Relative measurement quality means relative quality of the perception of truth. Thus, there can be a drastic difference in the perception of truth based on how well one can measure.

Almost all the answers found in practical life are within some proximity of the absolute truth. A glass is never completely full or empty in reality. The temperature of a liquid in a container is never exactly what the thermometer indicates. Examples are again countless. However, in practice, most uncertainties are tolerable, manageable, or negligible. A glass is not considered full if a drop of water is in it. It still qualifies as empty and functions as such. We do not assume that every molecule in a liquid has a thermal kinetic energy equivalent to 70 degrees Centigrade if a thermometer indicates so. We do not cancel our flight if the risk of crash is not absolutely zero. We live in a world of partial truth. So do the systems we build and operate.

As the primary tool for building and operating systems, science—in the traditional sense—has evolved around the idealism of mathematics, which sometimes falls short in dealing with the reality of life. Realism, referring to perception of partial truth, cannot be accurately expressed by true/false duality for obvious reasons. Historically, engineering design has been based on the idealistic theories of mathematics, physics, chemistry, and so on, although the system operations are subject to partial truth in practice. Computer science suffers from the same idealistic approach encouraged by the irresistible benefits of digital technology. Computer systems recognize the world by zeros and ones perfectly compatible with the true/false duality.

Referring to F. S. Marvin's quote, "The essence of science is to discover identity in difference," nowadays an engineer often asks, "How much identity in difference do I need in practice?" In this question, the level of identity in difference refers to the level of perception of truth, or the quality of partial truth. Fortunately, the traditional idealistic mathematical approach has been improved recently to accommodate partial truth by the introduction of fuzzy set theory invented by Professor Lotfi A. Zadeh. Unlike classical set theory, fuzzy set theory is flexible, and it focuses on the degree of being a member of a set. As discussed later, this simple notion leads to new concepts and ideas through which more realistic mathematical representations can be achieved in describing events observed with uncertainty.

Fuzziness primarily describes uncertainty or partial truth. There are a number of related concepts, mostly derived from daily language. Randomness, chaos, ambiguity, vagueness, undecidedness, inaccuracy, and imprecision are some of them. Although their interpretations in daily language may be different from their scientific definitions, we will briefly overview what has been academically referenced so far.

Randomness is the occurrence of an event without any recognizable pattern. A random noise—also characterized as white noise—can be a contributing factor to partial truth. Chaos is a highly complex iteration of order in which the principles of its evolution cannot be justified through repeated observations. Also known as nonlinear phenomena, chaotic behavior can be a disruptive factor in the perception of absolute truth. Ambiguity—as in ambiguous statements—is a definition of the state of an expression, rather than being a description of an event. Ambiguity suggests multiplicity in the perception of truth; thus, it is related to partial truth and to fuzziness. Vagueness is a mild form of ambiguity and a strong form of imprecision such that a statement can be either depending on the context. Like ambiguity, vagueness defines the state of an expression in describing truth. Undecidedness, very close to its meaning in daily language, defines the collision point of two or more equally justifiable decisions. In the ocean of possibilities, an undecided state of affairs may be a rare event, yet it somewhat implies fuzziness. Inaccuracy is a measure of distance between absolute truth and perceived truth caused by sheer lack of knowledge. It is a definition of the state of an outcome—such as inaccurate results—and is directly related to fuzziness. Imprecision is a measure of distance between absolute truth and measured truth caused by the limitations of the measurement vehicle. It is directly related to fuzziness.

Among the definitions stated above, none individually qualifies as the prime source of fuzziness or as the definition of fuzziness. That is because there is no way of determining that the cause of partial truth is purely imprecision, randomness, vagueness, or ambiguity in practical life. As experiential evidence suggests (i.e., successful applications of fuzzy logic in industry), it matters very little, if at all, to find a suitable definition of fuzziness in terms of its relatives. Once fuzziness is characterized at a reasonable level, fuzzy systems can perform well within the expected precision range. Encouraged by this fact, the technological design philosophy tends to depart from the academic quest for finding a formal definition of fuzziness and its constituents.

1.2 FOUNDATION OF FUZZY SYSTEMS

This section briefly examines the theoretical and practical foundations of fuzzy logic technology and explains how it is related to design and to this book. As mentioned earlier, the concept of partial truth characterized by fuzziness has launched the new theory of fuzzy sets, which has yielded a more accurate mathematical representation of the perception of truth than that of crisp sets. The term *perception* refers to the human brain and its way of observing and expressing reality. Thus, it also refers to natural and synthetic languages. Transition from crisp (true/false) mathematics to fuzzy mathematics by means of fuzzy set theory has allowed mathematical representations to become compatible with expressions in natural language. For example, the expression *he looks young* can now be mathematically represented by a fuzzy set *young* such that *he* belongs to this set in a fuzzy way.

In the transition from crisp sets to fuzzy sets, the key element is possibility theory and its extended interpretations. In *he looks young*, *he* belongs to the fuzzy set *young* with some possibility value. In a collective manner, possibilities are defined by a distribution function, often called a *membership function*. The membership function concept completes the mathematical formulation of an expression like *he looks young* and also of more complex expressions articulated in daily language. As a result, there is a way to compute with words using fuzzy set theory and possibility theory. *Computing with words* is a slightly futuristic phrase today since only certain aspects of natural language can be represented by the calculus of fuzzy sets.

One way to understand the relationship between fuzzy sets and fuzzy logic is to examine natural language. Expressions in natural language such as *he looks young* or *he works out a lot* are nothing but phrases describing an event or state of being. When they are put together in a sense-making order, a context is created that leads to reasoning. For example, the combination *he looks young, he works out a lot* creates a context in which looking young and working out become related. Such statements are called unconditional statements. One step further is the combination of simple expressions using some linguistic connectors (also called operators) such as in *If he works out a lot then he will look young*. Here, the connectors *if-then* modify the context and make it a conditional statement. When conditions are imposed, reasoning gets more restricted than in a simple relationship (such as that between looking young and working out), which leads to the subject of logical inference.

In the classical sense, logic—also referred to as Boolean Logic—consists of three elements: truth values, linguistic connectors, and reasoning types. In Boolean logic, truth values are either 1 or 0, which correspond to true/false duality. In fuzzy logic, truth is a matter of degree, thus truth values range between 1 and 0 in a continuous manner. This concept of continuum constitutes the most outstanding difference between classical logic and fuzzy logic. Linguistic connectors (or operators) in fuzzy logic function in the same way as in Boolean logic (union, intersection, negation). However, their properties are affected, as shown later in this book. Modes of reasoning such as syllogism or modus ponens are also the same in principle in both classical logic and fuzzy logic. Other forms of classical logic such as epistemic logic (involving belief), modal logic (necessity versus possibility), deontic logic (ought to be), tense logic (truth of past and future), and predicate calculus (logic using quantifiers) are all representable in fuzzy logic by the possibilistic interpretation of partial truth.

Historically, the language of logic and the mathematics of logic have been extensively used in computer technology. Therefore, by changing both of them from crisp to fuzzy, all computerized systems will be affected. Then, the question is: Why should such an avenue be pursued? There are several reasons, and the most important ones are addressed next.

> *The use of natural language in man–machine interface instead of using machine-specific language has great advantages in practice. Using fuzzy logic, it is possible to command a robot:* Turn left a little bit, then slowly speed up. *It is also possible to get a response from the robot such as* The obstacle looks like a cat sleeping on the floor. *The key element in this type of communication is the ability to address uncertainties that can be understood mutually.*
>
> *Since computation with words is possible, computerized systems can be built by embedding human expertise articulated in daily language. Also called a fuzzy inference engine or a fuzzy rule-base, such a system can perform approximate reasoning somewhat similar to (but much more primitive) than that of the human brain. Fuzzy logic is one of the most practical ways to mimick human expertise in a realistic manner. Computing with words also allows us to develop mathematical models of the events articulated in lan-*

guage only. For example, one can simulate interpretation of the law or simulate a crime based on witnesses' testimonies.

By fuzzifying crisp data obtained from measurements, fuzzy logic enhances the robustness of a system. Without fuzzification, systems designed to act at certain crisp input data points would not know what to do when data is somewhat corrupted. In one sense, a fuzzy system may be viewed as a crisp system designed conservatively by fuzzifying all variables based on the expectation that measurement corruption might occur. Therefore, fuzzy systems are inherently immune to uncertainties in measurements.

Representing a solution with fuzzy sets reduces the computational burden. Approximating a group of related data points by a few fuzzy categories serves this purpose. In some cases, fuzzy technology makes a solution possible that would otherwise be unthinkable due to cost of computing every single crisp data point.

By selecting the number of fuzzy representative sets, there is a way of adjusting (or controlling) the precision level of a solution. If more fuzzy sets are used in design, systems will require more memory, faster CPUs, and so forth. At the limit, the number of fuzzy sets becomes equal to the number of crisp data points. That represents the most precise and costly solution. Going back the other way, a solution can be tailored to a desired precision/cost criterion.

Nonfuzzy methods do not offer the practicality and cost-effectiveness of fuzzy methods in performing such functions in the computer domain. Therefore, the acceptance of fuzzy technology has been mostly based on economic reasons to date. Although a wider scope of benefits has been reported in the literature resulting from academic research, this book is focused on the list above due to the undisputable record of proof in industrial applications.

In our quest for designing a fuzzy system, the overall objective is to achieve one or more of the advantages listed above. In most cases, all the advantages become effective. Typically, a fuzzy system uses human friendly commands, embodies expert knowledge, yields robust performance, requires reasonable compution time and effort, and allows precision/cost adjustments. All these advantages are based on a simple foundation that converts the expression *he looks young* into a mathematical model and handles accompanying logical propositions by means of fuzzy set and possibility theories.

1.3 FUZZY SYSTEMS AT WORK

A system becomes a fuzzy system when its operations are entirely or partially governed by fuzzy logic or are based on fuzzy sets. Operations of systems are defined by several basic problems such as control, estimation (prediction, forecasting), modeling, pattern recognition (classification, clustering), optimization, and data compression as shown in Fig. 1.1. Problem definitions on the level of systems analysis include process control, image processing, signal processing, diagnostics, human factors, natural language processing, and so on. System analysis problems are also defined as a general decision-making problem for system operations. Some of the most prominent interdependencies are illustrated in Fig. 1.1.

A decision-making problem normally includes one of the analysis functions from the middle column. For example, a decision-making system in the field of medicine might only include a diagnostics function that uses a pattern recognition technique to recognize different illnesses, a learning algorithm to build its own memory, a modeling technique to simulate the progression of illnesses, and a data compression technique to speed up its computation time. Let's complete the definition of a fuzzy system with this picture in view. When fuzzy methods are applied to the leftmost column in the map of definitions of Fig. 1.1,[1] the system becomes a fuzzy system.

Examining fuzzy systems built to date, one finds that the majority of applications are in process control, signal processing, image processing, operations research, and in diagnostics. Most of the pioneering systems in these fields have emerged from engineering disciplines. With some time delay from the first generation of fuzzy systems, new applications have emerged in the fields of medicine, business, and social sciences. This new movement has put more emphasis on natural language processing capability using fuzzy logic and fuzzy set theory. We will discuss this issue later in this chapter.

---

[1] These concepts must be familiar to follow some of the discussions in this book.

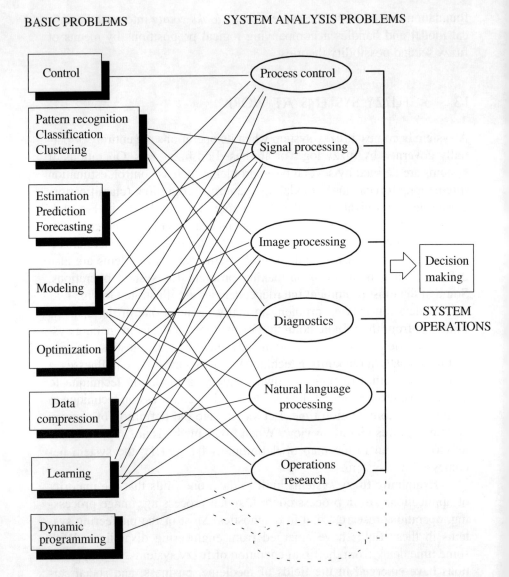

Figure 1.1 Interdependencies among basic problems and system analysis problems.

Let's continue our review of the first-generation fuzzy systems by looking at the examples listed in Table 1.1. This list, which includes a small portion of the existing applications, gives an idea about the diversity of fuzzy technology. For a more complete list of the existing applications and patents, two recent books are highly recommended [1, 2]. In Table 1.1, examples are briefly explained, including their origin. The product column indicates the status of the product. *Industrial* means a tailored design for a specific process, whereas *research* means that the system is under development or in a testing stage and might become a commercial product in the future.

An important observation is that most of the commercial and industrial products are quite simple in structure. A process controller, for example, only employs a fuzzy control law with no other connections, unlike what is shown in Fig. 1.1. Hybrid systems are still in the experimental stage. Some examples of hybrid systems include a fuzzy helicopter controller that uses a learning algorithm (neural network) [3],

Table 1.1 Some Successful Fuzzy Systems at Work

EXAMPLES	PRODUCT
Cement kiln control (Denmark)	Industrial
Blast furnace control (NKK Fukoyama, Japan)	Industrial
Automatic train operation (Sendai subway system, Japan)	Industrial
Nuclear reactor control (Art Fugen, Japan)	Industrial
Home air conditioner (Mitsubishi Heavy Industries, Japan)	Commercial
Washer machine (Matsushita Electric Industrial, Japan)	Commercial
Home heating system (Viessmann-INFORM, Germany)	Commercial
Fuzzy PLC (Klockner-Moeller, Germany)	Commercial
Helicopter control (LIFE, Japan)	Research
Fuzzy autofocus still camera (Sanyo, Japan)	Commercial
Photocopy machine (Sanyo, Japan)	Commercial
Fingerprint classification (NIST, USA)	Research
Smart sensors (Fisher-Rosemount, USA)	Commercial
Speech recognizer (NTT Communications, Japan)	Research
Fire detector (Cerberus, Switzerland)	Industrial
Health management (OMRON, Japan-USA)	Commercial
Fuzzy medical expert system (Univ. of California, USA)	Research
Autonomous robot control (SRI International, USA)	Research
Autonomous navigation of robots (ORNL, USA)	Research
Camera tracking (NASA, USA)	Industrial
Water treatment system (Fuji Electric, Japan)	Industrial
Target tracker in Patriot missile (MMES, USA)	Industrial
Used car selection (A used car center in Kansai, Japan)	Research

missile guidance based on the fuzzy Kalman filter [4], robot navigation using multiple simultaneous fuzzy solutions for sensing and controlling [5], and nuclear reactor control using both fuzzy logic and neural networks [6]. A second important observation is that most of the applications use simple fuzzy IF-THEN rules regardless of the type of problem.

Fuzzy technology—in the form it has been introduced to the marketplace—has already undergone testing by millions of consumers. From one application to another, the advantages are noted repeatedly regardless of the type of the problem or product. They are: high expressive power, greater generality, robustness, and adjustable precision. All these properties translate into two magic words compared to conventional or traditional systems of computation: practicality and cost-effectiveness.

Three important characteristics of fuzzy technology can be summarized as follows: Most of the successful fuzzy systems (1) are simple in terms of their objective and structure; (2) employ solutions articulated in daily language by means of fuzzy IF-THEN rules; and (3) meet their anticipated advantages regardless of the problem type. These three characteristics constitute a focal reference point in this book. Therefore, our design discussions will be centered around simple fuzzy systems using IF-THEN rules that yield practical and cost-effective implementation.

The question that has not been answered yet is: How does a basic problem become a fuzzy basic problem? For example, how does control or pattern recognition become fuzzy control or fuzzy pattern recognition, and so on? The mechanism of applying fuzziness to some method requires knowing the original method of solution because fuzzy set theory and fuzzy logic do not invent solutions, they modify solutions that already exist. Considering the characteristics of systems listed in Table 1.1, a method becomes a fuzzy method when its working principles are translated into fuzzy IF-THEN rules. The techniques of such a translation are described throughout the book as well as in a brief discussion in the next section.

The future of fuzzy technology lies in its potential use in every computation-based technology. There are two directions, as the current research topics suggest. The first one is the application of fuzzy sets to disciplines that use mathematics as the main form of expression; for example, geophysics, accounting, finance, navigation, atmospheric sciences, risk assessment, and so on. The second direction is the application of fuzzy inference to disciplines that use natural language as the main form of expression, such as law, political science, history, psychology, linguistics, anthropology, criminology, medicine, biology, economics, and so forth. Relatively more interesting progress is anticipated in those disciplines in which knowledge is articulated in daily language and mathematics is rarely used. This is because fuzzy inference offers a computational analysis option in these fields for the first time. At the limits of imagination, one may expect to see fuzzy systems analyzing legal contracts, policies, or procedures to detect inconsistencies, or analyzing the management of a big corporation to advise operational changes. Improved understanding of natural language and its more accurate computational models will also have a tremendous impact on how engineering systems operate. When such systems start to emerge, we will be talking about second-generation fuzzy systems—systems that are merely one step closer to human reasoning.

Another important area of progress has been signaled by the recent research on fuzzy computers. The ultimate fuzzy computer is one with hardware using continuous truth values compatible with multivalent logic instead of, or in addition to, binary values compatible with bivalent logic. Current research is focused on inventing a new analog–digital combination of hardware that can represent the calculus of fuzzy inference better than that of purely digital systems. The problem encountered in today's digital computers is the computation time caused by the time-consuming representations of the analog world via the digital world. Analog devices function in a continuum that is compatible with partial truth representation. Also, research on new fuzzy memory devices—where data of membership functions are stored—is under way. Fuzzy memory requires digital technology as one of the irreplaceable elements of fuzzy computers.

1.4   FUZZY SYSTEM DESIGN

In this section, we will briefly discuss the concept of design in general terms, including short answers to the questions:

>   What is fuzzy system design?
>   What are the design steps?
>   How are fuzzy systems implemented?
>   Who is the designer?
>   What are the design tools?

Design is a process of creating something new by implementing new concepts and/or new arrangements. Although every engineering system is subject to design, the number of books and courses on engineering design have unfortunately always been inadequate. The primary reason is that design requires creativity and imagination, which are impossible to teach in a systematic manner. The best a design book can offer is to highlight the most important theoretical aspects and to provide a framework through which a typical design activity is performed. This book follows a similar path for fuzzy system design.

What is fuzzy system design?

The essential part of fuzzy system design is the application of fuzzy sets and fuzzy logic to a solution, or to a method of solution, provided in the conventional, nonfuzzy form. In this context, it is important to focus on the meaning of *a solution*. Considering a control problem, there are three ways of obtaining a control solution:

- *Articulated expertise* of the human operator, such as explaining how to ride a bike
- *Recorded performance data* of a controller, such as in power plant operations
- *Mathematical formula,* such as the PID control law

It is generally valid that there are three sources of solution (as italicized above) for all kinds of problems. These three sources also indicate three domains of knowledge representation: natural language, numerical data, and closed-form mathematical formula. The design challenge is to translate the knowledge given in one of these forms into fuzzy IF-THEN form.

Expertise articulated in natural language is readily compatible with fuzzy IF-THEN rules. However, the calculus of fuzzy IF-THEN rules, as of today, cannot represent all expressions of language directly and accurately. Thus, a translation (also referred to as *knowledge acquisition*) is necessary from the natural language domain to the language of fuzzy inference. This type of design has been widely employed and continues to have great potential for the future.

Knowledge in the form of recorded performance (i.e., numerical data) represents a more difficult case for translation into fuzzy IF-THEN because numerical data is not expressive and is often misleading in terms of generalization. There are approximate methods for translation from the numerical data domain into IF-THEN language of fuzzy inference. This practice is known as rule (or knowledge) extraction from data, and it is explained in this book.

Translating a mathematical expression into fuzzy IF-THEN rules is also not very easy. Normally, implicitly embedded logical associations in a formula can be expressed by the language of fuzzy inference, either partially or completely. In some cases, however, a closed-form formula may contain knowledge that cannot be expressed by natural language. A typical example is an intrinsic function such as cosine of an angle, which does not have a linguistic equivalent. In such a case, the application of fuzzy set theory produces a fuzzy mathematical solution instead of a translation into fuzzy inference language. The application of fuzzy set theory yields cosine of fuzzy angles. The result becomes fuzzy and possibilistic rather than crisp and deterministic. Topics like fuzzy arithmetic and fuzzy graph theory operate in this manner without the involvement of logical inference expressed by language.

The discussions presented above are summarized in Fig. 1.2. Conversion from conventional method to fuzzy method by the application of fuzzy set theory and by the translation into fuzzy inference language yields two distinct forms of fuzzy systems. The first one—the main focus in this book—is the fuzzy inference by IF-THEN rules that involves language; the second is all other methods such as fuzzy arithmetics or fuzzy graphs that do not involve language. A good example to illustrate this difference is the pattern recognition problem. There are fuzzy inference engines developed for pattern recognition where the solution comes from articulated expertise (e.g. fingerprint analysts) and/or from a representative historical data set. There are also generalized mathematical fuzzy pattern recognition techniques such as the c-means clustering method that attempt to solve a class of pattern recognition problems by a piecemeal fuzzy strategy.

Fuzzy system design includes determining algorithmic elements of fuzzy inference which are of two distinct types: (1) well established in the literature and requiring very little creativity; and (2) not well established in the literature, such as uncertainty characterization (e.g., membership function design), thus requiring interpretation and intuition. This distinction also defines the design mentality suggested in this book. The well-established part is called *the basic fuzzy inference algorithm,* which is a skeleton of work flow to be filled with designable elements, sometimes

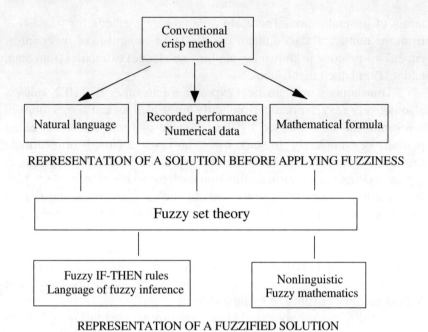

Figure 1.2 Converting a conventional method into a fuzzy method.

requiring selection from multiple options. The design philosophy is depicted in Fig. 1.3.

What are the design steps?

Fuzzy system design starts with obtaining the conventional method of solution. This is called the *knowledge acquisition* process. The second step is the translation, as shown in Fig. 1.2. The third step is designing the elements of the fuzzy inference algorithm, such as selecting certain logic operators. The second and third steps are sometimes interchangeable in their order or are performed simultaneously. Following these steps comes the final step of testing and simulation. Briefly, fuzzy system design steps include:

- *Knowledge acquisition*
- *Translation* into the language of fuzzy inference, composing fuzzy rules, and designing fuzzy variables

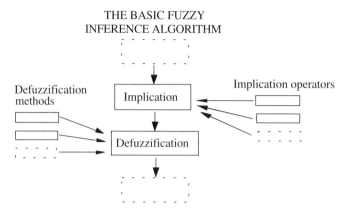

Figure 1.3 Fuzzy system design philosophy suggested in this book.

- *Developing the basic fuzzy inference algorithm,* filling the gaps by appropriate selections, and designing heuristic elements
- *Simulation and testing*

How are fuzzy systems implemented?

The fuzzy systems of Table 1.1 (and many others not listed) are implemented in a digital computer domain. Their design involves using special purpose software products that include design tools. Once a development is finalized, the fuzzy algorithm is converted into a computer program written in one of the programming languages, such as C, C++, Pascal, FORTRAN, PROLOG, Assembler, HC11, and so forth. Then the program is imported into the application environment. If this environment is a computer program or a software package, then fuzzy algorithms are embedded.[2] If the environment is digital hardware such as VLSI chips, then fuzzy algorithms are downloaded (or burned) on digital platforms. If the hardware is an analog device, then the fuzzy algorithm is implemented by circuitry design. Consequently, there are three forms of fuzzy system implementation:

---

[2] *Embedded programming* is a computer science term that refers to adding a subroutine or function written by someone other than the original programmer or by another computer program.

- *Embedded programming* for software applications
- *Digital hardware design* by downloading source codes
- *Analog circuitry design*

This phase of development is not considered within the framework of fuzzy system design in this book, since such developments require specialties not related to fuzzy logic.

Who is the designer?

The answer is, anybody with an appropriate background in fuzzy logic. The expertise required to develop a fuzzy system may be broken down to its components as follows. Examining the anticipated design steps listed previously, three different specialties are required to complete a working fuzzy system:

- *An expert* who knows the conventional non-fuzzy solution or method
- A *fuzzy logic specialist* who can perform the design steps
- A *software/hardware specialist* who can implement the design

Here, the fuzzy logic specialist, or the designer, plays a central role in obtaining the conventional solution from the expert, designing the fuzzy algorithm, and passing it on to the product specialist. As shown in the book, there are cases in which the requirement of an expert is relaxed such that the conventional solution can be readily obtained or fabricated using common sense.

What are the design tools?

Fuzzy system design requires a computer-aided design environment in which some of the tedious tasks are automated or already designed. There are several special purpose software products, also called *development shells,* on the market to aid the designer in transferring the conceptual design into a computer environment. Design tools are described in Chapter 4.

1.5 HOW TO USE THIS BOOK EFFECTIVELY

The elements of the basic fuzzy inference algorithm that are subject to heuristic interpretation or that require selection from multiple options are simply called *designable elements*. These elements, as mentioned earlier, are designed and placed in the empty slots of the skeleton of the basic fuzzy inference algorithm. Although the actual algorithm may be designed in a relatively small number of different ways, the basic fuzzy inference algorithm does not challenge one's creativity since it is a somewhat well-established work flow. This structure is presented in Chapter 3 and is used as a road map from that point on. The designable elements, on the other hand, are subject to interpretation, and there are multiple ways of designing them. A dozen design criteria are introduced in Chapter 4 that are type independent but context dependent for a given problem. In light of these criteria, several design principles are stated. Design principles do not concretely instruct the reader on what to do; instead they provide an insight into the resolution of the design issues identified in this book. In some cases, the design principles answer questions frequently asked by novice designers, such as "Why should I design membership functions with equal height?"

The book is organized in the following way. Chapter 2 presents a summary of theoretical subjects related to the design of simple fuzzy inference engines found in industrial applications. For more elaborate discussions on theory, the reader must refer to theoretical books, some of which are recommended throughout this book. Chapter 3 describes the skeleton structure of the basic fuzzy inference engine and addresses filling the empty slots by the most common methods employed today. These methods and alternative ones are described later in the book. The reader may employ the choices presented in Chapter 3 directly in his/her design work. A more detailed version of the basic fuzzy inference algorithm is included in the Appendix. Chapter 4 introduces the conceptual design phase in which subjects such as knowledge acquisition methods, general design criteria, problem types, and design tools are addressed. Design principles, which are presented in frames, also start from Chapter 4. Chapter 5 gives a detailed perspective on how to design a fuzzy variable—the heart of the design problem—with short and simple examples. This chapter focuses on decision boundaries, creating appropriate fuzzy categories, and linguistic modifiers. The design of membership functions from a simple perspective is discussed. Chapter 6, which is an extension of Chapter 5, presents membership function shape analysis. Issues such as membership function height, line style, and overlap are discussed. Chapter 7 includes logic operators, rule composition strategies, and the effect of membership function design with examples. Finally, Chapter 8 includes fuzzy implication methods, including the 10 most frequently used implication operators along with details of the calculus of the implication, aggregation, and defuzzification processes.

Compared to other types of books that deal with fuzzy logic, this book might serve better if read continuously from start to finish. In an application-oriented book, for example, this might not be necessary since each application might be presented in a self-sufficient context. Similarly, a theoretical book might be referenced from time to time to study certain aspects of theory without reading the entire book. The difference is mainly caused by the process of design such that every design subject becomes related to another one. This chain of relationships is best described in Chapter 3, where the basic fuzzy inference algorithm is explained as a road map.

As shown in Fig. 1.2, the application of fuzzy set theory to mathematical methods yields fuzzy methods such as fuzzy mathematical programming, fuzzy databases, fuzzy multiobjective programming, fuzzy statistical decision making, and so on. These subjects are not included in the book because their design characteristics are unique and do not match the design characteristics of fuzzy IF-THEN inference. On a higher level of complexity, the topic of fuzzy adaptive systems is also not discussed thoroughly. The reason is that adaptive fuzzy systems are not as widely applied in industry as simple nonadaptive systems, and their complex theory is not well established in the literature. A related subject called *neuro-fuzzy systems* is currently under intense research. A neuro-fuzzy system has a hybrid structure that includes a learning function provided by neural networks. This topic is also experimental and does not qualify well as a standard approach. In general, the reader should bear in mind that fuzzy systems design including all methods invented to date is a very broad subject to be examined in the limited space of one book.

1.6 TERMINOLOGY AND CONVENTIONS

Several terms used in fuzzy systems theory have special meaning beyond their dictionary meaning. Some of the terminology also overlaps or is used interchangeably in the literature. The following is a short list of fundamental terminology used throughout this book. It is recommended that the reader be familiar with the terminology at this point. The concepts behind the selected terminology are fully described in the following sections.

Antecedent. One or more variables that represent the conditions to be met before any conclusion can be made; same as an input variable.

Canonical form. A fuzzy rule is in canonical form when there is one term on both sides of the THEN operator, such as in *IF A THEN B*.

Composite rule. A fuzzy rule that is not in canonical form. A composite rule contains multiple statements on either or both sides of the THEN operator, such as in *IF A AND B THEN C*.

Composition. There are two distinct meanings of composition, which sometimes causes confusion. The first meaning refers to the formation of fuzzy rules (just like composing music). Tools of composition are AND, OR, NAND, NOR, THEN type logical operators. These operators are called *logic operators* in this book. The second meaning refers to the compositional inference, a principle in fuzzy systems theory, that describes a relation between two or more fuzzy sets (or variables) such as the max–min composition operator.

Consequent. One or more variables that represent the actions or conclusions; same as an output variable.

Decision boundary. A conceptual boundary on a product space that distinguishes decision categories. For example, if the drinking age is 21, then 21 is the boundary (on the universe of *age*) by which legal decisions are made. Decision boundary is a crisp concept but the actions resulting from decisions can be fuzzy based on some criterion, such as the distance to the boundary.

Degrees of commensurateness (DOC). A level of satisfaction or agreement to be imposed on the components of decision variables or consequents. Term used to analyze the right-hand side of a rule with respect to the THEN operator.

Degrees of fulfillment (DOF). A level of satisfaction or agreement achieved on the conditions or antecedents of a rule. DOF characterizes computation results of the entire left-hand side of a rule with respect to the THEN operator.

Hedges. A class of linguistic terms identifying degrees of strength such as *very*.

Implication. Implication is a special method of inference applied to fuzzy sets. Its calculus is well established. Implication is often associated with the THEN operator and denoted by an arrow.

Inference engine. An information processing system that draws conclusions based on given conditions or evidences. A fuzzy inference engine is an inference engine using fuzzy variables. Fuzzy inference refers to a fuzzy IF-THEN structure in this book. Note that this term traditionally emerged from expert systems technology and meant a special structure to obtain a solution. The term *inference engine* is used in a generalized manner in this book, and it does not refer to inference engines of expert systems.

LHS (left-hand-side). An abbreviation frequently used in this book that refers to the conditional part of a fuzzy rule before the THEN operator.

Linguistic variable. Same as fuzzy variable but its construction specifically requires linguistic values it can take. If *Temperature* is a linguistic variable, then its values (or predicates) might be *high, low, dangerous,* and so on. These values are always represented by a membership function (a fuzzy set) on the universe of discourse of the variable.

Membership function. A function that defines the degree of being a member of a fuzzy set. Membership functions are defined by fuzzy system designers based on possibility evaluation that might include probabilistic and heuristic measures.

Membership value. Single possibility value obtained after evaluating the membership function for a given input data.

Predicate. A linguistic phrase that is used as a semantic label for a membership function or a fuzzy category. *Low* is a predicate carrying semantic information describing the role of its membership function, so it is not an arbitrarily assigned label. A predicate is also known as linguistic value in the framework of a linguistic variable.

Premise. The left-hand side or the conditional part of a rule.

RHS (right-hand-side). An abbreviation frequently used in this book that refers to the consequent part of a fuzzy rule after the THEN operator.

Truth value. Same as membership value.

Universe of discourse. The numerical range of a fuzzy variable in which all of its membership functions reside.

Conventions Used in This Book

The following conventions are used in the presentation of fuzzy IF-THEN rules in this book. Variable names are denoted by uppercase first letters e.g., *Temperature*). The same applies to predicates such as *High*. Combined names are denoted by underscore signs, as in *Temperature_profile* or *Very_high*. Linguistic connectors including *is, is not, is less than, must be, was,* and so on, are all denoted by uppercase letters, as in *Temperature IS High*. The same consideration applies to logic operators such as AND, OR, and THEN. When we encounter these conventions in this book, we are encountering statements written in the language of fuzzy IF-THEN rules.

REFERENCES

[1] Terano, Toshira, Kiyoji Asai, and Michio Sugeno. *Applied Fuzzy Systems.* New York: AP Professional, 1994.

[2] Yen, John, Reza Langari, and Lotfi Zadeh. *Industrial Applications of Fuzzy Logic and Intelligent Systems.* Piscataway, NJ: IEEE Press, 1994.

[3] Yamaguchi, T. K. Goto, M. Yoshida, Y. Mizoguchi, and T. Mata,. "A Fuzzy Associative Memory System and Its Applications to Helicopter Control." Proc. 1st Int. Fuzzy Eng. Symp. pt. II, (1991): 770–779.

[4] Daniel McNeill and Paul Freiberger. *Fuzzy Logic.* New York: Simon & Schuster, 1993.

[5] Pin, F. G., and Y. Watanabe. "Navigation of Mobile Robots Using a Fuzzy Behaviorist Approach and Custom Designed Fuzzy Inferencing Boards." *Robotica* 12 (1990): 491–503.

[6] Berkan, R. C., et al. "Advanced Automation Concepts for Large Scale Systems." *IEEE Control Systems Magazine,* 11(no. 9): 1991.

Chapter 2

Theory

The objective in this chapter is to examine fuzzy systems theory at a level sufficient to understand the discussions related to fuzzy system design. Fuzzy systems theory is a broad subject including topics well beyond the intended scope of this book. Because our goal is to identify and resolve practical design issues for fuzzy inference systems of IF-THEN type, only closely related topics are examined. The reader should refer to the literature to build the sound theoretical background that is always necessary to become proficient in fuzzy system design.

> *The whole of science is nothing more than*
> *a refinement of everyday thinking.*
> Albert Einstein

2.1　CRISP VERSUS FUZZY SETS

A crisp set is a collection of distinct (precisely defined) elements. In classical set theory, a crisp set can be a superset containing other crisp sets. A superset will represent the universe of discourse if it defines the boundaries in which all elements reside. In any given situation, a new element can be tested to see whether it belongs to any set. This query is mathematically represented by the characteristic function. In classical set theory, an element either belongs to a set or not. If the set under investigation is A, testing of an element x using characteristic function χ is expressed as

$$\chi_A(x) = \begin{cases} 1 & x \in A \\ 0 & x \notin A \end{cases} \tag{2.1}$$

The numbers 1 and 0 constitute a valuation set (meaning x belongs to A and x does not belong to A, respectively) and are often shown inside curly brackets {1,0}. Using the curly brackets notation and Eq. (2.1), the crisp set A can be expressed as

$$A = \{(x, \chi_A(x))\} \tag{2.2}$$

where $(x, \chi_A(x))$ is called a *singleton*. Because the characteristic function only distinguishes between belonging or not belonging to a set, we can restate the crisp set A by $A = \{x\}$ where we only include the elements with characteristic function equal to one.

A fuzzy set is a collection of distinct elements with a varying degree of relevance or inclusion. The characteristic function test no longer has a trivial role because it determines the degree of relevance or inclusion. The characteristic function, which is known as the *membership function*, can take interval values between 1 and 0 and is often shown inside straight brackets [1,0]. Using the same notation as Eq. (2.2), the fuzzy set A can be expressed as

$$A = \{(x, \mu_A(x))\}, \quad x \in X \tag{2.3}$$

where μ denotes the membership function, and $(x, \mu_A(x))$ is a singleton. Another common way of representing a fuzzy set is

$$A = \bigcup_{x_i \in X} \mu_A(x_i)/x_i \tag{2.4}$$

Here, the fuzzy set A is the collection or union (denoted by $\cup$ in this book)[1] of all singletons $\mu_A(x_i)/x_i$. Note that "/" is not a division in this context. The difference between crisp sets and fuzzy sets can be examined by the difference between the two characteristic functions (membership functions) as shown in Fig. 2.1.

---

[1] In the fuzzy logic literature Σ and + are also used to denote union.

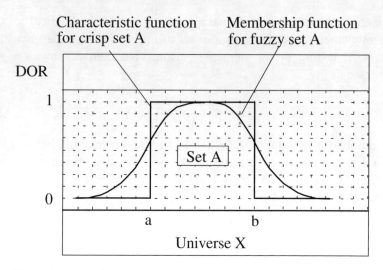

Figure 2.1 Degree of relevance (DOR) in identifying crisp and fuzzy sets using the membership function representation.

Set A, if it is a crisp set, is determined by the membership function that precisely identifies the boundaries of the set (a,b) on the universe of discourse. Set A, if it is a fuzzy set, is determined by the membership function that shows the distribution of degrees of relevance (or inclusion) across the universe of discourse.

EXAMPLE 2.1 FUZZY SETS VERSUS CRISP SETS

Assume that a basket of apples of various sizes is to be classified for packaging purposes. We also assume that size is determined by the weight of an apple.

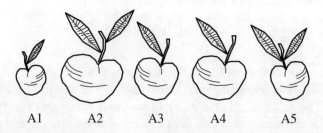

As shown in the graph below, a crisp set called *big apples* is determined by the characteristic function that represents a set boundary at 150 g for this particular example. Similarly, a fuzzy set called *big apples* is determined by the membership function that has no rigid boundary.

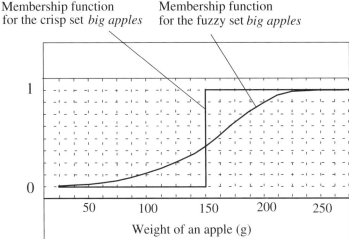

Suppose we have 5 apples labeled A1, A2, A3, A4, and A5 in the basket with weights, values of the characteristic function, and values of the membership function as shown in the Table.

ELEMENT	WEIGHT (g)	χ	μ
A1	44	0	0.01
A2	228	1	0.99
A3	148	0	0.45
A4	209	1	0.91
A5	151	1	0.47

Using the crisp set terminology, weights representing the *big apple* set are $A_c = \{228, 209, 151\}$. This is determined by the characteristic function (3rd column). Using the fuzzy set terminology, weights representing the *big apple* set are

$$A_F = \bigcup_{w_i \in W} \mu_A(w_i)/w_i$$
$$= 0.01/44 \cup 0.99/228 \cup 0.45/148 \cup 0.91/209 \cup 0.47/151$$

This is determined by the membership function (4th column). The fuzzy set representation can be translated such that weight 44 g has a degree of membership equal to 0.01 out of 1, weight 228 g has a degree of membership equal to 0.99 out of 1, and so forth. Note that both A_c and A_F are referring to the same information (set of big apples) but via different representations.

Two important properties of fuzzy sets should be noted. First, the elements (i.e., singletons) of a fuzzy set are, in one analogy, "aware" of each other by their relative distribution of the degree of membership. As shown in the example above, the membership value of 0.91 of weight 209 g is only meaningful due to the existence of another element in the set, say 148 g with a membership value of 0.45. Therefore, all elements carry twofold information: (1) the degree of inclusion to the set, and (2) the relative standing among others. This suggests that if there are more elements in a fuzzy set of certain size then there will be a more detailed relative standing profile. Thus, more singletons in a fuzzy set within a fixed universe of discourse indicates more information or more knowledge. Yet, it is another matter whether fuzzy sets with a high number of singletons are better mathematical tools than fuzzy sets with a small number of singletons. The usefulness of fuzzy sets and membership functions are discussed in more detail in Chapter 6, where design issues are presented. In crisp sets, on the other hand, the relative standing concept does not exist. Once an element belongs to a set, it does not matter where it stands in the set.

The second property to be examined is the error phenomenon. Referring back to Example 2.1, 148 g is not in the *big apple* set, whereas 151 g is, using the crisp set convention. When approaching a crisp set boundary defined by the characteristic function, the possibility of erroneous inclusion of an element to the set (or erroneous exclusion of an element from the set) due to measurement uncertainties exponentially increases. In an infinitesimally small interval around the exact boundary, the possibility of erroneous inclusion approaches 1, meaning an absolute indecision. In fuzzy sets, the possibility of error in the degree of inclusion is spread over the entire set somewhat uniformly. Thus, there is no localized point of absolute indecision in fuzzy sets.

2.2 FROM FUZZY SETS TO FUZZY EVENTS

Fuzzy set theory, compared to other mathematical theories, is perhaps the most easily adaptable theory to practice. The main reason is that a fuzzy set has the property of relativity, variability, and inexactness in the definition of its elements. Instead of defining an entity in calculus by assuming that its role is exactly known, we can use fuzzy sets to define the same entity by allowing possible deviations and inexactness in its role. This representation suits well the uncertainties encountered in practical life, which make fuzzy sets a valuable mathematical tool.

Uncertainties in life (or fuzzy events) are of many types. All measurement systems—man-made or biological—are subject to uncertainties. Therefore, every scientific discipline based on experiments and measurements can make use of fuzzy sets in mathematical modeling and in analytical solutions to improve the generality of the derived methods. The generality in this context means multiple solutions of varying possibilities instead of one crisp, so-called exact solution. Uncertainties in symbolic (i.e., nonnumerical) information can also be represented using fuzzy sets. This applies to natural and synthetic languages. Thus, the definition of an entity by means of language can be improved by equivalent (or analogous) mathematical representations using fuzzy sets, allowing us to compute with words. The "big apples" example in the previous section illustrates the point of departure in this direction. Finally, fuzzy sets can be used to model approximate reasoning, which governs our daily life. Examples would include a long list of activities such as driving a car, playing basketball, or controlling a power plant. A good source of information on fuzzy thinking and its remedies is available in the literature [1].

Transition from fuzzy sets to modeling fuzzy events requires a theory that would relate the membership function concept to the uncertainties encountered in different contexts as described above. This relation is also referred to as *uncertainty characterization* in this book. Possibility theory is the most widely accepted theory to describe uncertainty characterization. The reader should refer to the literature for elaborate description of possibility theory in mathematical terms [2]. We will describe the possibility theory briefly in the next paragraph. The remainder of this book includes practical interpretations of possibility theory for fuzzy system design.

Possibility theory originates from defining a possibility distribution function as a mapping of the singletons x in the universe X to the unit interval $s: X \rightarrow [0,1]$. Fuzzy set theory and possibility theory converge based on the conjecture that $s: X \rightarrow [0,1]$ represents membership function and the degree of elementhood is determined by possibility. Possibility theory can also be considered as a form of probability theory with relaxed constraints. Among the several interpretations (subjectivistic, axiomatic, and frequentistic), probability is mainly a measure of occurrence of an event (often a physically observable event). Possibility, on the other hand, is a measure of resemblance of an ideal sample, or a measure of capability to occur.[2] The fundamental point of departure between the two theories is that an ideal sample or capability of occurrence cannot always be characterized by likelihood. This happens in situations where there is no statistical evidence; instead, there is a consensus or expert opinion about the resemblance of an ideal sample. In another context, ideal sample is determined by choice, style, taste, or belief that has no direct connection to a probabilistically correct answer. In mathematical terms, the sum of possibilities can exceed 1, whereas the sum of probabilities cannot exceed 1. For such reasons, possibility is considered to be a broader concept than probability. Possibility also entails probability as one of its complementary elements. The probability/possibility consistency principle [3] is useful when both are used simultaneously for characterizing uncertainty. The basic axiom is that events that are not possible are not probable either, but events that are possible may not always be probable.

2.3 FUZZY LOGIC AND LINGUISTICS

An ideal medium to characterize fuzziness is natural languages. Just like a physicist improves his/her experimental knowledge by examining the physical world, we examine natural languages to improve our understanding of fuzziness and its extent. There are two basic objectives in examining natural language: (1) to explore the linguistic mechanisms of uncertainty characterization, and (2) to understand the logic mechanism in the presence of uncertainty. The underlying inspiration for many researchers has been to learn the principles of humanlike intelligence from the properties of language and logic and to apply those principles to practical problems.

The first objective has been a topic of extensive research with several proposed approaches to uncertainty characterization. According to one approach, uncertainty characterization can be based on *fuzzy measures* including the concepts of *possibility, belief, plausibility, necessity,* and *probability* [4]. More on this issue is discussed in the next section. In relation to natural logic, predicates (adjectives and adverbs) have been the focus of attention because of their function of characterizing and/or categorizing fuzziness. Fuzzy quantifiers exemplified by predicates such as *several, about,* and *much* provide an imprecise representation of the principality of one or more fuzzy or nonfuzzy sets [5]. Along the same lines, fuzzy quali-

---

[2] Capability to occur is different from likelihood of occurrence or chance.

fiers such as *best*, *terrible*, and *acceptable* fall into the class of heuristic measures. Three principle modes of qualification suggested by Zadeh are (1) truth qualification with predicates such as *not quite true*, (2) probability qualification with predicates such as *unlikely*, and (3) possibility qualification with predicates such as *almost impossible*. Examining the linguistic elements for the characterization of uncertainty has also yielded other concepts such as the linguistic hedges that further distinguish the degrees of qualification in the form of *intensification, dilution*, and *approximation* [6]. All these concepts are useful in fuzzy system design. However, their mathematical models found in the literature are not as deterministic as the mathematical models of physical events such as a heat transfer problem. Thus, the validity of the mathematical models of the linguistic concepts is a matter of reasoning rather than proof based on experimental data.

The second objective—examining natural logic and tautologies—has also been the subject of extensive research. Aristotle's bivalent logic comprising true and false duality has obvious shortcomings in handling practical problems involving uncertainty. Neither have advances in the classical theory of logic, such as multivalued or interval logic, yielded reasonable answers to paradoxical problems emerging from real-life situations. However, using the fuzzy measures and possibility theory, the applicability of the classical methods of inference (such as modus ponens and modus tollens) to a wider scope of decision-making problems is possible. In other words, a proposition such as *if the weather is bad then stay inside* can be modeled in mathematical terms using the membership function concept characterizing *bad weather*. The capability of representing *bad weather* mathematically, and all other similar fuzzy events for that matter, generates the idea of linguistic variables (i.e., *weather* is a linguistic variable just like a variable in calculus and it can take values such as *bad, mediocre, nice*, etc.). This elevates the classical inference concept embedded in pure symbolism to a new dimension in which, all of a sudden, a substantial portion of natural language can be represented by mathematical models.

2.4 PRACTICAL FUZZY MEASURES

The main idea behind fuzzy measures is to build an analytical bridge between mathematics (e.g., fuzzy sets, membership functions) and the uncertainty phenomenon encountered in practice. We have already discussed that possibility theory performs this role. We will see in this section that fuzzy measures theory as described in the literature comes to a similar conclusion with more detailed descriptions.

Note that measures of fuzziness is a different concept than fuzzy measures. Measures of fuzziness indicate the degree of fuzziness of a set. Entropy, distance, and similarity are suggested as useful concepts for such measurements. A comprehensive coverage of this topic can be found in the literature [7–9]. We use the concepts of entropy and distance in fuzzy variable design as described in Section 5.2, Chapter 5.

Uncertainty may be considered in two classes. The first class consists of uncertainty descriptions involving language only and the second class consists of uncertainty descriptions involving numbers and countable entities. The question *Does egg come from chicken or chicken come from egg?* implies uncertainty in the realm of language, whereas the statement *There is a crowd around the scene* implies uncertainty of the countable type. Note that the definition *countable* includes events that might not necessarily be countable or measurable in practice. *Ambiguity, vagueness*, and *undecidedness* articulate uncertainty caused by the mysterious notions of language and culture. *Randomness, chaos, imprecision*, and *inaccuracy* articulate uncertainty in countable events. Although these terms are used interchangeably in daily language, the distinction discussed above represents a complexity ranking. Uncertainties related to natural language and logic are quite complex[3] and well beyond the limits of the mathematical models in practice. Uncertainties related to countable events are somewhat more manageable. Imprecision and inaccuracy are the two concepts we are most familiar with historically. Most of the fuzzy systems built to date are centered around the notions of precision and accuracy. However, this fact does not constitute a precedent for the progress of fuzzy systems in the future.

A fuzzy measure is a mathematical tool that describes the *ambiguity, vagueness, undecidedness, randomness, chaos, imprecision*, and *inaccuracy* of a relation between crisp events, entities, or sets. A practical fuzzy measure only concerns the *imprecision* and *inaccuracy* of a relation between crisp events, entities, or sets. In Chapter 4, a set of linguistic design criteria is presented that is based on fuzzy measures to represent (or control) the level of imprecision and inaccuracy in today's fuzzy systems.

---

[3] Fuzzy logic, as implemented in practice today, refers to logic based on uncertain measurements, but not to uncertainties in logic.

In theory, belief and plausibility are the two fuzzy measures from which other fuzzy measures can be derived. A belief measure describes the degree of support, or evidence, for an element or subset to belong to a certain superset or universe. It is denoted as $bel(A)$ where A is the collection of elements. A plausibility measure is the complement of the disbelief measure and is given by $pl(A) = 1\text{-}bel(\bar{A})$ where $\bar{A}$ is the disbelief (i.e., complement of the belief measure). The intersection of the belief and plausibility measures yields a probability measure. When the evidence is nonconflicting, then belief and plausibility measures are referred to as necessity $\eta(A)$ and possibility $\pi(A)$ measures. It has been shown that these measures are dual relationships [10]; therefore, one can be derived from another by $\eta(A) = 1 - \pi(\bar{A})$, and $\pi(A) = 1 - \eta(\bar{A})$.

When designing a fuzzy system in practice, one of the main concerns is the level of compromise in achieving a technical objective. The compromise characterized by error (i.e., distance to the absolute truth) is due to the lack of two things: measurement quality and knowledge (of countable events). Measurement quality determines precision whereas knowledge determines accuracy. These definitions are shown in Fig. 2.2 for a hypothetical event. Note that truth is always unknown in Fig. 2.2.

Precision and accuracy have unrelated roots, but they interfere with each other, resulting in uncertainty comprising unknown proportions of lack of knowledge and measurement imperfections. Fuzzy measures characterize this phenomenon in a generalized and realistic manner, including all possible causes of degradation. A realistic interpretation of uncertainty caused by imprecision and inaccuracy is shown in Fig. 2.3. In this hypothetical drawing, the unknown truth was assumed to be nicely

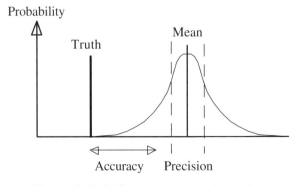

Figure 2.2 Definitions of accuracy and precision.

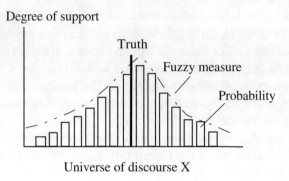

Figure 2.3 Representing truth based on fuzzy measures.

located within the expected domain of uncertainty characterization. This is never absolutely known in practice.

The main question for the fuzzy system designer is how to draw the fuzzy measure curve in Fig. 2.3. There are two answers.[4] The first answer applies to *truth representation* just like the case in Fig. 2.3. In this case, examples of truth may be the weight of an apple, the length of a bridge, or the speed of a bullet. The answer requires the collection of evidence either by measurements or via experts' opinion. The collected evidence, which is treated as the degree of belief (or the basic probability assignment), yields fuzzy measures as illustrated by several examples in the literature [11]. However, the requirement of the basic probability assignment as the initial step makes these methods dependent on the quality of evidence and on the uncertainties related to the method of evidence collection.

The second answer applies to *truth categorization,* in which truth may be the category of *high* room temperature, *short* people, *crowded* streets, or *valuable* stock. In this case, evidence becomes more subjective, involving personal experiences, cultural values, and style as well as deterministic elements such as data. Such a mixture of influences may potentially defy mathematical models based on basic probability assignment. Thus, fuzzy measures become heuristic measures. Note that fuzzy logic thrives in truth categorization problems since natural logic and lan-

---

[4] Answers suggested by mathematical models are not complete, and sometimes not adequate due to the complexity of the uncertainty phenomenon.

guage extensively involve truth categorization. Most of the fuzzy IF-THEN type inference engines embody knowledge of truth categorization as articulated by experts using natural language. The answer to how to draw the curve in Fig. 2.3 for this case is given in this book by means of the linguistic design criteria. These criteria are based on the objective patterns in designing a practical fuzzy system. The linguistic design criteria are described in Section 4.6.

Truth representation can be made for a combination (or a set) of truths. For example, weather temperature or flow rate of a river can have different truths for each molecule. Truth categorization is analogous to representing a combination of truths. However, categorization is flexible (matter of choice for the intended purpose), whereas, representing a combination of truths is not (no choice other than representing a property). The hypothetical drawing in Fig. 2.4 applies to both cases. The determination of the curve in Fig. 2.4 is made in the same way as discussed before.

Precision is often expensive. Accordingly, compromise with precision reduces cost. During the last two decades, the success of fuzzy technology in the field of engineering has largely depended on cost-effectiveness. The main theme has been to achieve tolerable imprecision by fuzzy approach which is cost-effective. In parallel, many applications previously considered too costly have become feasible using fuzzy technology. Along the same lines, accuracy is also expensive. Accuracy can be improved by increased knowledge that requires research, experimentation, and prototyping. Using a limited available knowledge, fuzzy inference approach facilitates the fabrication of commercial products without

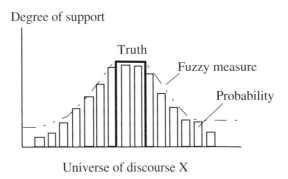

Figure 2.4 Representing a combination of truths based on fuzzy measures.

extensive research and development effort. A typical example is the new camcorder that can stabilize screen images during bumpy recording conditions [12]. A simple fuzzy algorithm employing common sense rules is able to stabilize screen images at a desirable level. The problem could also be solved by the advances in conventional (nonfuzzy) image-processing technology, however, for this particular product, fuzzy technology was more cost-effective.

2.5 FUZZY SET OPERATIONS

When we have two or more fuzzy sets describing the uncertainties of a given problem, analytical solutions often require operations among fuzzy sets. Fuzzy set operations is a widely studied subject in the literature with several new concepts and definitions. In the context of fuzzy system design as practiced today, we will only highlight those topics that are directly related to design.

2.5.1 Basic Definitions

Normal Fuzzy Set

A normal fuzzy set has a membership function that includes at least one singleton equal to unity. If there is no singleton equal to unity, the fuzzy set is called subnormal.

Support of Fuzzy Set

The support of a fuzzy set A is the crisp set of all $x \in X$ such that the membership values are nonzero, $\mu_A(x) > 0$.

Convex Fuzzy Set

A fuzzy set A is convex if its membership function is monotonically increasing and/or decreasing without any saddle point in the middle. Convexity is expressed by the following condition.

$$\mu_A(\lambda x_1 + (1-\lambda)x_2) \geq \min(\mu_A(x_1), \mu_A(x_2)), x_1, x_2 \in X, \lambda \in [0,1]$$
(2.5)

Complement of a Fuzzy Set

The complement of a fuzzy set is a new fuzzy set with a membership function representing the degree of exclusion or irrelevance. The new membership function is expressed by

$$\bar{\mu}(x) = 1 - \mu(x) \quad x \in X \tag{2.6}$$

where X is the universe of discourse, and 1 represents the maximum degree of inclusion.

Scalar Product of a Fuzzy Set

A fuzzy set can be multiplied by a scalar S. The membership function of the new fuzzy set is given as follows.

$$\mu(x) = S \cdot \mu_1(x) \quad x \in X \tag{2.7}$$

Power of a Fuzzy Set

A fuzzy set can be raised to a power m by raising its membership function to m. Here, m is a positive real number. After such an operation, the result is a new fuzzy set with a membership function given as follows.

$$\mu(x) = [\mu_1(x)]^m \quad x \in X \tag{2.8}$$

2.5.2 Set-Theoretic and Average Operators

There are several types of operations among fuzzy sets; to cover all of them might require the space of a whole new book. The objective behind these derivations is to apply classical set-theoretic operations such as intersection and union to fuzzy sets. Compared to classical set theory, fuzzy set theory offers a family of set operations due to the nature of fuzzy sets (e.g., partial truth and continuous valued membership functions). For example, the intersection between two crisp sets *people under age 21,* and *people with high school diplomas* will be quite deterministic when compared to the intersection between two fuzzy sets *young people* and *educated people*. The problem using fuzzy sets is who to include in the intersection set *young* and *educated,* which is a matter of degree expressed by membership functions.

The theoretical foundation is based on the generalized formulation of intersection and union, also known as *t-norms* and *t-conorms* (*s-norms*). Table 2.1 lists a collection of operators derived in the literature [13]. The family of intersection and union operators suggested in this table produce different outcomes. Minimum and maximum are the two extreme boundaries. Averaging operators, which are placed in the middle column in Table 2.1, result in possibility values away from both extremes. Some of these operators are parametrized, meaning that their behavior can be adjusted by a control parameter. Besides the linguistic design criteria described in Chapter 4, there is no concrete theory or experimental proof to tell us which operator is the exact mathematical recipe for a given problem.

These operators are also referred to as *aggregation operators* in the literature. However, we use the term *aggregation* in a special way in this book. The speciality is due to where it is used, but not how it is used. We use aggregation to combine two or more fuzzy sets that are outputs from each rule in an IF-THEN type fuzzy inference algorithm. Combination is necessary when all these fuzzy output sets represent a contribution to the same consequent fuzzy variable. Thus, we compute the total output by aggregation in such cases. As will be described later in this book, the aggregation mechanism in most of the fuzzy systems is either a minimum or maximum operator depending on the type of implication process.

In the realm of the basic fuzzy inference algorithm, some of these operators are used in several places, such as in computing the degrees of fulfillment of composite rules or in the implication process. Some of the more advanced concepts will be pursued in the later stages of the book. Four basic operations are described next.

Table 2.1 Set-Theoretic and Average Operators for Fuzzy Sets

INTERSECTION OPERATORS	AVERAGING OPERATORS	UNION OPERATORS
Minimum	Arithmetic mean	Maximum
Algebraic product	Geometric mean	Bounded difference
Bounded sum	Geometric mean	Bounded difference
Hamacher product[†]	Symmetric summation	Hamacher sum[†]
Einstein product	Symmetric differences	Einstein sum
Drastic product	Fuzzy AND, fuzzy OR[✦]	Drastic sum
Yager intersection[✦]	Compensatory AND[✦]	Yager union[✦]
Dubois intersection[✦]		Dubois union[✦]

[†] can also be parametrized.
[✦] only parametrized.

Union of Fuzzy Sets (Maximum)

The union of fuzzy sets (j of them) defined over the same universe of discourse is a new fuzzy set with a membership function that represents the maximum degree of relevance between each element and the new fuzzy set. The membership function of the new fuzzy set $\mu_\cup(x)$ is expressed by

$$\mu_\cup(x) = \mu_1(x) \vee \mu_2(x) \vee ... \vee \mu_j(x) \quad x \in X \tag{2.9}$$

where X is the universe of discourse, and $\vee$ is the maximum operation.

Intersection of Fuzzy Sets (Minimum)

The intersection of fuzzy sets (j of them) defined over the same universe of discourse is a new fuzzy set with a membership function that represents the minimum degrees of relevance between each element and the new fuzzy set. The membership function of the new fuzzy set $\mu_\cap(x)$ is expressed by

$$\mu_\cap(x) = \mu_1(x) \wedge \mu_2(x) \wedge ... \wedge \mu_j(x) \quad x \in X \tag{2.10}$$

where X is the universe of discourse, and $\wedge$ is the minimum operation. Note that the minimum operation applies to the corresponding singletons of each fuzzy set.

Algebraic Product

The product of fuzzy sets defined over the same universe of discourse is a new fuzzy set with a membership function on the universe X expressed as

$$\mu(x) = \mu_1(x) \cdot \mu_2(x) \cdot ... \cdot \mu_j(x) \quad x \in X \tag{2.11}$$

The algebraic product is the multiplication of the possibility values between each corresponding singleton.

Algebraic Mean

The algebraic mean of fuzzy sets over the same universe of discourse is a new fuzzy set with a membership on the universe of discourse X given by

$$\mu(x) = \frac{1}{j}\left[\mu_1(x) + \mu_2(x) + ... + \mu_j(x)\right] \quad x \in X \tag{2.12}$$

where j is the number of fuzzy sets. Computation given by Eq. (2.12) is the addition of the possibilities for each corresponding singleton divided by the number of fuzzy sets.

Summary of Operations on Fuzzy Sets

A summary of these operations is shown in Figs. 2.5 and 2.6 using two arbitrarily selected fuzzy sets called A and B.

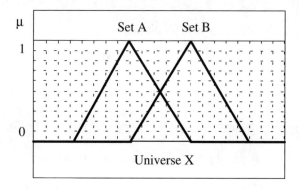

Figure 2.5 Two arbitrarily drawn fuzzy sets in the same universe of discourse X.

EXAMPLE 2.2 FUZZY SET OPERATIONS

To illustrate the operations between fuzzy sets, we select two fuzzy sets A and B as given below. The first fuzzy set A is the fuzzy set *big apple* in Example 2.1. The second fuzzy set is its complement, indicating a fuzzy set for *small apples* or *not big apples*.

$$A = \bigcup_{w_i \in W} \mu_A(w_i)/w_i$$
$$= 0.01/44 \cup 0.99/228 \cup 0.45/148 \cup 0.91/209 \cup 0.47/151$$

$$B = \bigcup_{w_i \in W} \mu_B(w_i)/w_i$$
$$= 0.99/44 \cup 0.01/228 \cup 0.55/148 \cup 0.09/209 \cup 0.53/151$$

The intersection of the two sets yielding a new fuzzy set C is computed by comparing each singleton in both fuzzy sets and selecting the one with minimum membership value. The membership function of the new fuzzy set is given by

$$\mu_C(w) = \mu_A(w) \wedge \mu_B(w) \quad w \in W$$

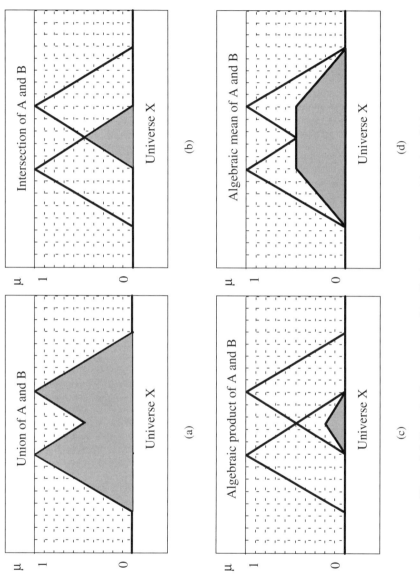

Figure 2.6 Operation between two arbitrarily drawn fuzzy sets: (a) union, (b) intersection, (c) algebraic product, and (d) algebraic mean.

where W is the universe of discourse. The new fuzzy set C is therefore

$$C = \bigcup_{w_i \in W} \mu_C(w_i)/w_i$$
$$= 0.01/44 \cup 0.01/228 \cup 0.45/148 \cup 0.09/209 \cup 0.47/151$$

The selected singletons are marked in the figure below. Note the difference between the continuous membership function (curves) and the discrete membership function (dots) representations. Obviously, the discrete fuzzy set representation using singletons has a lower resolution than that of continuous function representation. However, as will be shown later in the book, poor resolution in defining membership functions has very little effect in many practical applications as long as the characteristic features of the intended shape are represented adequately.

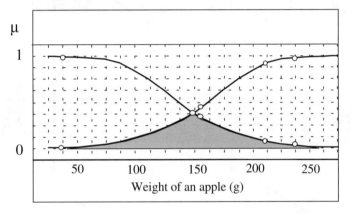

Consider the fuzzy sets selected in this example *big apples* and *not big apples*. Because the latter is the complement of the former, this example also shows the excluded-middle law for fuzzy sets. The intersection of two complementing sets in classical set theory would be a null set. This is explained next.

2.6 PROPERTIES OF FUZZY SETS

Using the intersection and union operators described in the last section, a class of properties between two fuzzy sets is described in this section. A detailed description of the properties of crisp and fuzzy sets can be found in the literature [14]. If A and B are two fuzzy sets, then the properties are given as follows.

$$(A \cup B) \cup C \equiv A \cup (B \cup C)$$
$$(A \cap B) \cap C \equiv A \cap (B \cap C)$$
Association

$(A \cup B) \cap C \equiv (A \cap C) \cup (B \cap C)$ $(A \cap B) \cup C \equiv (A \cup C) \cap (B \cup C)$	Distribution
$A \cap (A \cup B) \equiv A$ $A \cup (A \cap B) \equiv A$	Absorption
$A \cup A = A$ $A \cap A = A$	Idempotency
$A \cup B = B \cup A$ $A \cap B = B \cap A$	Commutativity
$\overline{(\overline{A})} = A$	Double Negation
$\overline{A \cup B} = \overline{A} \cap \overline{B}$ $\overline{A \cap B} = \overline{A} \cup \overline{B}$	De Morgan's Laws

All properties described above also apply to crisp sets. A difference between crisp sets and fuzzy sets is encountered for the excluded-middle laws. In fuzzy set theory, the union of a fuzzy set and its complement does not yield the universe. Also, the intersection of a fuzzy set and its complement is not null.

$$A \cup \overline{A} = X \quad \text{Crisp sets} \quad A \cup \overline{A} \neq X \quad \text{Fuzzy sets}$$
$$A \cap \overline{A} = \emptyset \qquad\qquad\qquad A \cap \overline{A} \neq \emptyset$$

These properties, although they are not directly useful to design a fuzzy inference system of IF-THEN type, satisfy the basic axioms of set theory, and hence provide a foundation for the derivation of more advanced tools that are used in fuzzy inference.

2.7 FUZZIFICATION TECHNIQUES

Fuzzification means adding uncertainty by design to crisp sets or to sets that are already fuzzy. This may be conceptually confusing to many readers who are new to the fuzzy logic topic because adding uncertainty seems like going backwards in reference to the classical school of thought in science. On the contrary, fuzzification is a useful concept and is not an entirely new idea. In practical terms, fuzzification is spreading the information provided by a crisp number or symbol to its vicinity so that the close neighborhood of the crisp number can be recognized by the computational tools. In the world of numbers, this is analogous to the interpolation method between two data points. Fuzzification corresponds to defining interpolation surfaces between two sets or two numbers by a controlled (designed) uncertainty distribution function.

Fuzzification, which seems like a dull exercise of repeating the interpolation concept in the world of numbers, is in fact very useful and unique when considering crisply defined logic, formulas, equations, relationships, and variables. The injection of "designed" uncertainty enables us to carry on computations in those gray areas not explicitly defined by crisp entities. For example, weather temperature = 70° F can be transformed into a fuzzy set called *nice weather* using a fuzzification technique so that temperatures around 70° F would have some association to the phrase *nice weather*. Then, this phrase may be used in a decision-making rule base.

The simplest form of fuzzification can be done using a fuzzifier function F that determines the degree of fuzziness in a set [15]. A fuzzifier function is identical to a membership function in principle except for the objective behind designing it. The objective is to be able to apply a controlled uncertainty distribution to the elements (e.g., inputs, outputs) of a fuzzy system in a consistent manner. For example, a fuzzifier given by

$$\mu(x) = \frac{1}{1+\left[\dfrac{x}{C_2}\right]^{-C_1}} \quad x \in X \quad (2.13)$$

produces an S-shaped curvature creating a generalized effect of *large*. C_1 and C_2 are called the exponential and denominational fuzzifiers, respectively. Note that by having controllable parameters such as C_1 and C_2, adaptive fuzzy algorithms can be developed where C_1 and C_2 will be updated based on certain criteria determined by selected measurements. Adaptive systems are explained in Sections 2.16 and 7.5.6. Using the fuzzifier given by Eq. (2.13), a fuzzy set can be written as the union (continuous) of singletons.

$$A \equiv \bigcup_x \left(\frac{1}{1+\left[\dfrac{x}{C_2}\right]^{-C_1}}\right) /x \quad x \in X \quad (2.14)$$

The effect of changing the controllable parameters is shown in Fig. 2.7. It is also apparent in these figures that this particular fuzzifier represents

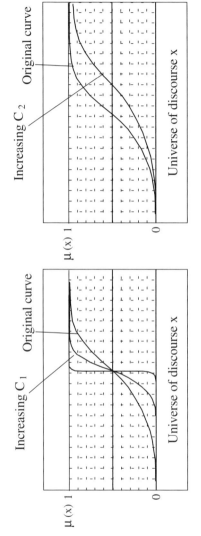

Figure 2.7 Effect of changing controllable parameters of a fuzzifier that models the linguistic phrase *large*.

large numbers on the universe. In the context of a rule-base system, such a fuzzifier can be used to represent large X where X is a context-specific fuzzy variable (e.g., temperature, flow, income, etc.).

Besides generating standardized membership functions via fuzzifiers, a fuzzifier can be imposed on crisp or fuzzy sets to change their attributes in a desired manner. Also known as the fuzzy kernel $K(x)$, the result produced by imposing a fuzzifier on another set is a new fuzzy set given by

$$F(A:K) = \bigcup_x \mu_A(x) \cdot \mu_K(x)/x \quad x \in X \quad (2.15)$$

where $F(A:K)$ denotes that the properties of set A are changed by fuzzy kernel K. Using the fuzzifier for *Large* as the kernel and a singleton 1.0/230, we can generate the membership function *big apples* given in Example 2.1. By storing the appropriate values of C_1, C_2, the kernel, and the singleton, we can reproduce membership functions of various sorts instead of storing all data points. Obviously, this comes in handy when developing software tools with memory restrictions.

Another common technique is called the extension principle. The extension principle suggests that fuzzifying the parameters of a function yields fuzzy outputs. Consider a function $y = f(x)$. If the input argument is a fuzzy set A on the universe of x, then the fuzzy output must obey the extension principle $B = f(A)$. Suppose set A is given by

$$\begin{aligned} A &= \bigcup_{i=1,n} \mu_A(x_i)/x_i \\ &= \mu_A(x_1)/x_1 \cup \mu_A(x_2)/x_2 \cup \ldots \cup \mu_A(x_n)/x_n \end{aligned} \quad (2.16)$$

then the fuzzy output set B is

$$\begin{aligned} B &= \bigcup_{i=1,n} \mu_A(x_i)/f(x_i) \\ &= \mu_A(x_1)/f(x_1) \cup \mu_A(x_2)/f(x_2) \cup \ldots \cup \mu_A(x_n)/f(x_n) \end{aligned} \quad (2.17)$$

where every single image of x under f becomes fuzzy to a degree determined by the membership value. In many-to-one mapping, which is often the case, the singleton with maximum membership value is selected.

In its most general form, a multivariate function $y = f(u,v,..,w)$ can be used to fuzzify the output space through the extension principle [14]. This is expressed as

$$\mu(y) = \vee \left(\left[\mu_u(u) \wedge \mu_v(v) \wedge \ldots \wedge \mu_w(w) \right] / f(u, v, \ldots, w) \right)_{U \times V \times \ldots \times W}$$

(2.18)

where fuzzy sets U, V,...,W are defined on multidimensional product space $U \times V \times \ldots \times W$. The underlying mechanism of the extension principle is therefore a supremum of pair-wise minima. The extension principle achieves generality at the expense of triviality. Earlier fuzzy systems used the extension principle for the formation of fuzzy sets representing the weighted output of fuzzy algorithms [16]. In such applications however, the extension principle tends to produce uniform distribution as the number of combined fuzzy sets increases.

2.8 ALPHA CUTS

A *level fuzzy set* [17] has membership values greater than α where α is between 0 and 1. This is expressed as

$$A^\alpha = \{x \, | \mu_A(x) > \alpha\}$$

(2.19)

which means that the new crisp set A^α contains elements with membership value higher than α. The new set can also be fuzzy depending on the computational objective.

In practical terms, an α-cut (also known as Lambda-cut) can be viewed as a lower-bound threshold for possibility. Accordingly, a membership function after being subjected to an α-cut represents a set smaller than its original size in terms of the number of singletons. This is illustrated in Fig. 2.8 for an arbitrarily drawn membership function and three different levels of α-cut. For each level cut, we have a different set.

The resolution principle offers an alternative way of representing fuzzy sets by means of α-cuts.

$$\mu_A(x) = \vee \left[\alpha \cdot \mu_{A\alpha}(x) \right] \quad x \in X$$

(2.20)

α is restricted between 0 and 1. Equation (2.20) indicates that the membership function of A is the union of all α-cuts, after each crisp set $\mu_{Aa}(x)$ is multiplied by α. Note that this approach yields an approximate representation of the membership function because some of the fuzzy singletons are converted to crisp singletons.

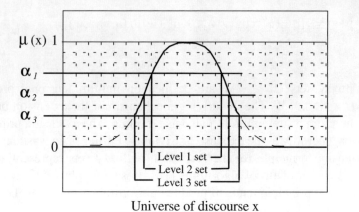

Figure 2.8 A membership function subjected to α-cuts.

2.9 RELATIONAL INFERENCE

We have seen various types of basic mathematical operations for fuzzy sets in Sections 2.6 and 2.7. Based on these basic operations, we will now discuss making something useful out of them. The objective is to apply fuzzy operators and operations to problem solving in the practical world. The most elementary step is to build models of decision making by defining relationships among fuzzy sets defined over different universes of discourse.

Establishing relationships among fuzzy sets, also referred to as *fuzzy relations*, is analogous to mapping between multiple-product spaces in mathematics, and is also formulated as fuzzy associate memories (FAMs) in another context [16]. The basic idea is to characterize the properties of connections between the elements of one fuzzy set and another. Very importantly, such relationships are not formed as a mathematical exercise; instead, they are formed as a model of inference (decision making) in a particular domain of interest. Relational inference among fuzzy sets defined over different universes can be viewed topologically as shown in Fig. 2.9. In the figure, every element (singleton) of fuzzy set A has one (or more) associations with a member in fuzzy set B where the cross-associations are also fuzzy. The direction of associations is important. The topology depicted in Fig. 2.9 is a bird's-eye view of a two-dimensional membership function similar to that of a hill.

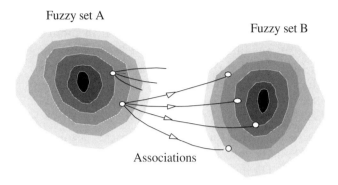

Figure 2.9 Relational inference between two fuzzy sets where every member in one set has an association to a member in another set.

Before elaborating on fuzzy relational inference, let's examine the mathematical properties of fuzzy relations using the definitions originally developed for crisp relations. Because fuzzy relations are also fuzzy sets, the fuzzy set properties described in Section 2.6 hold for fuzzy relations as well. Similarly, the difference between classical and fuzzy sets characterized by the excluded-middle law exists between fuzzy and crisp relations. There are more advanced associations applicable to fuzzy relations [17], as described next.

Consider a Cartesian product space $S = X \times Y$ in which x belongs to X and y belongs to Y. If R is a relation on S, then the relation is *reflexive* for xRx, and *irreflexive* if there is no x in S for which xRx is valid. A relation R is *symmetric* when xRy and yRx are both valid, and *asymmetric* if there is no x and y satisfying it. If xRy, yRz are valid, then a relation is *transitive* if xRz is valid. *Symmetry*, *reflexivity*, and *transitivity* are three fundamental properties of fuzzy relations. An important fuzzy relation called a similarity relation is symmetric, reflexive, and transitive. If a relation is only symmetric and reflexive, then it is called a *proximity* or *tolerance* relation. In terms of the membership function of a relation, these properties are expressed as follows:

$$\mu_R(x,x) = 1 \quad \text{reflexive} \tag{2.21}$$

$$\mu_R(x,y) = \mu_R(y,x) \quad \text{symmetric} \tag{2.22}$$

$$\mu_R(x,y) \geq \bigvee_z [\mu_R(x,z) \wedge \mu_R(z,y)] \quad \text{max-min transitive} \tag{2.23}$$

The relational inference shown in Fig. 2.9 can be represented either by fuzzy sets of the associations or by fuzzy sets in product spaces. The membership functions of fuzzy sets A and B can be obtained given the membership function of the associations, or vice versa. However, such transformations result in generalization at the expense of losing some information. This is explained next.

Consider a hypothetical case where the body shape of 5-year-old children is to be identified for medical purposes. Assume that a category of body shape called *chubby* is to be determined based on the weight and height of a child. The membership function of the association between weight and height spaces is shown in Fig. 2.10. The surface in this figure represents expert knowledge (or commonsense reasoning) of what is considered as chubby body shape.

This representation is the simplest form of relational inference. Given data of height and weight, the child's body shape (its resemblence to chubbiness) will be determined by the value of membership. This type of inference is mainly used in diagnostic problems. Although it looks simple, this type of decision making can become quite complex when there are other fuzzy associations in the same product space (e.g., body shape slim, normal, overweight, skinny, healthy, etc.). Problems arise when fuzzy sets overlap, creating undecided zones. More on this type of problem is discussed in Chapter 5 under the topic of decision boundary analysis.

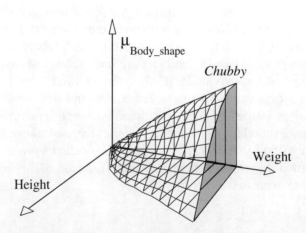

Figure 2.10 An example of relational inference where the associations between the product spaces *Height* and *Weight* are represented by a two-dimensional membership function for the category *Chubby*.

Because the association between the product spaces is represented by a (two-dimensional or higher) membership function, the associations can also be represented by a new fuzzy set. The new fuzzy set is expressed as the collection (union) of singletons defined over the n-dimensional product space $X \times Y \times \ldots \times Z$, given by

$$A^R = \bigcup_{X \times Y \times \ldots \times Z} \mu(x, y, \ldots, z)/(x, y, \ldots, z) \qquad (2.24)$$

where A^R is the fuzzy set of associations. Now, let's try to construct membership functions on the product spaces given by A^R. This process is called the *projection* principle (also called *marginal fuzzy restrictions*) and it resembles the geometrical projection of A^R on product spaces. The projected membership function on a product space X is given by

$$\mu(x) = \vee_\perp [\mu^R(x, y, \ldots, z)] \quad x \in X, \ y \in Y, \ldots, z \in Z \qquad (2.25)$$

which is equivalent to finding the supremum among membership values that correspond to each single point on X. Note that the projection is from the direction normal to the product space. This is denoted by $\perp$ in Eq. (2.25). Considering the ongoing example, the projections of the fuzzy set A^R on the *Height* and *Weight* product spaces are shown in Fig. 2.11.

The second form of building relational inference is to utilize the membership functions given on each product space separately to construct the membership function of the associations. This is the reverse of the *projection* process described above. One of the methods applicable to such a problem is called the *Cartesian product*, which is a pair-wise minima computation in the entire product space. The membership function of the associations is given by the following expression:

$$\mu(x, y, \ldots, z) = \bigwedge_{X \times Y \times \ldots \times Z} [\mu(x), \mu(y), \ldots, \mu(z)] \quad x \in X, y \in Y, \ldots, z \in Z$$

(2.26)

Referring back to the chubbiness example, the Cartesian product that forms the membership function for the associations is shown in Fig. 2.12. In this case, we are given one-dimensional membership functions for each product space separately as shown in Fig. 2.12(a), and we are to form the membership function of the associations. The result shown in Fig 2.12(b) is different from 2.11(b) due to the generalization nature of the pair-wise minima method. Membership functions that do not retain complete information after projection or Cartesian product operations are called interac-

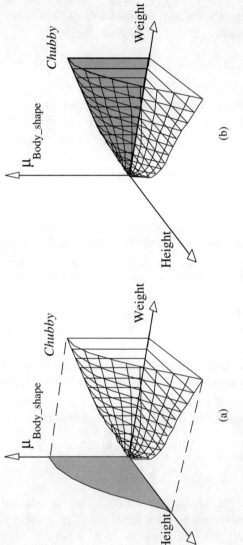

Figure 2.11 Projections (shaded areas) on the product space *Height* (a) and on the product space *Weight* (b).

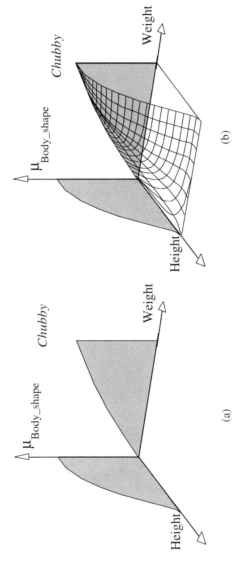

Figure 2.12 Cartesian product rule using the two separate one-dimensional membership functions (a) to form the membership function of the associations (b).

tive. Piece-wise-linear membership functions are noninteractive. When membership functions are interactive, building relational inference based on Cartesian product is not preferable compared with building the association surfaces directly by expert knowledge. In cases where association is unknown, then Cartesian product may be useful. There are also several other mathematical approaches, known as similarity methods, in the literature to form fuzzy relations.

2.10   COMPOSITIONAL INFERENCE

A more complex form of inference is called the *compositional inference,* where the associations among fuzzy sets are of concern in addition to associations among the elements of each fuzzy set. This is symbolically illustrated in Fig. 2.13 where fuzzy associations represent system structure, as an adaptive clustering algorithm would discover or as an expert would articulate linguistically [16].

Compositional inference provides a medium for defining linguistic variables (fuzzy variables) that leads to the formation of fuzzy logic. Consider a domain of fuzzy sets as shown in Fig. 2.14. Assume that the fuzzy sets in this domain belong to a variable called *investment risk* that is characterized by the product space of interest rates and depreciation.

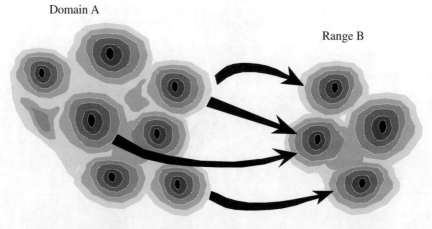

Figure 2.13 Compositional inference illustrated symbolically by a mapping-like association between a domain of fuzzy sets (input space) and a range of fuzzy sets (output space).

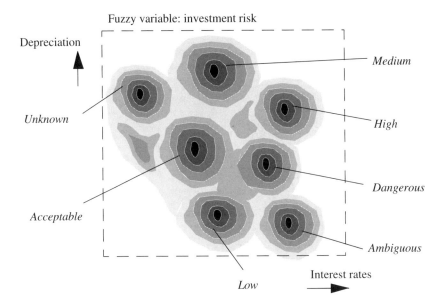

Figure 2.14 Definition of a linguistic variable (fuzzy variable).

In this example, each fuzzy set indicates a different risk region labeled as *Low, Medium, Dangerous, Acceptable, High,* and *Unknown*. Obviously, the formation of fuzzy sets (membership functions) reflects a context-specific knowledge (either acquired from an expert or from a data set). Now consider fuzzy associations between two fuzzy variables as shown in Fig. 2.15.

As it can be seen in Fig. 2.15, compositional inference is the essence of translating natural language into mathematics, and converting the accompanying logic into mathematical inference computations. The linguistic equivalence of one of the associations in Fig. 2.15 is *Don't invest our money when risk is dangerously high*. As we shall see later, compositional inference gets more complicated when there are multiple fuzzy antecedents (variables in the input space) related to multiple fuzzy consequents (variables in the output space).

Some of the important operations in compositional inference are the operations between fuzzy associations. Consider three fuzzy sets as symbolically shown in Fig. 2.16. In this example, we assume that the association *a* indicates the relationship between lion population (predator) and hyena population (scavenger), whereas *b* indicates the relationship be-

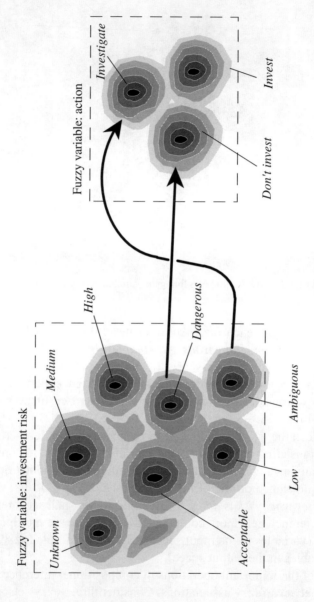

Figure 2.15 Compositional inference between two linguistic (fuzzy) variables.

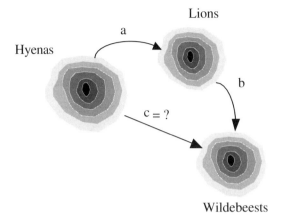

Figure 2.16 A fundamental compositional problem: given the fuzzy associations a and b, what is the association c?

tween the wildebeest (large prey) population and the lion population. The question under investigation is c, the relationship between the scavengers and large prey. In a more generalized format, the question is the similarity or elementhood of one fuzzy set to another where only their associations to an intermediate fuzzy set are known. Composition operators that can handle this type of problem are discussed next.

Max–Min Composition

The max-min composition applied to two associations $a(h,l)$ and $b(l,w)$ defined over Cartesian products $H \times L$ and $L \times W$ yields a new association c defined over the product space $H \times W$. The new association, which is also a fuzzy set, is expressed by

$$c = \bigcup_{H \times W} \vee_l [\mu_a(h,l) \wedge \mu_b(l,w)]/(h,w) \quad h \in H, l \in L, w \in W$$

(2.27)

where symbols $\wedge$ and $\vee$ correspond to min and max operations of fuzzy sets, respectively. Note that the singleton in Eq. (2.27) indicates the membership function of the new association produced by max-min composition.

$$\mu_c(h,w) = \vee_l [\mu_a(h,l) \wedge \mu_b(l,w)] \quad \in H, l \in L, w \in W \quad (2.28)$$

The subscript "*l*" under the max operator sign (∨) indicates the common product space L. The composition operations are very similar to multiplication in matrix algebra (where the first operator is the summation of elements and the second operator is the multiplication). Accordingly, max-min operation (denoted by ∘) can be represented in a matrix format

$$\begin{bmatrix} a_{1,1} & a_{1,2} \\ a_{2,1} & a_{2,2} \end{bmatrix} \circ \begin{bmatrix} b_{1,1} & b_{1,2} \\ b_{2,1} & b_{2,2} \end{bmatrix} = \begin{bmatrix} [a_{1,1} \wedge b_{1,1}] \vee [a_{1,2} \wedge b_{2,1}] & [a_{1,1} \wedge b_{1,2}] \vee [a_{1,2} \wedge b_{2,2}] \\ [a_{2,1} \wedge b_{1,1}] \vee [a_{2,2} \wedge b_{2,1}] & [a_{2,1} \wedge b_{1,2}] \vee [a_{2,2} \wedge b_{2,2}] \end{bmatrix}$$

$H \times L \qquad L \times W \qquad\qquad\qquad\qquad H \times W$

(2.29)

where only four elements per association are represented for simplicity. Example 3.6 in Chapter 3 illustrates this computation in more detail.

Max-* Composition

This type of composition, also called *max-star composition,* is a generalized formula for maximization operations where * can be replaced by multiplication, summation, average, or some other binary operation. Whatever the second operand is, the membership function of the new association is expressed as in Eq. (2.29) and the computation is done in the same manner as in Eq. (2.30).

$$\mu_{A_3}(h,w) = \vee_l [\mu_{A_1}(h,l) * \mu_{A_2}(l,w)] \quad \in H, l \in L, w \in W \qquad (2.30)$$

2.11 LINGUISTIC VARIABLES AND LOGIC OPERATORS

As mentioned earlier, compositional inference allows heuristic interpretation (articulated in linguistic form) to be part of mathematical representation, which leads to the linguistic/fuzzy variable concept. Fuzzy variables are the basic building blocks of compositional inference. Because they directly connect numbers to words, the compositional inference in fuzzy logic is unique among all other mathematical methods. A fuzzy variable has the following hierarchical information structure:

- Fuzzy variable name
- Predicates linguistically identifying different regions of the universe of discourse

- Membership function for each fuzzy set labeled by one predicate
- Universe of discourse

Fuzzy variables in this book are denoted by X and the predicates (also called *fuzzy values*) are denoted by P. In cases where more than one fuzzy variable is required to describe an event, the predicates that belong (by design) to a particular fuzzy variable are denoted by subscripts. Accordingly, the N number of predicates of the jth fuzzy variable are denoted by $P_{j,i}$ where $i = 1,..., N$. This representation leads to the definition of fuzzy statements (also known as *fuzzy propositions*) that are the basic linguistic elements of rule description. Accordingly, the kth fuzzy statement S_k will be expressed as

$$S_k \equiv X_j \bullet P_{j,i} \tag{2.31}$$

where the connector (denoted by $\bullet$) corresponds to the linguistic premise IS. Similarly, this statement can be made negative using another connector (denoted by $\neg$) that corresponds to IS NOT. The index k is not related to j. The reader must be careful not to confuse these connectors with the logic operators (such as AND, and OR denoted by Θ) described in the next section, or with the composition operators (such as max-min often denoted by $\circ$) described in the previous sections. Also note that there are two more basic linguistic connectors in the classical computer science terminology as shown in Table 2.2.

Although the greater than and less than operations have not been widely studied in the context of fuzzy theory, we present their equivalence in fuzzy systems later in the book because of their importance in compositional inference. This discussion is in Section 5.4.2 of Chapter 5.

Table 2.2 Linguistic Connectors in Classical and Fuzzy Logic Inference

CLASSICAL THEORY		FUZZY THEORY	
Equal to	=	IS	$\bullet$
Not Equal to	≠	IS NOT	$\neg$
Less than	<		
Greater than	>		

EXAMPLE 2.3 LINGUISTIC VARIABLE CONCEPT

Consider a typical case of categorizing body temperature. Either by common sense or by an expert's opinion, it is possible to categorize different regions of the temperature universe using predicates such as *Hypothermia, Normal, Fever,* and *High Fever*.

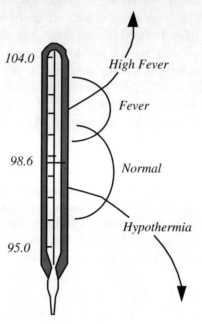

The distributions (the shape of the membership functions) shown above are determined according to the underlying general knowledge about body temperature. Membership functions can be designed and labeled in any way as long as it serves the intended purpose.

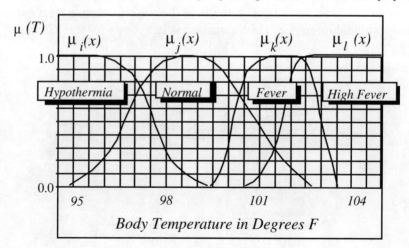

Considering the body temperature example given here, the first fuzzy statement might be *Temperature is Normal* where *Temperature* is a fuzzy variable (*X*) and *Normal* is one of its predicates (*P*) represented by the membership function $\mu_j(x)$ either in continuous functional form or by fuzzy sets.

Classical logic operators (logic gates) are useful to combine fuzzy statements to form a conditional statement or rule. Three of the most commonly used operators are AND, OR, and THEN. In fuzzy systems theory, the AND operation is represented by minimum operation (intersection) denoted by ∧, and an OR operator is represented by maximum operation (union) denoted by ∨. The THEN operator performs a mapping-like function, which is modeled by the implication process explained in the next section. By using such operators along with fuzzy

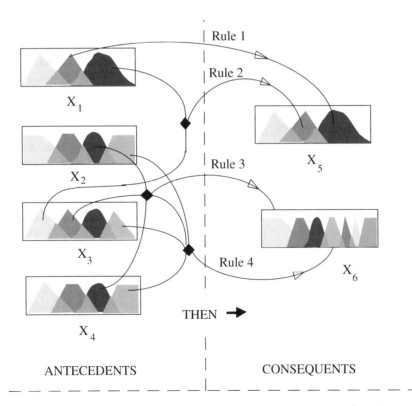

Figure 2.17 An illustration of the compositional inference process for a hypothetical case (with arbitrarily selected numbers of variables, predicates, and rules).

statements, it becomes possible to make an IF-THEN type inference model in a form similar to that shown in Fig. 2.17.

In Fig. 2.17, the number of fuzzy variables (X), predicates (membership functions), and rules are selected arbitrarily for illustration purposes. The connecting points (marked with small tetragons) among associations (lines) indicate logical operators such as AND and OR. If viewed as a mapping process, the compositional inference depicted in Fig. 2.17 can be broken into input and output spaces with antecedent and consequent fuzzy variables. When an IF-THEN rule format is used, the center line corresponds to the THEN operator where the mapping operation is symbolized.

2.12 INFERENCE USING FUZZY VARIABLES

The basic principles of inference in fuzzy logic are the adaptations of classical principles to the fuzzy domain. Now that we have formally introduced the linguistic/fuzzy variable concept, we can examine the properties of relational and compositional inference of fuzzy sets in view of classical inference methods. The methods of inference listed below are accompanied by "on-the-spot" linguistic examples to familiarize the reader with the knowledge acquisition process described in Chapter 4. An important part of the knowledge acquisition process consists of translating daily language into fuzzy rules in a format suitable for inference engine development. The linguistic examples are subject to interpretation and they do not constitute unique answers.

Entailment Principle

Consider two membership functions *Warm* and *Nice* where the definition of *Warm* is included in the definition of *Nice*. In terms of fuzzy sets, *Warm* is a subset of a larger set *Nice*. Then the entailment principle says that

$$\frac{\text{Weather is Nice}}{\text{Nice is (means) Warm}} \quad \frac{X \bullet P}{P \subset A}$$
$$\text{Weather is Warm} \quad X \bullet A$$

where the conclusion is stated under the horizontal line. In this example, the entailment denoted by $\subset$ can be modeled using linguistic terms such as *means* or *includes*.

Conjunctive Inference

Consider two membership functions *Warm* and *Cold* where they belong to different fuzzy variables (different universes). Suppose that the first fuzzy variable is the temperature of liquid A, and the second fuzzy variable is the temperature of liquid B. The conjunctive inference (also known as *Cartesian product*) principle states that

Temperature of Liquid A is Warm	$X \bullet W$
Temperature of Liquid B is Cold	$Y \bullet C$
Temperature of Liquid A and B is between Warm and Cold.	$(X,Y) \bullet W \times C$

Mixing liquids is a good example to visualize this principle. However, what is mixed above is the possibilities, not the liquids in the physical world. The membership function of the relation $W \times C$ is given by

$$\mu_{W \times C}(w,c) = \mu_W(w) \wedge \mu_C(c) \quad w \in W, c \in C \qquad (2.32)$$

In the accompanying example above, the term *between* is figuratively used, and it does not refer to geometric positioning. Suppose the same two membership functions *Warm* and *Cold* are defined over the same universe. In this case, we have one variable. The conjunctive inference is then as follows:

Temperature of Liquid A is Warm	$X \bullet W$
Temperature of Liquid A is also Cold	$X \bullet C$
Temperature of Liquid A is in between Warm and Cold	$X \bullet (W \cap C)$

The membership function of this relation is given by

$$\mu_{W \cap C}(t) = \mu_W(t) \wedge \mu_C(t) \quad t \in T \qquad (2.33)$$

where T is the universe of discourse.

Projection

Now consider an association called *Body_shape* and a membership function *Chubby*. Two fuzzy variables *Weight* and *Height* are assumed to determine the association *Body_shape*.

Body_shape is Chubby	$(W,H) \bullet C$
Weight must be Excessive	$W \bullet E$

where E (called *Excessive* in the example) is the projection defined by

$$\mu E(w) = \bigvee_{W \times H} [\mu_C(w,h)] \quad w \in W, h \in H \quad (2.34)$$

and the maximum (supremum for continuous valued functions) is taken over $w \in W$.

Extension Principle

The extension principle provides a mechanism to compute induced constraints. Consider a fuzzy variable called *Temperature* with a membership function *High*. Suppose another variable *Pressure* is a function of *Temperature*, $P = f(T)$. The extension principle states that

Temperature is High	$T \bullet H$
Pressure is affected by High Temperature	$f(T) \bullet f(H)$

with a membership function of $f(H)$ given by

$$\mu_{f(H)}(p) = \sup_t [\mu_H(t)] \quad p \in P, t \in T \quad (2.35)$$

subject to the condition $p = f(t)$. Note that an inference such as *Pressure is High* cannot be readily drawn because the function f can be anything. Thus, the linguistic equivalence *affected by* is used in the example above to indicate this fact.

Compositional Rule of Inference

In Section 2.10, we introduced an example for compositional inference using fuzzy sets and associations. Now that we have linguistic variables, the compositional inference can be restated in a more expressive manner.

Temperature is Dangerous	$T \bullet D$
Pressure-Temperature relation is (indicates) Unsafe_operation	$(T,P) \bullet U$
Pressure is composed of Dangerous and Unsafe characteristics	$P \bullet (U \circ D)$

where $(U \circ D)$, the composition of the binary relation U and unary relation D is defined by

$$\mu_{U \circ D}(u) = \vee_u [\mu_U(u) \wedge \mu_D(t)] \quad t \in T, u \in U \quad (2.36)$$

Again, the linguistic terms *composed of* and *indicates* are used to maintain the linguistic integrity. Note that the membership function in Eq. (2.36) is the max-min composition operation described previously.

Generalized Modus Ponens

Generalized *modus ponens* (GMP) is one of the most suitable inference methods in articulating expert knowledge in terms of IF-THEN rules. Consider two fuzzy variables: A (*Air-temperature*) and H (*Heater switch*). Suppose that Air-temperature takes linguistic values of *Low* and *Warm*, both defined by fuzzy sets on the temperature universe. Similarly, the fuzzy variable *Heater switch* takes linguistic values, one of which is called *Crank up* defined by a fuzzy set on the universe of heater switch values.

If Air-temperature is Low Then Crank up the Heater	$A \bullet L \rightarrow H \bullet C$
Air-temperature is Warm now	$A \bullet W$
Then Heater must be adjusted accordingly.	$\rightarrow H \bullet \phi(C,W,L)$

The linguistic statement *must be adjusted accordingly* describes what is known as the implication process denoted by $\phi(C,W,L)$ or more formally by $\phi[\mu_C(s), \mu_W(t), \mu_L(t)]$ where $t \in T$, and $s \in S$. The membership function of the implication relation is not unique; different authors have suggested different models in the literature. These options are discussed in the next section.

Generalized Modus Tollens

Generalized *modus tollens* (GMT) is the reverse of GMP in the sense that the objective is to find inputs (reasons) causing the known action. Using the same example shown above, GMT can be stated as follows:

When Air-temperature is Low Heater is Cranked-up	$A \bullet L \rightarrow H \bullet C$
Heater is running Half-way	$H \bullet Hw$
Air-temperature must be inferred accordingly	$A \bullet \phi(C,L,Hw)$

The linguistic value (membership function) called *Half-way* also resides on the universe of heater switch values. In comparison to GMP, the linguistic term *must be inferred accordingly* describes an implication process in the reverse direction. When the number of antecedents and consequents is more than one—that is, analogous to many-to-one (or many) mapping—the GMT inferencing scheme cannot be applied casually due to the uniqueness problem. In other words, there may be more than one combination of inputs yielding the specified output. In such a case, there will be no basis to judge which combination of inputs is the answer. Note that more than one answer may be acceptable in certain control applications whereas multiple answers are often not acceptable in diagnostics applications.

Contraposition

This type of inference deals with the truth of negatives or exclusions.

| If Temperature is Mild Then Weather is Nice | $T \bullet M \rightarrow W \bullet N$ |
Temperature is NOT Mild	$T \neg M$
Then Weather is NOT Nice	$\rightarrow W \neg N$

Here, the negation NOT is denoted by the symbol ($\neg$), and the relationship is expressed by the complement of the corresponding membership function.

2.13 FUZZY IMPLICATION

Fuzzy implication $P \rightarrow Q$ (P implies Q) is a mechanism for GMP inference applicable to a chain of conditional propositions. Because conditional reasoning can also be applied to unconditional cases (by relaxing the constraints), the implication process is widely applicable to many problems. The implicational inference is the main focus in this book. Now, consider the following propositions.

Unconditional

I have lots of money, I will buy a nice car.

Conditional

If I have lots of money then I will buy a nice car.

When I have lots of money, I will buy a nice car.

In unconditional reasoning, the relationship of interest is the association between the composed elements (fuzzy sets). The association in this example is between the two events: having lots of money and buying a nice car. Regardless of its strength (weak, none, or strong), the association stands by itself as the entity of interest and a vehicle for reasoning. In the conditional case however, the first event constitutes a requirement for the second event to occur. In other words, conditional statements impose a condition (degrees of fulfillment) on the first event that determines the strength of action (degrees of commensurateness) on the second event.

Deductive reasoning[5] represents a case where conditionality holds regardless either in an implicit or explicit manner. As illustrated in the example below, implicitly conditional propositions often expressed using terms such as *must be, therefore,* and *because of* indicate the existence of a condition or association that leads to a conclusion.

Deductive, Implicitly Conditional	**Deductive, Explicitly Conditional**
I have lots of money, I must be rich	If I have lots of money then I must be rich

Just like physical objects with equal mass may have different centers of gravity depending on geometry, propositions using similar linguistic terms can have different levels of conditionality depending on the context-related meaning. The fuzzy implication process, however, is suitable for a variety of linguistic structures including conditional, deductive, and inductive[6] reasoning.

A membership function that defines the implication relation can be expressed in a number of ways, as suggested by different authors in the literature. The main question often asked by fuzzy system designers is which one to use under which conditions. This issue is discussed in Chapter 8 along with all other design-related discussions. Now, we will list those useful implication operators by assuming that we have the following simple conditional proposition (canonical rule):

$$If\ X\ is\ A\ Then\ Y\ is\ B \qquad X \bullet A \rightarrow Y \bullet B \qquad (2.37)$$

The implication relation is defined by

$$R(x,y) = \bigcup_{x,y} \mu(x,y)/(x,y) \qquad (2.38)$$

---

[5] Deduces the validity of a logical proposition under different circumstances.

[6] Generalizes the validity of a logical proposition from specific examples.

where linguistic/fuzzy variables X and Y take the values of A and B, respectively, and $\mu(x,y)$ is the membership function of the implication relation. The membership function of interest is denoted by

$$\mu(x,y) = \phi[\mu_A(x),\mu_B(y)] \quad x \in X, \ y \in Y \tag{2.39}$$

The most widely used implication operators are given in Table 2.3. The symbols $\wedge$ and $\vee$ correspond to intersection and union operations, respectively.

Table 2.3 Fuzzy Implication Operators

Implication Operators	$\phi[\mu_A(x),\mu_B(y)] =$
Mamdani	$\mu_A(x) \wedge \mu_B(y)$
Larsen	$\mu_A(x) \cdot \mu_B(y)$
Lukasiewicz	$1 \wedge (1 - \mu_A(x) + \mu_B(y))$
Kleen-Dienes	$(1 - \mu_A(x)) \vee \mu_B(y)$
Bounded Product	$0 \vee (\mu_A(x) + \mu_B(y) - 1)$
Zadeh	$(\mu_A(x) \wedge \mu_B(y)) \vee (1 - \mu_A(x))$
Standard	$1 \quad \text{if} \quad \mu_A(x) \leq \mu_B(y)$ $0 \quad \text{if} \quad \mu_B(x) > \mu_A(y)$
Drastic Product	$\mu_A(x) \quad \text{if} \ \mu_B(y) = 1$ $\mu_B(x) \quad \text{if} \ \mu_A(y) = 1$ $0 \qquad \text{if} \ \mu_A(y) < 1 \ \mu_B(y) < 1$
Gougen	$1 \qquad \text{if} \quad \mu_A(x) \leq \mu_B(y)$ $\dfrac{\mu_B(y)}{\mu_A(x)} \quad \text{if} \quad \mu_B(x) > \mu_A(y)$
Godelian	$1 \qquad \text{if} \quad \mu_A(x) \leq \mu_B(y)$ $\mu_B(y) \quad \text{if} \quad \mu_B(x) > \mu_A(y)$

2.14 FUZZY SYSTEMS AND ALGORITHMS

A fuzzy system is defined as a system (mechanical, electrical, etc.) with operating principles based on fuzzy information processing and decision making. Designing a fuzzy system refers to developing mechanisms for fuzzy information processing and decision making within a digital platform and a soft computing environment. The theoretical elements introduced so far constitute the main building blocks of such a development. An organized collection of the theoretical elements implemented in a computer program is also known as *fuzzy algorithms*. In general, a fuzzy algorithm can employ relational, compositional, or implicational inference methods. However, the implicational inference in the form of conditional IF-THEN rules includes theoretical features with all levels of complexity (including relational and compositional inference). Accordingly, in the next chapter, we have introduced *the basic fuzzy inference algorithm* that implements fuzzy IF-THEN rules in dealing with a large domain of practical problems such as control, classification, pattern recognition, diagnostics, modeling, prediction, forecasting, and general decision making. An important question to consider before examining the basic fuzzy inference algorithm is whether there is any theory that would systematically explain how to integrate theoretical tools for a given objective. Some of the useful theoretical aspects are discussed in this section to serve this purpose. Nevertheless, a complete theory is not yet available.

There are three basic domains of information in a fuzzy algorithm, as shown in Fig. 2.18: input data, output data, and design data. Input and output data are dynamically updated whereas design data is permanently stored in memory. Design data embodies the knowledge of a specific solution. If we denote the input, output, and design domain data by $I(t)$, $O(t)$, and D respectively, then a typical fuzzy algorithm can be expressed in its most generalized form by

$$O(t) = \psi[D, I(t)] \qquad (2.40)$$

where ψ is an algorithmic function of the fuzzy system. The independent variable t indicates an instance (or step) for output generation, and it is not necessarily the clock time. The design data D consists of IF-THEN rules composed of fuzzy statements ($S \equiv X \bullet P$) such as *Temperature IS High*, logic operators (Θ) such as AND, OR, and the implication operator ($\rightarrow$) THEN. As a result, a collection of rules forming a fuzzy algorithm is expressed by

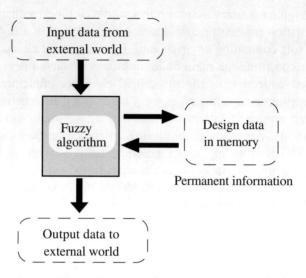

Figure 2.18 The information structure of a fuzzy algorithm.

$$\begin{array}{c} S_{1,1} \, \Theta_{1,1} \, S_{1,2} \, \Theta_{1,2} \ldots S_{1,i} \rightarrow S_{1C} \\ S_{2,1} \, \Theta_{2,1} \, S_{2,2} \, \Theta_{2,2} \ldots S_{2,j} \rightarrow S_{2C} \\ \ldots\ldots \\ S_{n,1} \, \Theta_{n,1} \, S_{n,2} \, \Theta_{n,2} \ldots S_{n,k} \rightarrow S_{nC}. \end{array} \qquad (2.41)$$

Every row corresponds to one rule and there are n rules. Subscripts i, j, and k indicate that the number of fuzzy statements and logic operators before the THEN operator (consequents) can vary in an arbitrary manner depending on the specific solution they represent. The subscript C which appears after the THEN operator indicates that the statements are consequents or action variables. Equation (2.40) represents domain data (D) residing in memory in an algorithmic format.

To illustrate the input data absorption and output data generation process, we need to rewrite Eq. (2.41) in a more explicit manner. Here, we have expanded the statements using the $S \equiv X \bullet P$ notation where X is a linguistic variable and P is one of its predicates (membership functions):

$$[X_x \bullet P_{x,y}]_{1,1} \Theta_{1,1} [X_x \bullet P_{x,y}]_{1,2} \Theta_{1,2} \ldots [X_x \bullet P_{x,y}]_{1,i} \to [X_x \bullet P_{x,y}]_{1C}$$
$$[X_x \bullet P_{x,y}]_{2,1} \Theta_{2,1} [X_x \bullet P_{x,y}]_{2,2} \Theta_{2,2} \ldots [X_x \bullet P_{x,y}]_{2,j} \to [X_x \bullet P_{x,y}]_{2C}$$
$$\ldots\ldots$$
$$[X_x \bullet P_{x,y}]_{n,1} \Theta_{n,1} [X_x \bullet P_{x,y}]_{n,2} \Theta_{n,2} \ldots [X_x \bullet P_{x,y}]_{n,k} \to [X_x \bullet P_{x,y}]_{nC}$$
(2.42)

The subscripts x and y indicate that variables and predicates can be any selection from a list residing in memory. In other words, what goes inside the parentheses represents a context-related solution and it can be in any order. The same thing applies to logical operators denoted by (Θ). To illustrate the input absorption (evaluation) process that is part of $\psi[D, I(t)]$ in Eq. (2.40), we first look at the statements before THEN operator, often called the *left-hand-side* (*LHS*) *statements*.

$$[X_x \bullet (P_{x,y} \circ I(t)_x)]_{1,1} \Theta_{1,1} [X_x \bullet (P_{x,y} \circ I(t)_x)]_{1,2} \Theta_{1,2} \ldots$$
$$[X_x \bullet (P_{x,y} \circ I(t)_x)]_{1,i} \to$$
$$[X_x \bullet (P_{x,y} \circ I(t)_x)]_{2,1} \Theta_{2,1} [X_x \bullet (P_{x,y} \circ I(t)_x)]_{2,2} \Theta_{2,2} \ldots$$
$$[X_x \bullet (P_{x,y} \circ I(t)_x)]_{2,j} \to$$
$$\ldots$$
$$[X_x \bullet (P_{x,y} \circ I(t)_x)]_{n,1} \Theta_{n,1} [X_x \bullet (P_{x,y} \circ I(t)_x)]_{n,2} \Theta_{n,2} \ldots$$
$$[X_x \bullet (P_{x,y} \circ I(t)_x)]_{n,k} \to$$
(2.43)

Instantaneous inputs $I(t)$ are evaluated on each membership function space denoted by $(P_{x,y} \circ I(t)_x)$ where the symbol ∘ indicates a composition operator such as max-min. Note that all inputs in Eq. (2.43) have the subscript x in accordance with the linguistic variable to which it belongs. In other words, when X_x is *Temperature* then $I(t)_x$ is a temperature input; when X_x is *Flow* then $I(t)_x$ is a flow input, and so forth. The LHS computations after input absorption produce values for each rule called the degrees of fulfillment $r(t)$.

The other operations symbolized by $O(t) = \psi[D,I(t)]$ consist of using the degrees of fulfillment values to produce outputs through the implication process. We will express this process by

$$\to [X_x \bullet \phi(P_{x,y}, r_1(t))]_{1C} \equiv O_1(t)$$
$$\to [X_x \bullet \phi(P_{x,y}, r_2(t))]_{2C} \equiv O_2(t)$$
$$\ldots\ldots$$
$$\to [X_x \bullet \phi(P_{x,y}, r_n(t))]_{nC} \equiv O_n(t)$$
(2.44)

where there are n degrees of fulfillment (one for each rule) for each evaluation step t. The output from each rule, which is a fuzzy set, is obtained through the implication operation denoted by ϕ between the membership function (labeled by predicate P) and the corresponding $r(t)$ value. The actual computations are based on a geometrical interpretation of the implication operators listed in Table 2.3. The working principles of this process are explained in Chapter 3 in more detail. This process becomes simpler for canonical rules (of the form *IF A THEN B*) with single statements on both sides of the THEN operator. Now we will discuss some theoretical issues related to output processing.

2.15 DEFUZZIFICATION

Aggregating two or more fuzzy output sets (or membership functions) yields a new fuzzy set (or a new membership function) in the basic fuzzy inference algorithm. In most cases, a result in the form of a fuzzy set is converted into a crisp result by the *defuzzification* process. Defuzzification is especially necessary for hardware applications, because conventional systems operate based on crisp data exchange. However, selecting a single value from a fuzzy set that represents the information contained within a fuzzy set is a challenging task. Among several methods suggested in the literature, the most widely used methods are listed in Table 2.4. Note that these methods definitely produce different results as the outcome of the inference computations. Unfortunately, there is no theory to justify their behavior other than commonsense reasoning such that the defuzzified output must represent a weighted, voted, or most suitable solution.

Table 2.4 Defuzzification Methods

CENTROID METHODS:
Center of Gravity
Center of Weights
Center of Largest Area
Center of Mass of Highest Intersected Region
MAXIMA METHODS:
Mean of Maximums
Maximum Possibility
Left–Right Maxima

As illustrated in Table 2.4, there are two basic mechanisms: centroid and maxima. The centroid methods are based on finding a balance point of a property that can be the total geometric figure, the weight (area) of each fuzzy set, the area of the largest fuzzy set, or the area of the highest intersection. The maximum possibility method searches for the highest peak whereas the left-right maxima method searches for the peaks in a selected direction. The mean of maxima method may also be considered as one of the centroid techniques since mean and center practically refer to the same property. These techniques are also discussed in Chapter 3, within the domain of the basic fuzzy inference algorithm, and in Chapter 8 (including their design properties). Their properties are summarized as follows [18].

The center of gravity method is calculated by

$$x' = \frac{\int \mu(x) x \, dx}{\int \mu(x) \, dx} \qquad (2.45)$$

where $\mu(x)$ is the output fuzzy set after the aggregation of individual implication results. A practical and approximate way to solve Eq. (2.45) is given by

$$x' = \frac{\sum_{i=1}^{N} x_i \mu_o(x)}{\sum_{i=1}^{N} \mu_o(x)} \qquad (2.46)$$

which consists of dividing the weighted possibilities by total possibilities. Approximation embedded in Eq. (2.46) is fully justified when the membership function is defined as a fuzzy set comprising singletons. Otherwise, (that is, when a continuous membership function is used), the approximation is determined by the number of summations (i.e., the sampling frequency of the continuous function). The membership function $\mu_o(x)$ represents the fuzzy set of the final output (either aggregated or not) of one linguistic/fuzzy variable, and x is the location of each singleton on the universe of discourse. Suppose that a linguistic variable X has three predicates, which appear on the right-hand side of three rules. Also assume that the membership functions corresponding to the predicates were subject to an implication process where the three outputs before aggregation are shown in Fig. 2.19.

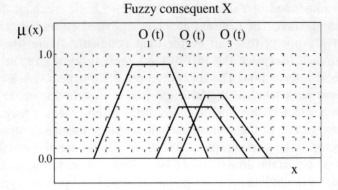

Figure 2.19 Three hypothetical implication results that belong to consequent X.

The result of the center of gravity method is shown in Fig. 2.20. The center of weights (also known as *composite moments*) method is expressed as

$$x' = \frac{\sum_{i=1}^{N} w_i \, C_i \, A_i}{\sum_{i=1}^{N} w_i \, A_i} \qquad (2.47)$$

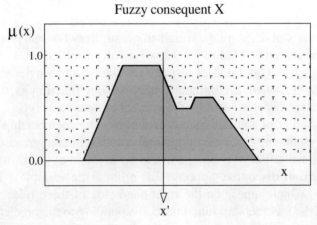

Figure 2.20 Union aggregation and center-of-gravity method for defuzzification.

where w, C, and A are the rule importance weight, center of gravity, and area of each individual implication result, respectively. Because each implication result is used, this method eliminates the aggregation process. Note that rule importance weights are not the reason for calling this technique *center of weights method;* rather, the term *weight* refers to the weight of each implication result (area A) in a balancing beam analogy.

The center of the largest area method evaluates each individual implication result, finds the one with the largest area, then computes its center of gravity by Eq. (2.46). For symmetric consequent membership functions, the center of gravity of each output fuzzy set can be stored in memory that eliminates the computations. However, this approach will produce discrete output values often undesirable in practice. Thus, this method requires asymmetric membership functions for continuous output. The center of the highest intersected region is a method that uses Eq. (2.45) or (2.46) on the area produced by intersection aggregation of the output fuzzy sets obtained from individual rules.

The maxima methods are based on the maximum possibility point on the universe of discourse as the answer to defuzzification.

$$x' = x_i \quad \mu_o(x_i) > \mu_o(x_j) \, j = 1,..N \quad j \neq i \qquad (2.48)$$

If the maximum point is nonsingular (or a plateau), then the defuzzified output is the average of maximums or the center of maximums. The right-left maxima method selectively searches for the maximum possibility point either at the far left or far right on the universe of discourse.

2.16 ADAPTIVE FUZZY SYSTEMS AND ALGORITHMS

An adaptive system by definition has the ability to adjust its working structure in accordance with changes in the external world. This broad definition is subject to interpretation because there is no theory to draw a clear line between what is a structural element and what is a variable element. In control systems, for example, controller gain is considered as a structural element fixed by design and tuning. When tuning is automated by utilizing an on-line measurement, the system is considered adaptive. A more sophisticated adaptive controller has the ability to change its control law according to the changing process conditions [19]. At the top of the sophistication pyramid, a true adaptive system not only changes its working principles according to varying circumstances, but also learns from the experience and anticipates future events. In application to fuzzy systems, the definition of *adaptive* is as ambiguous as in any other field. Consider the generalized representation of a fuzzy algorithm given by Eq. (2.49).

$$O(t) = \psi[D,I(t)] \tag{2.49}$$

The structural element is D, where design information is stored in the memory permanently. All of the elements of D may be considered nonstructural (variable) and adaptable at the extreme levels of complexity. More methodically, let's first consider a simple form where some part of D is adaptable.

$$O(t) = \psi[D,D_a(t),I(t)] \tag{2.50}$$

The adaptable part of D is indicated by subscript a. The objective of adaptive systems in general is to improve system performance to meet specified objectives through adaptation that would not be met otherwise. The mechanism of changing $D_a(t)$ using external input $I_a(t)$ is done by a new set of adaptive IF-THEN rules or by means of another fuzzy/non-fuzzy method.

$$D_a(t) = \psi_a[D',I_a(t)] \tag{2.51}$$

Described by Eq. (2.51), the adaptive inference engine computes the necessary changes $D_a(t)$ by using an input set $I_a(t)$ and a new design data D' that includes adaptive rules. Note that the input data set $I_a(t)$ need not be the same as $I(t)$, and it can also include some error criteria generated by the difference between the desired and actual system response. An adaptive fuzzy system is shown in Fig. 2.21. The simplest form of an adaptive fuzzy system is the one in which the original rule structure is modified by changing the shapes of the membership functions [20]. Referring back to Eq. (2.50), the membership functions labeled by predicates are updated at each initiation step t. The left-hand side of the rules are then given by

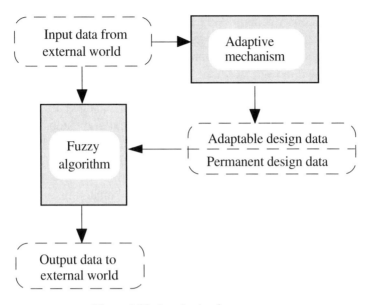

Figure 2.21 An adaptive fuzzy system

$$[X_x \bullet (P_{x,y}(t) \circ I(t)_x)]_{1,1} \Theta_{1,1} [X_x \bullet (P_{x,y}(t) \circ I(t)_x)]_{1,2} \Theta_{1,2} \cdots$$
$$[X_x \bullet (P_{x,y}(t) \circ I(t)_x)]_{1,i} \rightarrow$$
$$[X_x \bullet (P_{x,y}(t) \circ I(t)_x)]_{2,1} \Theta_{2,1} [X_x \bullet (P_{x,y}(t) \circ I(t)_x)]_{2,2} \Theta_{2,2} \cdots$$
$$[X_x \bullet (P_{x,y}(t) \circ I(t)_x)]_{2,j} \rightarrow$$
$$\cdots$$
$$[X_x \bullet (P_{x,y}(t) \circ I(t)_x)]_{n,1} \Theta_{n,1} [X_x \bullet (P_{x,y}(t) \circ I(t)_x)]_{n,2} \Theta_{n,2} \cdots$$
$$[X_x \bullet (P_{x,y}(t) \circ I(t)_x)]_{n,k} \rightarrow$$

(2.52)

where $P(t)$ indicates the adaptive feature. Similarly, the right-hand side of the rules are expressed as follows.

$$\rightarrow [X_x \bullet \phi(P_{x,y}(t), r_1(t))]_{1C} \equiv O_1(t)$$
$$\rightarrow [X_x \bullet \phi(P_{x,y}(t), r_2(t))]_{2C} \equiv O_2(t)$$
$$\cdots$$
$$\rightarrow [X_x \bullet \phi(P_{x,y}(t), r_n(t))]_{nC} \equiv O_n(t)$$

(2.53)

Note that these equations describe the main fuzzy inference engine. The adaptive rules or mechanism may be described by Eqs. (2.52) and (2.53) such that the consequents modify the adaptive membership functions of the main algorithm. One simple way of updating membership functions is to change the coordinates (if membership functions are defined by breakpoints) as shown in Fig. 2.22.

In this hypothetical drawing, the coordinates of the membership function labeled by predicate *high* have been changed by an external adaptive mechanism for the three consecutive initiation steps. More complicated forms of adaptation may include updating the number of membership functions, creating new fuzzy variables, employing more than one implication operator, and changing the defuzzification methods, all based on some adaptive principle evaluating the changing circumstances. The essential components and working principles of the basic fuzzy inference algorithm presented in the next chapter will help the reader understand these issues better. After the next chapter, more on adaptive fuzzy algorithms can be found in Section 7.5.6 of Chapter 7.

The most complicated form of an adaptive fuzzy inference engine is the one where new rules are discovered by examining the external world (see Fig. 2.23). In an automated manner, such an adaptive system relies on data (often numerical). When on-line data retrieval and decision making are automated in a fast manner, there is no time for knowledge acquisition and rule composition by human expertise. In a relatively slower pace, an interactive version of the algorithm would allow human inter-

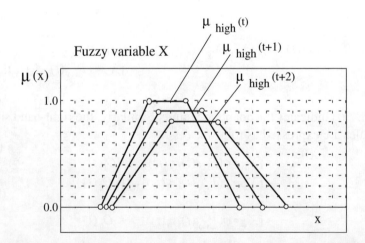

Figure 2.22 A hypothetical adaptive fuzzy algorithm with a membership function of varying coordinates determined by an external adaptive mechanism.

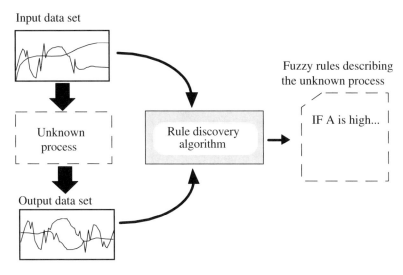

Fig. 2.23 A rule discovery algorithm generating fuzzy rules using input/output data of an unknown process.

vention and control over new rule generation. A rule discovery algorithm is analogous to artificial neural networks and supervised learning algorithms.

In general, there is no piecemeal strategy for composing adaptive rules that would work for a class of problems. Every adaptive system is uniquely built, which requires context-dependent adaptive solutions to be invented by the designer. The only exception is the rule discovery algorithm where an adaptive process invents its own rules. However, this process suffers from several problems as its theory has not been fully developed yet.

2.17 EXPERT SYSTEMS VERSUS FUZZY INFERENCE ENGINES

Expert systems and fuzzy systems are closely related. A fuzzy inference system based on IF-THEN rules is practically an expert system if the rules are developed via expert knowledge (e.g., a computerized system that diagnoses an illness). However, if the rules are based on common sense reasoning (e.g., a rule base system for steering a car automatically based on common knowledge of driving) then the term *expert systems* does not apply. Today's fuzzy expert systems are quite different from classical expert systems of two decades ago. One of the notable differences is that fuzzy inference has been extensively deployed in hardware applications due to its simple mathematical equivalence, whereas expert systems (or fuzzy expert systems) have long been software oriented.

Examples of the pioneering expert systems include MYCIN, DENDRAL, PROSPECT, and others, most of which were developed independently for mainframe computers. Earlier systems worked in a manner similar to a database search. Once the given conditions match a particular knowledge in the database, the corresponding action is retrieved as the answer. More like a lookup table, this technology is still in use today. Later, the emergence of special computer languages such as LISP, OPS5, and PROLOG, along with expert system development shells such as EMYCIN has broadened the application spectrum. These tools allow list-oriented programming and advanced knowledge base structures. In classical expert systems, a knowledge base is traditionally represented by production rules (also used are the frames and semantic nets) by which the programmer defines (1) the search space by the left-hand-side conditions and (2) the action space by the right-hand-side propositions. These tools also offer Boolean logic operators (AND, OR) such that composite and nested statements can be formed in the search space. The term *inference engine* has emerged from structures similar to this; practically it means a search engine (such as the forward-chaining, backward-chaining, and breadth-width search engines) that operates within the knowledge base. Advances in the field of uncertainty management have yielded expert system solutions with confidence levels when imprecise data is presented as input. In most of the classical expert systems, Bayesian approach is the basic mechanism to estimate belief propagation and to manage conditional probabilities.

One of the most important characteristics that defines classical expert systems is its path-dependent search mechanism on the antecedent domain to reach a solution. This is illustrated in Fig. 2.24 using a simple rule. In the figure, four arbitrary conditions are tested. The hypothetical results (1 for match and 0 for no match) are also displayed under each operation. The first input set has produced 1 for $X = A$, 0 for $Y = B$, 1 for $W = C$, and 1 for $Z = D$. The second set has produced 0, 1, 1, 0, and so forth. Notice the propagation of intermediate results through AND and OR gates towards the action domain. Out of four input sets, we have 1, 0, 0, 0 at the output indi-

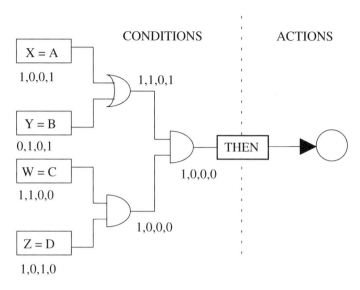

Figure 2.24 Propagation of inputs in a traditional expert system.

cating that only one of the four input sets will produce an answer. If there is some uncertainty-handling mechanism in this inference engine, the result (the only result) will have some confidence level associated with it. The rest of the three input cases are null. Within the classical expert systems framework, inference engines work in this manner.

A fuzzy inference algorithm operates in a quite different manner. Figure 2.25 shows the fuzzy equivalent of the same rule in a fuzzy algorithm. We again applied four hypothetical input sets that produced truth values as displayed under each operation in Fig. 2.25.

After AND (min) and OR (max) operations, we obtain an output value for each input set. In general, a fuzzy inference engine always produces a number of outputs equal to the number of input cases. Obviously, this is due to the membership function concept and the propagative nature of logic operators in fuzzy logic. Note that this may seem ambiguous in the sense that a fuzzy inference engine always finds a solution (before the THEN operator) as if there is no search process at all. The reader may also wonder: What happened to searching for a solution? In fact, one of the most characteristic properties of fuzzy inference engines is that the search for a solution is not confined to the antecedent domain; instead the solution is formed in the consequent domain by means of aggregation and defuzzification. Therefore, Fig. 2.25 depicts only half of the whole process. This

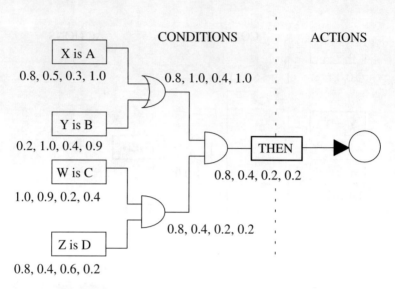

Figure 2.25 Propagation of inputs in a fuzzy inference engine.

constitutes one of the many—perhaps the most outstanding—difference between a traditional expert system and a fuzzy inference engine. The fact that fuzzy inference engines evaluate all the rules simultaneously and do not search for matching antecedents on a decision tree makes them perfect candidates for parallel processing computers. Therefore, the emergence of parallel computers in the future will enhance their speed and efficiency.

As a direct result of the difference stated above, we come to another comparison. Consider the right-hand side of the THEN operator. The state of an action in expert systems is either true or false (by some certainty factor or confidence level). Such a duality (i.e., true/false) somewhat restricts the evaluation of multiple composite conditions by writing two or more rules with the same consequent. The question is: What happens if there is more than one true answer? The manner in which multiple true answers are amalgamated in some advanced expert systems is quite different from that of fuzzy inference engines. In expert systems, multiple true answers are evaluated by their confidence levels, which originated from the uncertainty levels of the antecedents; then rule importance weights are taken into account to decide the result. Also, several belief propagation methods exist in the literature based on the Bayesian approach to manage joint probabilities. In fuzzy inference engines, there are no multiple true answers; instead, there are many partially true answers, all of which are determined by the possibility evaluation of the consequents and antecedents. The fusion of partial answers in fuzzy systems includes the aggregation and defuzzification processes that do not deal with the management of joint probabilities.

There are many other differences in the practical approach and conceptual foundation between expert systems and fuzzy inference engines. However, the expert system technology has been converging towards fuzzy inference quite rapidly within the last few decades. This convergence is explained in detail in a book written by Kandel with several examples [21]. Some of the latest fuzzy expert system developments include ASK (credit evaluation), CADIAG (medical expert system), EXPERT (rheumatology), PROTIS (treatment of diabetes), SYNTEX (hospital management), SPERIL (earthquake damage assessment), SPHINX (medical diagnostics), and ES/KERNEL (fuzzy expert shell) [22]. Also note that the classical tools of expert systems such as PROLOG have recently incorporated fuzzy elements, indicating the recent trend that the two technologies are converging.

REFERENCES

[1] Kosko, Bart. *Fuzzy Thinking*. New York: Hyperion, 1993.

[2] Dubois, D., and H. Prade. *Possibility Theory—An Approach to Computerized Processing of Uncertainty*. New York: Plenum Press, 1987.

[3] Zadeh, Lotfi A. "The Role of Fuzzy Logic in the Management of Uncertainty in Expert Systems." *Fuzzy Sets and Systems* 11 (1983): 199–227.

[4] Sugeno, Michio. "Theory of Fuzzy Integral and Its Applications." Ph.D. Diss., Tokyo Institute of Technology, 1974.

[5] Zadeh, Lotfi A. "Fuzzy Logic." *IEEE Computer Magazine,* IEEE, (April 1988): 83–92.

[6] Cox, E. *The Fuzzy Systems Handbook*. New York: AP Professional, 1994.

[7] de Luca, A., and S. Termini. "A Definition of a Nonprobabilistic Entropy in the Setting of Fuzzy Sets Theory." *Information and Control* 20 (1972): 301–312.

[8] Kaufman, A. *Introduction to the Theory of Fuzzy Subsets*. Vol 1. New York: McGraw-Hill, 1975.

[9] Yager, R. R. "On the Measure of Fuzziness and Negation, Part I: Membership in the Unit Interval," *Int'l. Jour. Gen. Syst.* 5, 1979.

[10] Klir, G. J., and T. A. Folger. *Fuzzy Sets, Uncertainty, and Information*. Englewood Cliffs, New Jersey: Prentice Hall, 1988.

[11] Ross, T. J. *Fuzzy Logic with Engineering Applications.* New York: McGraw-Hill, Inc., 1995.

[12] Egusa, Y., H. Akahori, A. Morimura, and N. Wakami. "An Application of Fuzzy Set Theory for an Electronic Video Camera Image Stabilizer." *IEEE Trans. on Fuzzy Systems* 3 (no. 3): 1995.

[13] Zimmerman, H. J. *Fuzzy Set Theory and Its Applications.* Boston: Kluwer Academic Publishers, 1990.

[14] Dubois, D., and H. Prade, *Fuzzy Sets and Systems: Theory and Applications.* Boston: Academic Press, 1980.

[15] Mamdani, E. H. "Application of Fuzzy Logic to Approximate Reasoning Using Linguistic Systems," *IEEE Trans. on Computers,* 26 (1977): 1182–1191.

[16] Kosko, B. *Neural Networks and Fuzzy Systems.* Englewood Cliffs, NJ: Prentice Hall, 1992.

[17] Tsoukalas, L. H., and R. E. Uhrig. *Fuzzy and Neural Approaches in Engineering,* New York: John Wiley & Sons, 1996.

[18] Gupta, M. M. and T. Yamakawa. *Fuzzy Computing.* Amsterdam: North Holland, 1991.

[19] Tsoukalas, L. H., R. C. Berkan, B. R. Upadhyaya, and R. E. Uhrig. "Expert System Driven Fuzzy Control Application to Power Plants." *Proc. 2nd Int'l. Conf. on Tools for Artificial Intelligence,* IEEE (1990): 445–450.

[20] Terano, Toshira, Kiroji Asai, and Michio Sugeno. *Applied Fuzzy Systems.* New York: AP Professional, 1994.

[21] Kandel, A. *Fuzzy Expert Systems,* Boca Raton: CRC Press, 1991.

[22] McNeill D., and P. Freiberger. *Fuzzy Logic.* New York: Simon & Schuster, 1993.

Chapter 3

The Basic Fuzzy Inference Algorithm

> The basic fuzzy inference algorithm is a generalized work flow through which fuzzy IF-THEN computations take place in a sequence, as seen in many industrial applications. The algorithm includes elements subject to design in fuzzy logic as well as the elements of peripheral computations such as input-output processing. This chapter presents the basic structure of fuzzy IF-THEN calculus along with the most widely used design options. The design issues addressed later in this book mainly refer to this structure; therefore, understanding the basic fuzzy inference algorithm must be considered as the first crucial design step.

3.1 INTRODUCTION

The basic fuzzy inference algorithm introduced in this chapter is a generalized implementation of fuzzy logic for a variety of problems. It is an algorithm that solves the problems expressed in the basic IF-THEN rule format and formulated as the generalized *modus ponens* inference in the literature. Fuzzy logic applications that do not fit this scheme are often formulated based on a simpler relationship than IF A THEN B, and correspond to one of the main building blocks of the basic fuzzy inference algorithm. However, the application of fuzzy set theory is not confined to this structure as suggested by examples found in the literature.

The algorithm, in the form presented in this book, consists of the summary of events (computations or other forms of information processing) arranged in the most logical order. The overall layout is invariant, but the order and type of computations shown in this book can be altered in a number of ways. Designing a fuzzy system starts from understanding the basic fuzzy inference algorithm. As illustrated later in this chapter, the design effort consists of selecting appropriate methods at the different stages of the basic algorithm that depend on the properties of the problem at hand. Abundance of options along with the designer's interpretation make every fuzzy system different and unique. Therefore, the overall design process is hardly a standard operation.

The basic fuzzy inference algorithm is mostly employed in today's control applications. However, the inference mechanism is the same for all other types of applications in which approximate reasoning plays a major role. The primary requirement is the availability of a solution articulated in the IF-THEN form. Once this requirement is met, then the basic fuzzy inference algorithm can be designed for any problem regardless of the problem type (e.g., control, classification, estimation, forecasting, diagnostics, modeling, etc.).

This chapter includes three equivalent representations: (1) block diagram representation, (2) mathematical representation, and (3) linguistic equivalence. The block diagram representation provides a visual environment for easy understanding. The details of the algorithm are included in the Appendix mostly in this format. The mathematical representation is a computational platform where the actual solution is implemented. The solutions are represented both in standard calculus (also called the geometric approach) and in fuzzy set terminology. The linguistic equivalence is the translation of fuzzy computations into fuzzy IF-THEN statements. These are all essential elements of fuzzy systems. However, the equivalent representations are used only when necessary in this book.

We have expanded the concept of *algorithm* in this book by including fuzzy elements and by creating a more generalized domain of operations than that used in the traditional binary computer terminology. Fuzzy elements, which are denoted by dashed lines, are the complementary operations that involve fuzzy decision making (or fuzzy computation), and are subject to design.

3.2. OVERALL ALGORITHM

A fuzzy inference engine has a simple input-output relationship as shown in Fig. 3.1 (a). Input data collected from the external world is processed by the fuzzy inference engine to produce output data to be used back in the external world. The events taking place in this process are outlined in Fig. 3.1 (b) to which we refer as the basic fuzzy inference algorithm. The details of each block are explained later in this chapter and also in the Appendix.

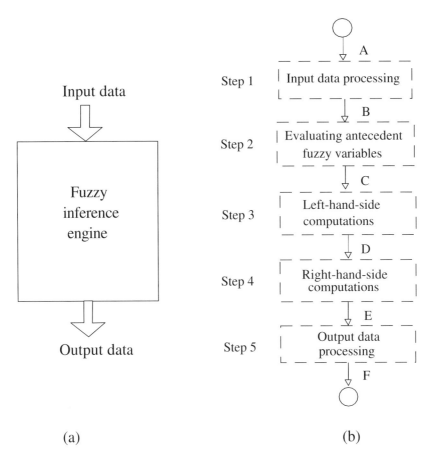

Figure 3.1 Generalized fuzzy computing: (a) inference engine, (b) the basic fuzzy algorithm.

The partitioning of the overall algorithm into five steps as shown in Fig. 3.1 (b) does not necessarily represent one universal philosophy. The same work flow may also be represented in a different number of steps. The purpose of the five-step representation proposed here is to provide a simplified work flow from the design point of view. Letters A– F indicate terminal points. The outputs of each block are inputs to the next one. Thus, these five steps cannot be shuffled. The basic fuzzy algorithm is invariant in that respect. On the other hand, all of the five steps drawn with dashed lines involve fuzzy computations that are subject to design.

Steps 3 and 4 in Fig. 3.1 refer to the left- and right-hand side of a general rule with respect to the THEN operator. The splitting of the computations into left and right provides a simple algorithmic approach to handle composite rules with several statements before and after the THEN opera-

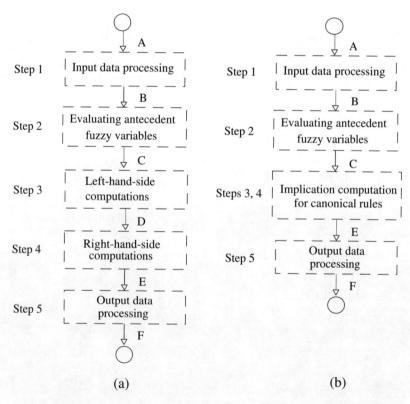

Figure 3.2 Computation steps of the basic fuzzy inference algorithm including (a) composite and (b) canonical rules.

tor, combined with logic operators such as AND and OR. An alternative approach is to translate all of the composite rules into their canonical form (only one statement on both sides, such as IF A THEN B). This replaces steps 3 and 4 with a single step computation as indicated in Fig. 3.2.

Transforming composite rules into canonical forms, which is explained in Chapter 7, simplifies computations at the expense of complicating the rule handling steps and compromising the understandability of the original representation. The examples included in this chapter explain these aspects in detail.

The overall algorithm describes standard IF-THEN inference based on the generalized *modus ponens* (GMP) inference paradigm. GMP and other forms of inference paradigms are explained in Chapter 2. In general, the GMP algorithm is considered as a transformation from the LHS degrees of fulfillment to the RHS degrees of commensurateness by means of a selected implication or composition operator.

3.3 INPUT DATA PROCESSING

In most of the existing applications, input data received from the external world is analyzed for its validity (in syntax, format, and range) before it is propagated into a fuzzy inference engine. Most of the time, an input data processing step is included within the peripheral computations, and it is not considered as part of the fuzzy inference engine owing to its trivial function. However, this important step cannot be taken without design knowledge.

A fuzzy inference engine can process mixed data. Mixed data in this context refers to the mixture of numerical and symbolic (linguistic) data. The capability of processing mixed data is based on the membership function concept by which all input data are eventually transformed into the same unit (possibility) before the inference computations. This transformation process is explained in the next section under the title, "Evaluating Fuzzy Variables." Accordingly, the goal of input data processing is to ensure that input data is in an appropriate form for transformation. The input data processing is depicted in Fig. 3.3. A block diagram representation of input data processing is illustrated in the Appendix. This algorithm applies only to antecedent (input) fuzzy variables.

A fuzzy inference engine normally includes several antecedent fuzzy variables. If the number of antecedent variables is k then there will be k information collected from the external world. This is referred to as

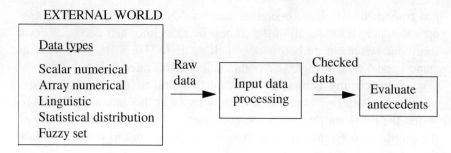

Figure 3.3 Input data processing.

the *input data set* in this book and denoted by $[I_1\ I_2 \ldots I_k]$. Note that it is a crisp set containing crisp and fuzzy elements. Furthermore, each element in the set may contain more than a single data point, namely a set of distribution points or a fuzzy set.

The input data processing step performs a verification task to ensure that the input data set is acceptable for transformation to the possibility units. The acceptability criterion, which is subject to design as explained later, is based on the fact that the antecedent variables of the fuzzy inference engine are only functional in certain intervals (universe of discourse) by design. In addition, nonnumerical data (linguistic statements) must be recognized by the inference engine, again by design. As one would expect, a substantial portion of the memory of the fuzzy system (where all design knowledge is embedded) will be accessed during input data processing.

Another important property is that when an input data set is partially ambiguous or unacceptable, a fuzzy inference engine may still produce reasonable answers. This property, often called *robustness* against missing data, requires a unique treatment during input data processing. For instance, all acceptable and unacceptable input data may be recorded to make a decision about whether to accept the entire input data set. Depending on the design, the acceptance criterion can be made strict or tolerant.

When time or sequence is the independent variable, the concept of the input data set may become confusing. As mentioned earlier, some elements in the input data set may contain more than one data point (a subset). The definition of the input data set is that every element in the set, regardless of whether it is a subset or a single data point, belongs to the same initiation step (sequence, iteration, or time). Thus, $[I_1\ I_2 \ldots I_k]$ means $[I_1(t)\ I_2(t) \ldots I_k(t)]$ where t indicates an initiation step. For instance, $I_2(t)$ may contain more than one data point. The next initiation step denoted by $t + 1$ corresponds to a new data set $[I_1(t + 1)\ I_2(t + 1) \ldots I_k(t + 1)]$, and so forth.

When considering all the possible forms of data collected from the external world, some form of data transformation may become necessary either within the design envelope of an inference engine or within the peripheral computations (such as graphical user interfaces). Fuzzification and normalization are the two typical transformations. In some particular cases, the relationship among each input data point can be important, so that special transformations are needed. For example, suppose that the temperature profile of a furnace along the vertical axis is a fuzzy variable in a particular design. The profile data, which contains many data points, will be one input containing more than a single data point. Starting from the highest elevation to the lowest, assume that this data is represented by a vector $[T_1\ T_2\ T_3 \ldots T_N]$. Because the order of data means something, this type of data is called *sequentially correlated*. Most of the dynamic data are sequentially correlated with time as the independent variable. Now assume that a similar data vector $[M_1\ M_2\ M_3 \ldots M_N]$ indicates the mass amounts of N chemicals mixed to form a special plastic compound. Because the order of data in this vector means nothing, this data set is called *cross section data*. The following example illustrates how to transform sequentially correlated and cross section data into a form that can be processed in the basic fuzzy inference algorithm.

EXAMPLE 3.1 DATA PROCESSING OF MIXED INPUTS

Suppose that a fuzzy inference engine is designed to compute the heart attack risk of each health insurance applicant. A segment of the input questions is shown below.

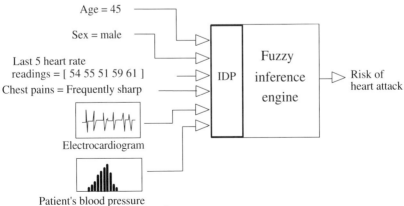

In this example, the input data set has mixed data components: (1) a crisp number for the patient's age, (2) linguistic data for the sex category, (3) a crisp set indicating the last five heart rate readings, (4) linguistic data for patient's complaints, (5) a set of numerical data from the patient's last electrocardiogram, and (6) statistical data for the blood pressure.

The input data processor (IDP) checks each numerical data (such as 45) to make sure that the entry is within the universe of discourse of the fuzzy variable (*age*). It may be the case that this fuzzy inference engine is designed for the age group of 35 to 75. Therefore, an entry such as *age* = 6 will be out of the domain of the embedded expert knowledge in a region where the embedded risk assessment principles do not apply. Symbolic (linguistic) input data *male* and *frequently sharp* must match one of the fuzzy set definitions residing in the memory of the fuzzy inference engine. For the variable *chest pains* for example, there may be a list of frequency definitions (*never, seldom, regular, frequent*) along with class definitions (*sharp, mild, light*) each represented by a fuzzy set. If the symbolic data is not recognized, this information cannot be used by the inference engine. Other complex forms of input data such as distribution points (*electrocardiogram*) may require data transformation from a high dimensional space to a lower dimensional space. However, the universe of discourse test applies to all numerical data regardless of the type of data transformation. That is, all the points in the *electrocardiogram* must be within the universe of discourse. The same principle applies to statistical data.

Assume that this inference engine was designed to assess the heart attack risk even in the absence of some of the input data. In this example, it may be the electrocardiogram data which is not available for all applicants. Then, the input data acceptability criterion can be made flexible for this particular antecedent variable so that the absence of an electrocardiogram data would not halt the risk assessment process.

The next issue to discuss in this example is timing. Regardless of how long it takes to collect one set of data, they all belong to the same initiation step. The last five heart rate readings may have been taken within the last five hours or five days. All five readings belong to one initiation step when implementing this fuzzy inference engine. The risk assessment process may be repeated with some (but not necessarily all) updated data from the same patient. This will be the second initiation step. Therefore, the parameter t in $[I_1(t)$ $I_2(t) \ldots I_k(t)]$ is not the time of data collection procedure, but it is the initiation step of the inference engine.

There are basically three ways to handle the sequentially correlated input data depending on how this input variable is used in rule composition during design. The first way is to rotate all data points vertically (a figure of speech in reference to the schematic above) and define new input variables for each time instance. Application of this method to the electrocardiogram data is shown below. This can be a very costly exercise since there must be as many variables as data points. In addition, all these variables must be used in fuzzy rules in one or more places, which becomes very tedious and sometimes impractical for large data sets.

The second way is to reduce the data to a practical size by a data compression technique or by a nonlinear filter such that the characteristics of the original data are retained. This is depicted next.

Chap. 3 The Basic Fuzzy Inference Algorithm

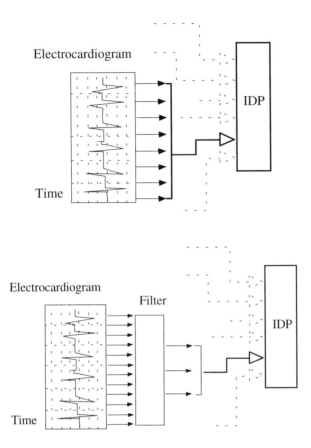

Note that the nonlinear filter must be sensitive to the order of data points. In other words, if we change the order without changing the data points, the filter must respond differently. The filter can be a pattern recognition technique—or a linguistic classifier—to recognize previously defined categories of heart performance. In such a case, the classifier converts numerical data into linguistic data and the fuzzy inference engine is designed to receive linguistic input from the electrocardiogram channel. One of the simplest methods of designing such a classifier is to employ the image matching technique by minimizing the total differential distance between the data points of any two images.

The third way is to dissect the frequencies occurring in the data. This is only applicable to a periodic data set such as the electrocardiogram. There are many time-to-frequency conversion methods in signal processing. Applying such a technique can reduce hundreds of time points to a handful of frequency intervals. Each frequency interval must be a fuzzy variable in the inference engine by design.

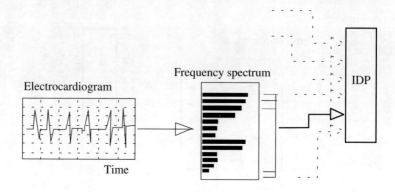

Now, we will examine how to handle cross section data. The blood pressure measurements taken within the last seven days are shown as a list below. Statistical data of this type is cross section data because the order of measurements is not important or known, and the objective is to acquire an average value that describes the patient's unique regular blood pressure. If the input data is not in a histogram format, cross section data must go through such a conversion as the first step. Then the frequency histogram is converted into a possibility diagram, which is an acceptable input for a fuzzy inference engine.

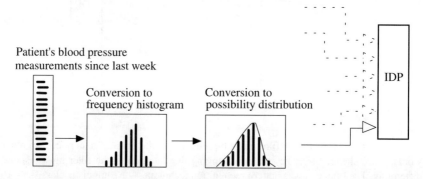

Transition from frequency of occurrence to possibility distribution can be done by a simple normalization of the frequency data between 1 and 0.

3.4 EVALUATING ANTECEDENT FUZZY VARIABLES

Following the natural flow of the basic fuzzy inference algorithm, we will now discuss the absorption of the input data by a fuzzy system. This step is the evaluation of antecedent fuzzy variables for given input data (input from the external world).

A fuzzy inference engine is based on possibility computations only. Therefore, the units of the input data to the inference computations must be possibility units. In practical life, however, data is often in units such as Kg, m/sec, or dollars. The transformation from the practical units to possibility units (also known as transformation from the measurement space to the possibility space) is referred to as evaluating fuzzy variables in this book. Obviously, such a transformation is not possible using a fixed formula; instead, it is made by means of transformers (or membership functions) modeled by designers.

A fuzzy variable may be considered as a crisp set of fuzzy sets $X = \{\mu_1(x), \mu_2(x), \ldots, \mu_N(x)\}$. When evaluating an antecedent fuzzy variable for a given input data $I_k(t)$, all membership functions are evaluated. The result of the evaluation is a vector (or a set) of membership values $\{\mu_1, \mu_2, \ldots, \mu_N\}$ with each element indicating the possibility produced by this input data entry. This process is applied to all antecedent variables at the same initiation step.

The overall fuzzy variable evaluation event at a given initiation step is illustrated in Fig. 3.4. The input data set includes N elements for N antecedent variables. Each input data, which can either be a single point or a distribution of points, is applied to the data transformation process via membership functions. Depending on the design, the number of membership functions (i.e., the length of the membership value vector) may be

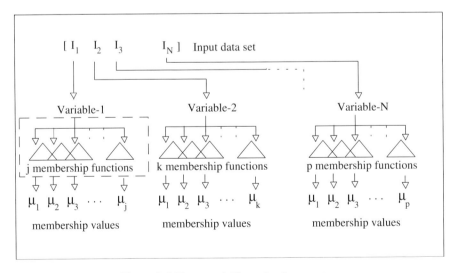

Figure 3.4 Fuzzy variable evaluation event.

different. In Fig. 3.4, this is denoted by *j* and *k* for the first and second antecedent variables.

The evaluation of *variable-1* marked with a dashed square in Fig. 3.4 is outlined in Fig. 3.5 in terms of block diagrams. Threshold filtering in Fig. 3.5 is discussed in Section 3.4.3. Because the input data can be either a single point or a distribution of points, there is an alternative operation for each type. This process will be applied to all of the antecedent variables of the fuzzy inference engine. The block diagram for the entire process includes a data loop through which all antecedent variables are evaluated as shown in the Appendix. Remember that we are analyzing the generalized working principles of a fuzzy system that is already designed. Accordingly, the membership function to be evaluated is assumed to be known (designed).

How to design such membership functions is the subject of Chap-

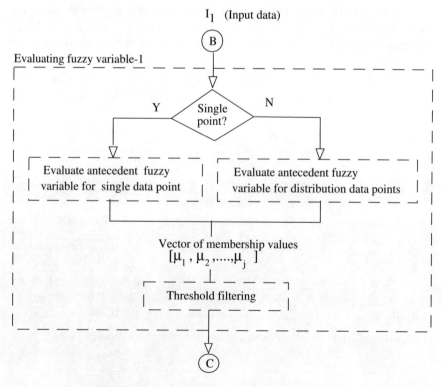

Figure 3.5 Evaluating an antecedent fuzzy variable.

3.4.1 Evaluation Based on Single Input Data

When evaluating a membership function, single input data (also considered a crisp number) is used in a manner identical to that of evaluating a function in calculus. As shown in Fig. 3.6, the crisp numerical data a has a membership value of b in this evaluation. This also means that the possibility of a being a member of the fuzzy set $\mu_j(x)$ is b. The fuzzy set $\mu_j(x)$ is a category of the antecedent fuzzy variable X.

This process may be expressed in calculus by assuming continuum for the membership function. Then, a membership value is computed by

$$\mu_a = \mu_j(x)\big|_{x=a} = b \qquad (3.1)$$

where a and b are real numbers. In fuzzy set terminology, a membership value is expressed in terms of a singleton $\mu_j(x)/x$ or b/a indicating that the crisp number a is the member of the fuzzy set $\mu_j(x)$ with the possibility b. A single element of data can also be represented by a fuzzy set in which all the singletons are zero except one that corresponds to the input value. Then, the evaluation process is undertaken by the max-min operation as shown below.

$$\mu_a = \vee[\mu_j(x) \wedge I(x)] = b \qquad (3.2)$$

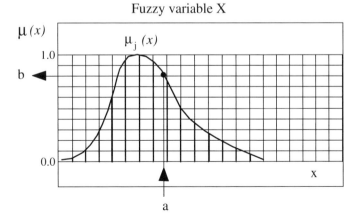

Figure 3.6 Evaluating a membership function for single input data.

Here, $I(x)$ is the input fuzzy set including one nonzero singleton. The trouble with this operation in practice is that inputs are often continuous; thus, an input data point may not coincide with the singletons of the fuzzy set representing the membership function. This requires approximate solutions to form an input set acceptable for the fuzzy set computations. The continuous form of Eq. (3.1) is more practical; thus, this chapter includes continuous solutions as well as fuzzy set solutions.

The linguistic equivalence of this process is trivial. Suppose that the fuzzy variable X is called *temperature*, and the membership category $\mu_j(x)$ is called *low*. The membership function will be represented as $\mu_{low}(Temperature)$ in such a case. For a = 27 and b = 0.8, the linguistic equivalence of input absorption corresponding to the ongoing discussion is:

Temperature is 27 degrees.
[The possibility of 27 degrees to be considered as low temperature is 0.8.]

The first statement expresses the crisp input data and it connects the fuzzy algorithm to the external world. Linguistic statements of this nature are the basic elements of a rule-based system. The second statement is the translation of the membership evaluation process into daily language, and it is entirely within the domain of fuzzy algorithm. Such statements are not part of the rule-based representation, and they are indicated by square brackets in this book. The importance of these statements is that they can be useful during design to check the logical inconsistencies.

EXAMPLE 3.2 EVALUATION OF A MEMBERSHIP FUNCTION

Suppose we will evaluate the membership function $A(x)$ given a single input data, 2. The membership function $A(x)$ is shown below.

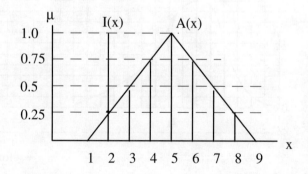

Using the fuzzy set operations, the fuzzy set $A(x)$ is given by

$A(x) = 0/1 \cup 0.25/2 \cup 0.5/3 \cup 0.75/4 \cup 1/5 \cup 0.75/6 \cup 0.5/7 \cup 0.25/8 \cup 0/9$

The input 2 must be converted into a fuzzy set $I(x)$ as follows.

$I(x) = 0/1 \cup 1/2 \cup 0/3 \cup 0/4 \cup 0/5 \cup 0/6 \cup 0/7 \cup 0/8 \cup 0/9$

The max-min operation between $A(x)$ and $I(x)$ is the result.

$\mu = \vee[(0 \wedge 0)(0.25 \wedge 1)(0.5 \wedge 0)(0.75 \wedge 0)(1 \wedge 0)(0.75 \wedge 0)(0.5 \wedge 0)(0.25 \wedge 0)(0 \wedge 0)]$
$= \vee[0 \quad 0.25 \quad 0 \quad 0 \quad 0 \quad 0 \quad 0 \quad 0 \quad 0] = 0.25$

Note that if the input were 2.3 instead of 2, the operation above would not be possible. On the other hand, the formulation using continuous functions can handle any input. The membership function $A(x)$ is now defined by a function.

$$A(x) = \begin{pmatrix} 0.25(x-1) & 1 \le x \le 5 \\ -0.25(x-9) & 5 < x \le 9 \end{pmatrix}$$

Its evaluation for $x = 2$ yields $A(2) = 0.25$. It is also possible to evaluate $A(x)$ for $x = 2.3$ or any other value in the universe of discourse. The difficulty in this approach is to represent the membership function in analytical form, which can have more complex appearance than what is shown above. However, simple geometric shapes are commonly used with success even in the most complex applications.

Because there is normally more than one membership function per variable, the evaluation process described above is applied to all of them. As shown in Fig. 3.7, single input data often produces more than one membership value. Most of the time, there are two or three nonzero elements in the membership value vector per input data. Obviously, the number of nonzero elements depends on the degree of overlap among membership functions that is subject to design. If there are N membership functions defined in variable X, then there will be N membership values computed as follows.

$$\begin{aligned} \mu_{I_1} &= \mu_1(x)\big|_{x=I_1} = b_1 \\ \mu_{I_2} &= \mu_2(x)\big|_{x=I_2} = b_2 \\ &\cdots \\ \mu_{I_N} &= \mu_N(x)\big|_{x=I_N} = b_N \end{aligned} \qquad (3.3)$$

The results are often put in a vector format $[b_1 \, b_2 \, \ldots \, b_N]$ for convenience. After the fuzzy variable evaluation stage, these vectors will be

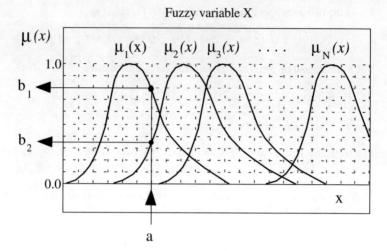

Figure 3.7 Single input data evaluation of all membership functions.

used in the evaluation of fuzzy statements that appear in the rule base. Figure 3.7 illustrates evaluation of a fuzzy variable given a single input data point.

Suppose that the membership categories $\mu_1(x)$ and $\mu_2(x)$ are called *low* and *medium*, respectively. Then, the linguistic equivalence of input absorption will have two statements. Assume that a = 27, b_1 = 0.8, and b_2 = 0.35.

Temperature is 27 degrees.
[The possibility of 27 degrees to be considered as low Temperature is 0.8.]
[The possibility of 27 degrees to be considered as medium Temperature is 0.35.]

For this particular case, all other possibilities are zero. Thus, the membership value vector of the fuzzy variable *Temperature* is [0.8 0.35 0 0 .... 0] for the input data of 27 degrees.

3.4.2 Evaluation Based on Distribution Data

The distribution $A(x)$ as the input data and the membership function $\mu_j(x)$ as a category of the antecedent variable X are both fuzzy sets as shown in

Fig. 3.8. As one can infer, every singleton of the fuzzy set $A(x)$ can potentially produce a membership value if all the singletons were treated point by point in the same manner as explained in the previous section. Thus, the result is also another fuzzy set.

To find the correct implementation in calculus, we first examine the fuzzy set approach by applying the max-min composition operator. The new fuzzy set is shown by the shaded area in Fig. 3.8 and the maximum possibility value in this set is the resultant membership value (real number b for the case in Fig. 3.8).

The analytical representation of this process using the fuzzy set terminology is given by

$$\mu_a = \vee[\mu_j(x) \wedge A(x)] = b \qquad (3.4)$$

where $A(x)$ is the fuzzy set equivalent of the distribution data. Note that this equation is identical to Eq. (3.1) except that $I(x)$ is replaced by $A(x)$. Again, the distribution data $A(x)$ must have all of its distinct values coincide with the singletons of the fuzzy set $\mu_a(x)$, otherwise max-min operation cannot be performed. Instead of approximating the $A(x)$, the continuous solution in calculus is more appropriate. The solution is the maximum possibility value among the intersecting points between the two continuous functions. This is expressed by

$$\mu_a = b = \mu(x_i^*); \quad x_i^* = x_i \quad for \quad \mu(x) - A(x)|_{x_i} = 0 \qquad (3.5)$$

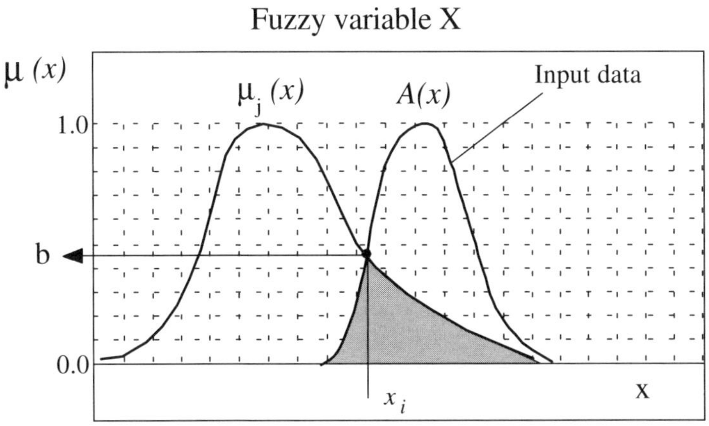

Figure 3.8 Input data superimposed on a membership function.

where the superscript asterisk (*) means the intersection point that yields the maximum value. Similar to the case of single input data, the distribution data evaluation will be performed on all of the membership functions of the antecedent variable. The hypothetical case illustrated in Fig. 3.9 includes three membership functions in addition to the input data.

Compared to the case of single input data, distribution input data will produce more nonzero elements in the membership value vector. If there are N number of membership functions defined for the antecedent variable and $A(x)$ is the input data, then this process is expressed by

$$\begin{aligned}
\mu_{a1} &= b_1 = \mu_1(x_i^*) \quad ; \quad x_i^* = x_i \quad \mu_1(x) - A(x)\big|_{x_i} = 0 \\
\mu_{a2} &= b_2 = \mu_2(x_i^*) \quad ; \quad x_i^* = x_i \quad \mu_2(x) - A(x)\big|_{x_i} = 0 \\
&\dots \\
\mu_{aN} &= b_N = \mu_{N1}(x_i^*) \quad ; \quad x_i^* = x_i \quad \mu_{N1}(x) - A(x)\big|_{x_i} = 0
\end{aligned} \quad (3.6)$$

The resultant membership value vector is expressed as

$$\bar{\mu}_{XA} = [b_1 \; b_2 \; \dots \; b_N] \quad (3.7)$$

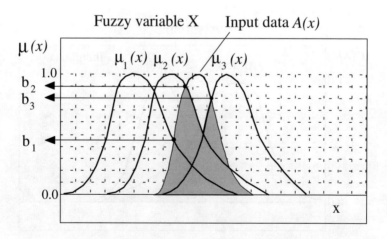

Figure 3.9 Distribution data evaluation of all membership functions.

In the ongoing discussion, suppose that $A(x)$ is called *Moderate*, the membership categories $\mu_2(x)$ and $\mu_3(x)$ are called *Medium* and *High*, respectively. Then, the linguistic equivalence of input absorption (for $b_1 = 0.47$, $b_2 = 0.9$, and $b_3 = 0.8$) will be:

Temperature is Moderate
[The possibility of the moderate temperature distribution to match the low temperature category is 0.47 at maximum]

[The possibility of the moderate temperature distribution to match the medium temperature category is 0.9 at maximum]
[The possibility of the moderate temperature distribution to match the high temperature category is 0.8 at maximum]

All other possibilities are zero. Thus, the membership value vector of the fuzzy variable *Temperature* is [0.47 0.9 0.8 0 . . . . 0] for the input data distribution called *Moderate*.

3.4.3 Evaluation With Thresholding

The need for thresholding arises from the fact that low possibilities (small membership values) are sometimes undesirable because they trigger unnecessary computations. In fuzzy logic terminology, a small membership value is called residual truth [1], emphasizing that the results produced by small possibilities are not designed. This is illustrated in Fig. 3.10 where the input data produces a small membership value.

A small membership value, if not desired, can be avoided by applying a number of different thresholding methods. Selecting an appropriate method is subject to design and is explained later in the book. It is also important to remember that the thresholding of membership functions of antecedent variables yields different results from that of consequent variables.

The block diagram representation in Fig. 3.11 shows the interior of the thresholding block. When a vector of membership values are received, each membership value undergoes this filtering process. Terminal points B' and C' denote that this block diagram is a part of the overall B to C flow described previously.

A threshold function or filter is applied to each membership value in the basic fuzzy inference algorithm. The result is questioned to decide

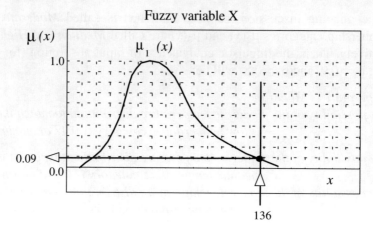

Figure 3.10 Input data producing a membership value too small to consider.

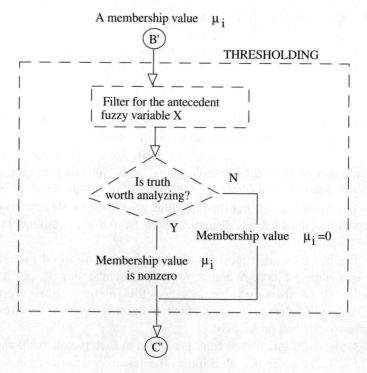

Figure 3.11 Thresholding operation.

whether to let the membership value propagate into fuzzy computation. The questioning process denoted by the dashed diamond may involve fuzzy computations and it is also subject to design. In Fig. 3.11, if the truth is not worth analyzing then the membership value is zero. Otherwise, the membership value is unchanged from its original value.

EXAMPLE 3.3 THRESHOLD OPERATION USING α-CUTS

Consider the case in Example 3.2 with an α-cut threshold of 0.3.

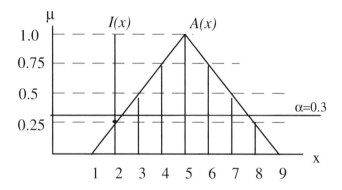

Applying an α-cut of 0.3 to fuzzy set $A(x)$ yields a new crisp set A^α as shown below.

$$A(x) = 0/1 \cup 0.25/2 \cup 0.5/3 \cup 0.75/4 \cup 1/5 \cup 0.75/6 \cup 0.5/7 \cup 0.25/8 \cup 0/9$$
$$A^\alpha = \{x \mid \mu_A(x) > 0.3\} = \{3 \quad 4 \quad 5 \quad 6 \quad 7\}$$

Then, the input $x = 2$ is checked if it is included in the crisp set A^α. In this case, the evaluation of this input will not produce any possibility value. If the input value was 4, then the input evaluation steps would be pursued. Note that for continuous inputs (i.e., such as 2.7), the boundaries of A^α (3 and 7) can be used for the same purpose. In actual implementation, either the crisp set A^α or its boundaries can be stored in memory to speed up the evaluation process.

3.5 LEFT-HAND-SIDE COMPUTATIONS

Step 3 between the terminal points C and D in Fig. 3.2 consists of the left-hand-side (LHS) computations of a rule with respect to the THEN operator. Because the membership values are obtained in the previous step, LHS computations reduce to calculating the effects of logic operators on the LHS that yield the degree of fulfillment (DOF) for each rule.

This process is symbolically illustrated in Fig. 3.12. Before describing the LHS computation, let's recall some of the fundamental definitions from Chapter 2.

Fuzzy Statement

A fuzzy statement denoted by S consists of a fuzzy variable connected to one of its predicates. If X is a fuzzy variable, then a fuzzy statement is expressed, in general, as

$$S_j \equiv X_k \bullet P_{k,m} \qquad (3.8)$$

where $\bullet$ is the linguistic operator IS.[1] The complement IS NOT is denoted by $\neg$. The indices j, k, and m in Eq. (3.8) indicate that the jth fuzzy statement is composed with the kth fuzzy variable and with one of its predicates (the mth one). The indices j and k are not related; thus, a fuzzy statement can be composed of any fuzzy variable. However, the fuzzy predicate denoted by k,m belongs to the kth fuzzy variable. Another form of fuzzy statement is expressed as

$$S_j \equiv \lambda_j (X_k \bullet P_{k,m}) \qquad (3.9)$$

where λ is an auxiliary parameter that modifies the fuzzy statement, similar (but not equivalent) to a function in calculus. Auxiliary parameters are used to bring extra modifications to a fuzzy statement, more specifically to the membership function. These parameters, being part of the design, are associated with fuzzy statements. Therefore, the indices are the same.

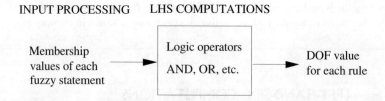

Figure 3.12 Inputs and outputs of LHS computations in the basic fuzzy inference algorithm.

---

[1] The operator $\bullet$ must not be confused with the max-min composition operator $\circ$.

The two primary effects that can be represented by λ are linguistic hedges and adaptive solutions. The way in which such effects are incorporated into the calculus of fuzzy inferencing takes different forms, unlike the functional appearance in Eq. (3.9). A linguistic hedge such as *often* modifies a predicate such as *cold* by changing the shape of the membership function for *cold*. This is discussed in Chapter 5 in more detail.

EXAMPLE 3.4 LINGUISTIC HEDGE EFFECT

Suppose a fuzzy variable, *Temperature*, has three membership functions that correspond to the three predicates *low, medium, high*. The components of this representation are

$$X_1 \equiv Temperature$$
$$P_{1,1} \equiv Low$$
$$P_{1,2} \equiv Medium$$
$$P_{1,3} \equiv High$$

A fuzzy statement $S_j \equiv X_k \bullet P_{k,m}$ may be in one of the following forms:

Temperature IS Medium
Temperature IS NOT High

Using linguistic hedges, a fuzzy statement $S_j \equiv \lambda_i(X_k \bullet P_{k,m})$ may also be in the following form:

Usually (Temperature IS Medium)
Frequently (Temperature IS NOT High)

where λ_i is defined by *Usually* or *Frequently*. As seen in this example more clearly, the hedges modify the entire statement and therefore carry the same indice *j*.

Logic Operators

In the rule base of a fuzzy inference engine, where the specialized knowledge is embedded, fuzzy statements are combined with logic operators. One complete proposition is referred to as a *fuzzy rule*. As a result, a fuzzy rule is represented by

$$S_1 \Theta_1 S_2 \Theta_2 \ldots \Theta_{z-1} S_z \qquad (3.10)$$

where the symbol Θ denotes logic operators such as AND, OR, ELSE, and THEN. A detailed discussion on logic operators is given in Chapter 7. The rule shown above consists of *z* fuzzy statements and *z*-1 logic operators. These operators perform mathematical operations in accordance with their linguistic function. Note that the logic operators denoted by Θ are not related to the linguistic operators denoted by • and ¬.

Among the logic operators, THEN has a different role in the basic fuzzy inference algorithm. From the algorithmic solution point of view, operations taking place on the left-hand side of THEN are different from those on the right-hand side. Thus, this operator conceptually divides every rule into two distinct parts. In this book, the left-hand-side (LHS) and right-hand-side (RHS) abbreviations are frequently used. The THEN operator is also referred to as the implication operator in different contexts and is denoted by an arrow →. Note that nested rules containing more than one THEN operator are always reduced to a form in which only one of the THEN operators functions as the implication operator in the basic fuzzy inference algorithm. This is discussed in Chapters 7 and 8.

LHS of a Fuzzy Rule

First, let's restate the representation of single rule expressed by Eq. (3.10) using the fuzzy variable and fuzzy predicate representation of Eq. (3.9).

$$(X_k \bullet P_{k,m}) \Theta_1 (X_s \bullet P_{s,f}) \Theta_2 \ldots \Theta_{z-1} (X_r \bullet P_{r,u}) \quad (3.11)$$

The significance of this representation is hidden in the indices. The indices $k, m, s, f, r,$ and u indicate that each statement can contain any fuzzy variable and any predicate belonging to that fuzzy variable. This freedom in structure makes some definitions necessary, such as the antecedent predicate matrix and membership value matrix. These matrices are used for algorithmic solutions presented in the Appendix.[2] The indices in Eq. (3.11) are in reference to the order of variables, predicates, and membership values in those matrices always kept in the memory of the fuzzy system. Before defining the indices exactly, let's introduce the THEN operator using the arrow convention and split the rule in Eq. (3.11) into the LHS and RHS portions. The THEN operator is shown below as the nth logic operator.

$$(X_k \bullet P_{k,m}) \Theta_1 (X_s \bullet P_{s,f}) \Theta_2 \ldots \Theta_{n-1} (X_t \bullet P_{t,y})$$
$$\rightarrow (X_p \bullet P_{p,q}) \Theta_{n+1} \ldots \Theta_{z-1} (X_z \bullet P_{z,u}) \quad (3.12)$$

---

[2] Memory of a fuzzy system can be organized in a number of ways while the relationship among the indices are kept invariant.

Chap. 3   The Basic Fuzzy Inference Algorithm

An example of representation Eq. (3.12) is shown below.

If *Temperature IS High* OR *Flow IS Low* THEN *Valve IS Open* AND *Operation IS Correct*
   $(X_1 \bullet P_{1,3})$ Θ_1 $(X_2 \bullet P_{2,1})$ $\rightarrow$ $(X_3 \bullet P_{3,2})$ Θ_3 $(X_4 \bullet P_{4,1})$

In this example, *Temperature, Flow, Valve,* and *Operation* are the 1st, 2nd, 3rd, and 4th fuzzy variables in the memory of the fuzzy inference engine, respectively. Accordingly, the predicates *high, low, open* and *correct* are 3rd, 1st, 2nd, and 1st membership functions of the fuzzy variables *Temperature, Flow, Valve,* and *Operation,* respectively. The operators Θ_1, Θ_2, and Θ_3 are OR, THEN, and AND operators, respectively. Therefore, the IF-THEN rules can be viewed as the map of the memory where the design information resides. The LHS portion of a fuzzy rule with respect to the operator THEN is expressed, in general, as follows

$$(X_k \bullet P_{k,m}) \; \Theta_1 \; (X_s \bullet P_{s,f}) \; \Theta_2 \; ... \; \Theta_{n-1} \; (X_t \bullet P_{t,y}) \qquad (3.13)$$

All the statements (*n* of them) and logic operators (*n-1* of them) belong to the LHS. Now, the indices *k, s,* and *t* are the row numbers in the antecedent predicate and membership value matrices which reside in the memory of the fuzzy system. Similarly, the indices *m, f,* and *y* correspond to the column numbers in these matrices. Based on the dimensional equivalency between the membership value matrix and the antecedent predicate matrix, each individual membership value, which is computed during the input absorption process, can be extracted from the memory using these indices, and they can be plugged into Eq. (3.13) for the LHS computations. Thus, the expression in Eq. (3.13) reduces to

$$DOF = (\mu_{k,m}) \; \Theta_1 \; (\mu_{s,f}) \; \Theta_2 \; ... \; \Theta_n \; (\mu_{t,y}) \qquad (3.14)$$

in the basic fuzzy inference algorithm, where DOF is a scalar known as the *degree of fulfillment*. Using the auxiliary parameters, the LHS of a fuzzy rule is expressed in a similar manner.

$$DOF = \lambda_k \; (\mu_{k,m}) \; \Theta_1 \lambda_s \; (\mu_{s,f}) \; \Theta_2 \; ... \; \Theta_n \lambda_t \; (\mu_{t,y}) \qquad (3.15)$$

Above, the indices of λ follow the statement numbers. The two symbolic expressions given by Eqs. (3.14) and (3.15) describe the basic LHS operations. The indices and the type of the logic operators (and fuzzy probabilities if dispositional logic is employed) are fixed via design, whereas the membership values inside the parentheses change from iteration to iteration by the input absorption process. Next, we will illustrate the computation based on Eqs. (3.14) and (3.15).

EXAMPLE 3.5 DOF COMPUTATIONS

Assume that we have two statements on the LHS of a rule expressed by

$$S_k \; \Theta_k \; S_{k+1}$$

where $S_k \equiv \lambda_k (X_l \bullet P_{l,j})$ and $S_{k+1} \equiv \lambda_{k+1} (X_m \bullet P_{m,n})$. The membership functions corresponding to the predicates $P_{l,j}$ and $P_{m,n}$ are shown below along with hypothetical input data (a_1 and a_2) for the antecedent fuzzy variables X_l and X_m.

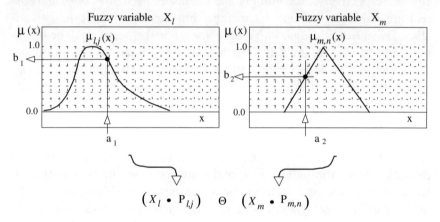

Assuming that auxiliary parameters λ_k, λ_{k+1} are not specified, the computation between these two fuzzy statements yields a DOF value of b_1 or b_2 depending on the type of logic operator:

$$S_k \; \Theta_k \; S_{k+1} \equiv \mu_{l,j}(a_1) \; \Theta_k \; \mu_{m,n}(a_2) = b_1 \; \Theta_k \; b_2 = \begin{cases} b_1 \vee b_2 = b_1 & \Theta_k \equiv \text{OR} \\ b_1 \wedge b_2 = b_2 & \Theta_k \equiv \text{AND} \end{cases}$$

In general, there are other logic operators besides AND and OR that may be used in a fuzzy inference engine. These operators and their functions are described in Chapter 7. Note that the logic operators AND and OR are implemented as intersection and union (minimum and maximum). In fuzzy systems theory, intersection and union formulations for AND and OR logic operators are approximations based on rigorous derivations [2].

Now, let's examine the effect produced by auxiliary parameters. Consider the previous fuzzy statements $S_k \equiv \lambda_k(X_l \bullet P_{l,j})$ and $S_{k+1} \equiv \lambda_{k+1}(X_m \bullet P_{m,n})$. Here, the auxiliary parameters λ_k, λ_{k+1} modify the membership functions corresponding to the predicates $P_{l,j}$ and $P_{m,n}$ by an external input. For simplicity, let's assume that the λ_k, λ_{k+1} pair modifies the membership functions in the manner shown below.

$$S_k \equiv \lambda_k (X_l \bullet P_{l,j}) \equiv (X_l \bullet P_{l,j}^s)$$
$$S_{k+1} \equiv \lambda_{k+1}(X_m \bullet P_{m,n}) \equiv (X_m \bullet P_{m,n}^s)$$

Here, the superscript s denotes the external input, which changes the power of the exponent and therefore the shapes of the membership functions. As discussed in Chapter

Chap. 3 The Basic Fuzzy Inference Algorithm

5, $s = 2$ corresponds to the linguistic hedge *very*. As a result, we can externally change the LHS of the fuzzy rule from

Temperature IS High AND Flow IS Low

to

Temperature IS (very)High AND Flow IS (very)Low

by the input of $s = 2$. This effect may also be supplied externally by an adaptive algorithm such that the exponent s is updated based on some criteria. Let's iterate the $s = 2$ case one step further. Referring back to the previous example, we now have new shapes for the membership functions.

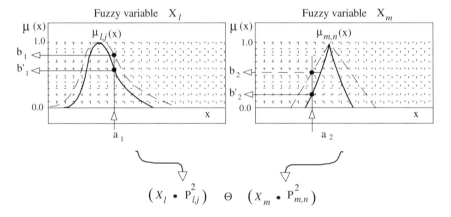

The computation between these two fuzzy statements yields a DOF value of b'_1 or b'_2 depending on the type of logic operator:

$$S_k \,\Theta_k\, S_{k+1} \equiv \mu_{l,j}(a_1) \,\Theta_k\, \mu_{m,n}(a_2) = b'_1 \,\Theta_k\, b'_2 = \begin{cases} b'_1 \vee b'_2 = b'_1 & \Theta_k \equiv \text{OR} \\ b'_1 \wedge b'_2 = b'_2 & \Theta_k \equiv \text{AND} \end{cases}$$

3.6 RIGHT-HAND-SIDE COMPUTATIONS

The fourth step in the basic fuzzy inference algorithm consists of the implication computations. The implication relation has been defined in different forms in the literature and each one constitutes a legitimate design option as explained in Chapter 8. The principle implication relation is formed by imposing the LHS result vector, which includes the DOF of each composed rule, on the RHS fuzzy statements located after the

THEN operator. The input/output relation of this computational step is shown in Fig. 3.13.

As with the previous case of LHS computations, some definitions are introduced first to facilitate the understanding of the RHS computation algorithm.

RHS of a Fuzzy Rule

Referring back to the generalized representation of a fuzzy rule given by Eq. (3.12):

$$(X_k \bullet P_{k,m})\Theta_1(X_s \bullet P_{s,f})\Theta_2 \ldots \Theta_{n-1}(X_t \bullet P_{t,y})$$
$$\rightarrow (X_p \bullet P_{p,q})\, \Theta_{n+1} \ldots \Theta_{z-1}(X_z \bullet P_{z,u}) \quad (3.16)$$

we can now highlight the right-hand side (RHS) of the rule after the THEN operator in its most generalized form.

$$\rightarrow (X_p \bullet P_{p,q})\Theta_{n+1} \ldots \Theta_{z-1}(X_z \bullet P_{z,u}) \quad (3.17)$$

In its most frequently used form (based on the generalized *modus ponens* inference), the RHS statements are reduced to one statement per rule by interpreting the logic operators in different ways. As a result, each rule has one RHS statement expressed as

$$\rightarrow (X_p \bullet P_{p,q}) \quad (3.18)$$

where the other RHS statements may be restated by repeating the LHS statements. The treatment of multiple RHS statements is subject to design and is explained in Chapter 8. In the basic fuzzy inference algorithm, we will pursue a single RHS statement expression Eq. (3.18) per rule by assuming that fuzzy rules are converted into this form either during the design or by the original composer. The RHS fuzzy variables are also called consequent (or action) variables.

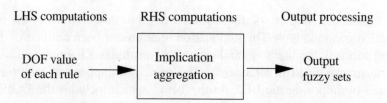

Figure 3.13 RHS computations of the basic fuzzy inference engine.

Chain Rule Syllogism

We also assume that the antecedent and consequent fuzzy variables are separated to the LHS and RHS of the rules, respectively, either by design or by the original rule composition. This means that no consequent appears on the LHS and no antecedent appears on the RHS of any rule. If the original form of rule composition involves a syllogism (where consequents become antecedents), the design can deal with it by separation. Using two simple fuzzy rules, this type of syllogism is expressed as

$$(X_k \bullet P_{k,m}) \rightarrow (X_p \bullet P_{p,q})$$
$$(X_p \bullet P_{p,q}) \rightarrow (X_r \bullet P_{r,s}) \quad (3.19)$$

where the RHS consequent fuzzy statement in the first rule reappears on the LHS of the second rule as antecedent. This set of rules is equivalent to the set below where the LHS statement in the first rule is repeated.

$$(X_k \bullet P_{k,m}) \rightarrow (X_p \bullet P_{p,q})$$
$$(X_k \bullet P_{k,m}) \rightarrow (X_r \bullet P_{r,s}) \quad (3.20)$$

This transformation may be better understood using the linguistic equivalents of Eqs. (3.19) and (3.20) in terms of a hypothetical set of rules.

IF Temperature IS Low THEN Flow IS High
IF Flow IS High THEN Power IS ON
⇔
IF Temperature IS Low THEN Flow IS High
IF Temperature IS Low THEN Power IS ON

As it can easily be inferred from above, the fuzzy statement *Flow IS High* functions as an intermediate condition linking two actual conditions *Temperature IS Low* and *Power IS ON*. The intermediate condition stated by the first rule may be kept in the inference engine if the fuzzy variable *Flow* must be monitored. The basic fuzzy inference algorithm is based on removing the chain rule syllogism to simplify computations. The chain rule syllogism does not apply to the following pair of rules.

IF Temperature IS Low THEN Flow IS Medium
IF Flow IS High THEN Power IS ON

Here, the consequent *Flow* in the first rule becomes an antecedent in the second rule, but they modify the different classes *Medium* and *High*. In this case, the antecedent in the second rule only receives an input from the rule above. Thus, it is a cascade arrangement as described in Chapter 7.

3.6.1 Implication Computation

In implication computations, the information from the LHS and RHS of a rule are input to an implication operator (often denoted by ϕ). The choice of the implication operator is subject to design and is explained in Chapter 8. Among them, one of the most commonly used implication operators is the Mamdani operator, and the implication operation is simply expressed as

$$\mu(x,y) = \phi[\mu_A(x),\mu_B(y)] = \mu_A(x) \wedge \mu_B(y) \quad (3.21)$$

$$\mu'_B(y) = I(x) \circ \mu(x,y) \quad (3.22)$$

where $\mu'_B(y)$ is the implication result. The equation above is based on a simple rule of the canonical form *IF X is A Then Y is B,* where X and Y are the antecedent and consequent fuzzy variables, and A and B are the predicates represented by the membership functions $\mu_A(x)$ and $\mu_B(y)$. Example 3.6 illustrates this process using the Larsen implication operator.

In general, the LHS of a rule is not restricted to one statement like *X is A;* instead, it may contain composite statements involving logic operators. Splitting fuzzy computations into RHS and LHS parts facilitates the implication computations for rules having composite LHS structures. Accordingly, the implication computations given by Eqs. (3.21) and (3.22) must be modified. Example 3.7 shows one practical way to handle this case.

The implication computations based on fuzzy sets cannot be easily implemented when the inputs are continuous (i.e., the inputs do not match the singletons of the antecedent membership functions). To handle such cases, the implication process can be formulated in the functional form of calculus. For example, the clipping property of the Mamdani implication operator can be represented by the area under the DOF line as the implication result. Example 3.8 illustrates this technique, which is also known as the geometrical approach.

As the most generalized case, a technique using both the geometric and fuzzy set representations, which can handle composite rules and continuous inputs, is presented in Example 3.9. Note that this process can be designed in a number of ways; however, they should all yield the same results.

In Fig. 3.14, the indices show that a LHS result does not clip all of the membership functions of a fuzzy variable; only the one that is identified by its predicate, such as *Y is medium* where medium is designed as $\mu_j(x)$.

Chap. 3 The Basic Fuzzy Inference Algorithm   113

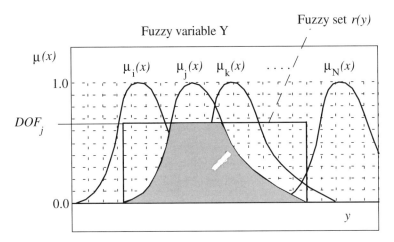

Figure 3.14 Clipping a consequent membership function by the LHS result DOF at step τ.

The fuzzy set $r(y)$ shown in Fig. 3.14 is explained in Example 3.7. Note that using a different implication operator, Eq. (3.21) will be in a different form; so will the type of clipping. The clipped membership function denoted by the shaded area is the result of one implication process. If discrete values are used to represent a membership function, then the clipped membership function $\mu_j(y)$ is a set $\{\mu_1 \; \mu_2 ... \mu_M\}$ including M data points (Fig. 3.15). Thus, the intermediate result stored in the implication algorithm can be a fuzzy set for convenience.

The RHS implication results are computed for every rule modifying

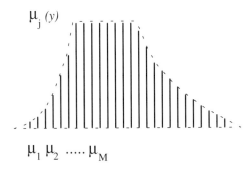

Figure 3.15 Discrete values representing a clipped membership function in the memory.

the same consequent fuzzy variable. This is illustrated in Fig. 3.16 for a hypothetical case with three rules

(Rule I)......*THEN Y IS Medium*
(Rule J)......*THEN Y IS High*
(Rule K).....*THEN Y IS Low*

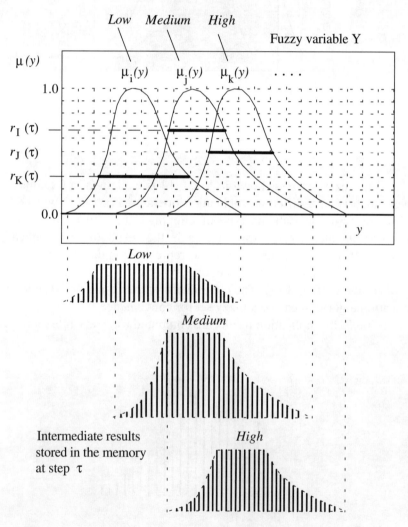

Figure 3.16 Implication results stored in the memory to be used in the next step.

where the predicates *Low, Medium,* and *High* are represented by membership functions $\mu_i(y)$, $\mu_j(y)$, and $\mu_k(y)$, respectively. The rule indices *I, J,* and *K* indicate that these rules may have been aligned in the rule base (memory) in any manner. The same freedom applies to the membership functions with indices *i, j,* and *k*, such that they do not have to be located in an increasing order in the memory. The bottom portion of Fig. 3.16 illustrates the discretized form of the clipped membership functions in the memory if the fuzzy set representation is employed in the algorithm.

EXAMPLE 3.6 IMPLICATION COMPUTATIONS FOR RULES IN THE CANONICAL FORM

Suppose we have the following fuzzy sets for the predicates of the rule *IF X is A THEN Y is B*.

$A(x) = 0/1 \cup 0.25/2 \cup 0.5/3 \cup 0.75/4 \cup 1/5 \cup 0.75/6 \cup 0.5/7 \cup 0.25/8 \cup 0/9 \cup 0/10$
$B(x) = 0/1 \cup 0.25/2 \cup 0.5/3 \cup 0.75/4 \cup 1/5 \cup 1/6 \cup 0.75/7 \cup 0.5/8 \cup 0.25/9 \cup 0/10$

Assume that this rule receives the following input from the external world:

$I(x) = 0/1 \cup 0./2 \cup 0/3 \cup 1/4 \cup 0/5 \cup 0/6 \cup 0/7 \cup 0/8 \cup 0/9 \cup 0/10$

This input, which is represented by a fuzzy set with its fourth singleton equal to 1.0, is equivalent to a scalar input 4. Applying the Larsen implication operator, Eq. (3.21) is modified to

$$\mu(x,y) = \phi[\mu_A(x), \mu_B(y)] = \mu_A(x) \cdot \mu_B(y)$$

$$\mu(x,y) = \begin{bmatrix} 0 \\ 0.25 \\ 0.5 \\ 0.75 \\ 1 \\ 0.75 \\ 0.5 \\ 0.25 \\ 0 \\ 0 \end{bmatrix} \cdot [0 \; 0.25 \; 0.5 \; 0.75 \; 1 \; 1 \; 0.75 \; 0.5 \; 0.25 \; 0]$$

$$\mu(x,y) = \begin{bmatrix} 0 & 0 & 0 & 0 & 0 & 0 & 0 & 0 & 0 & 0 \\ 0 & 0.06 & 0.13 & 0.19 & 0.25 & 0.25 & 0.19 & 0.13 & 0.06 & 0 \\ 0 & 0.13 & 0.25 & 0.38 & 0.50 & 0.50 & 0.38 & 0.25 & 0.13 & 0 \\ 0 & 0.19 & 0.38 & 0.56 & 0.75 & 0.75 & 0.56 & 0.38 & 0.19 & 0 \\ 0 & 0.25 & 0.50 & 0.75 & 1.00 & 1.00 & 0.75 & 0.50 & 0.25 & 0 \\ 0 & 0.19 & 0.38 & 0.56 & 0.75 & 0.75 & 0.56 & 0.38 & 0.19 & 0 \\ 0 & 0.13 & 0.25 & 0.38 & 0.50 & 0.50 & 0.38 & 0.25 & 0.13 & 0 \\ 0 & 0.06 & 0.13 & 0.19 & 0.25 & 0.25 & 0.19 & 0.13 & 0.06 & 0 \\ 0 & 0 & 0 & 0 & 0 & 0 & 0 & 0 & 0 & 0 \\ 0 & 0 & 0 & 0 & 0 & 0 & 0 & 0 & 0 & 0 \end{bmatrix}$$

The implication result is found by Eq. (3.22):

$$\mu'_B(y) = I(x) \circ \mu(x,y)$$

$\mu'_B(y) =$
$[0\ 0\ 0\ 1\ 0\ 0\ 0\ 0\ 0\ 0] \circ \begin{bmatrix} 0 & 0 & 0 & 0 & 0 & 0 & 0 & 0 & 0 & 0 \\ 0 & 0.06 & 0.13 & 0.19 & 0.25 & 0.25 & 0.19 & 0.13 & 0.06 & 0 \\ 0 & 0.13 & 0.25 & 0.38 & 0.50 & 0.50 & 0.38 & 0.25 & 0.13 & 0 \\ 0 & 0.19 & 0.38 & 0.56 & 0.75 & 0.75 & 0.56 & 0.38 & 0.19 & 0 \\ 0 & 0.25 & 0.50 & 0.75 & 1.00 & 1.00 & 0.75 & 0.50 & 0.25 & 0 \\ 0 & 0.19 & 0.38 & 0.56 & 0.75 & 0.75 & 0.56 & 0.38 & 0.19 & 0 \\ 0 & 0.13 & 0.25 & 0.38 & 0.50 & 0.50 & 0.38 & 0.25 & 0.13 & 0 \\ 0 & 0.06 & 0.13 & 0.19 & 0.25 & 0.25 & 0.19 & 0.13 & 0.06 & 0 \\ 0 & 0 & 0 & 0 & 0 & 0 & 0 & 0 & 0 & 0 \\ 0 & 0 & 0 & 0 & 0 & 0 & 0 & 0 & 0 & 0 \end{bmatrix}$

This operation is similar to matrix multiplication, where multiplying each element is replaced by finding the pair-wise minimum, and adding all the elements is replaced by finding the maximum.

$$\mu'_B(y) = [0\ \ 0.19\ \ 0.38\ \ 0.56\ \ 0.75\ \ 0.75\ \ 0.56\ \ 0.38\ \ 0.19\ \ 0]$$

To complete the fuzzy set representation, this vector is expressed as the union of singletons on the corresponding universe of discourse.

$\mu'_B(y) = 0/1 \cup 0.19/2 \cup 0.38/3 \cup 0.56/4 \cup 0.75/5$
$\cup 0.75/6 \cup 0.56/7 \cup 0.38/8 \cup 0.19/9 \cup 0/10$

EXAMPLE 3.7 IMPLICATION COMPUTATIONS FOR COMPOSITE RULES

Now, the same computation will be performed using the alternative formulation based on the DOF values. This method can be applied to rules with composite LHS statements. Applying Eq. (3.2), the membership value is computed by

$\mu_a = \vee[\mu_j(x) \wedge I(x)]$
$\mu_a = \vee[(0 \wedge 0)(0.25 \wedge 0)(0.5 \wedge 0)(0.75 \wedge 1)(1 \wedge 0)(0.75 \wedge 0)(0.5 \wedge 0)(0.25 \wedge 0)(0 \wedge 0)]$
$= 0.75$

The DOF value of the entire LHS of the rule is given by Eq. (3.14).

$$DOF = (\mu_{k,m})\Theta_1(\mu_{s,f})\Theta_2 \ldots \Theta_n(\mu_{t,y})$$

However, in this particular example we have only one LHS statement. Thus,

$$DOF = (\mu_a) = 0.75$$

Now, we modify the implication relations described by Eqs. (3.21) and (3.22). We create a new fuzzy set $r(y)$ on the same universe of $\mu_B(y)$ using the DOF value

$r(y) = DOF \cdot \mu_{B,n}(y) = 0.75 \cdot [1\ 1\ 1\ 1\ 1\ 1\ 1\ 1\ 1\ 1]$
$r(y) = [0.75\ \ 0.75\ \ 0.75\ \ 0.75\ \ 0.75\ \ 0.75\ \ 0.75\ \ 0.75\ \ 0.75\ \ 0.75]$

The implication result is computed by modifying Eq. (3.21) for the Larsen operator and using the new fuzzy set $r(y)$.

Chap. 3 The Basic Fuzzy Inference Algorithm

$\mu'_B(y) = \phi[r(y), \mu_B(y)] = r(y) \cdot \mu_B(y)$
$\mu'_B(y) = [(0.75 \cdot 0)(0.75 \cdot 0.25)(0.75 \cdot 0.5)$
$\qquad (0.75 \cdot 0.75)(0.75 \cdot 1)(0.75 \cdot 1)(0.75 \cdot 0.75)(0.75 \cdot 0.5)(0.75 \cdot 0.25)(0.75 \cdot 0)]$
$\mu'_B(y) = [0 \;\; 0.19 \;\; 0.38 \;\; 0.56 \;\; 0.75 \;\; 0.75 \;\; 0.56 \;\; 0.38 \;\; 0.19 \;\; 0]$

In this process, the application of Eqs. (3.2) and (3.14) provides us a practical method to deal with the composite rule structures involving logic operators such as AND and OR on the LHS.

EXAMPLE 3.8. IMPLICATION COMPUTATIONS USING THE GEOMETRICAL APPROACH

Because the inputs of the basic fuzzy inference algorithm in the practical world are often continuous, the implication computation illustrated in the previous examples cannot be used exactly as stated. We will illustrate the geometrical approach for the same problem examined in Examples 3.6 and 3.7. In this technique, every fuzzy set is converted into its functional form in calculus. Thus, $\mu_A(x)$ and $\mu_B(y)$ are given by

$$\mu_A(x) = \begin{pmatrix} 0.25(x-1) & 1 \leq x \leq 5 \\ -0.25(x-9) & 5 < x \leq 9 \\ 0 & 9 < x \end{pmatrix} \qquad \mu_B(y) = \begin{pmatrix} 0.25(y-1) & 1 \leq y \leq 5 \\ 1.0 & y = 6 \\ -0.25(y-10) & 6 < y \leq 10 \end{pmatrix}$$

Note that the notations do not refer to fuzzy sets anymore. The evaluation of the membership function using Eq. (3.2) is

$$\mu_a = \mu_A(x)|_{x=4} = 0.75$$

for the scalar input 4. The DOF value of the entire LHS of the rule is given by Eq. (3.14).

$$DOF = (\mu_{k,m}) \Theta_1 (\mu_{s,f}) \Theta_2 \dots \Theta_n (\mu_{t,y})$$

However, in this particular example we have only one LHS statement. Thus,

$$DOF = (\mu_a) = 0.75$$

The geometrical interpretation of the Larsen operator is

$$\mu'_B(y) = DOF \cdot \mu_B(y)$$

which yields the following implication result:

$$\mu'_B(y) = \begin{pmatrix} 0.187(y-1) & 1 \leq y \leq 5 \\ 0.75 & y = 6 \\ -0.187(y-10) & 6 < y \leq 10 \end{pmatrix}$$

This result is the functional equivalent of the result obtained in the previous example. Note that we have applied the same scalar input (4) in this example to illustrate the equivalency of each method. However, if the input were 4.23 or 3.88, this method would handle it in the same manner.

EXAMPLE 3.9 IMPLICATION COMPUTATION BY COMBINING THE GEOMETRICAL APPROACH AND THE FUZZY SET OPERATIONS

This example combines the last two methods given in Examples 3.7 and 3.8 so that the implication process can handle both composite rules and continuous inputs. We start from the geometrical approach by converting the fuzzy sets into the functional form in calculus only for the antecedent membership functions.

$$\mu_A(x) = \begin{pmatrix} 0.25(x-1) & 1 \leq x \leq 5 \\ -0.25(x-9) & 5 < x \leq 9 \\ 0 & 9 < x \end{pmatrix}$$

The evaluation of the membership function using Eq. (3.2) is

$$\mu_a = \mu_A(x)\big|_{x=4} = 0.75$$

for the scalar input 4. The DOF value of the entire LHS of the rule is given by Eq. (3.14).

$$DOF = (\mu_{k,m}) \Theta_1 (\mu_{s,f}) \Theta_2 ... \Theta_n (\mu_{t,y})$$

However, in this particular example we have only one LHS statement. Thus,

$$DOF = (\mu_a) = 0.75$$

At this point, we have already dealt with continuous inputs. The DOF value could be anything corresponding to any continuous input. Now, we will use the DOF value to carry on fuzzy set operations for the rest of the implication computation. The DOF value is used to create a new fuzzy set $r(y)$.

$r(y) = DOF \bullet \mu_{B,n}(y) = 0.75 \bullet [1\ 1\ 1\ 1\ 1\ 1\ 1\ 1\ 1]$
$r(y) = [0.75\ 0.75\ 0.75\ 0.75\ 0.75\ 0.75\ 0.75\ 0.75\ 0.75]$

The implication result is computed by modifying Eq. (3.21) for the Larsen operator and using the new fuzzy set $r(y)$.

$\mu'_B(y) = \phi[r(y), \mu_B(y)] = r(y) \bullet \mu_B(y)$
$\mu'_B(y) = [(0.75 \bullet 0)(0.75 \bullet 0.25)(0.75 \bullet 0.5)(0.75 \bullet 0.75)(0.75 \bullet 1)$
$\qquad\qquad (0.75 \bullet 1)(0.75 \bullet 0.75)(0.75 \bullet 0.5)(0.75 \bullet 0.25)(0.75 \bullet 0)]$
$\mu'_B(y) = [0\ 0.19\ 0.38\ 0.56\ 0.75\ 0.75\ 0.56\ 0.38\ 0.19\ 0]$

3.6.2 Aggregation

An arbitrary collection of fuzzy rules does not necessarily constitute a fuzzy inference algorithm unless the relationship among them is defined. The relationship of interest determines how much each fuzzy rule contributes to the output of the fuzzy inference engine. Let's consider an arbitrary set of fuzzy rules with the same consequent fuzzy variable.

Chap. 3 The Basic Fuzzy Inference Algorithm

$$S_{1,1}\Theta_{1,1}S_{1,2}\Theta_{1,2}...\Theta_{1,n-1}S_{1,n} \to (X_p \bullet P_{p,q})$$
$$S_{2,1}\Theta_{2,1}S_{2,2}\Theta_{2,2}...\Theta_{2,k-1}S_{2,k} \to (X_p \bullet P_{p,r})$$
$$.....\qquad\qquad\qquad ......$$
$$S_{Z,1}\Theta_{Z,1}S_{Z,2}\Theta_{Z,2}...\Theta_{Z,m-1}S_{Z,m} \to (X_p \bullet P_{p,s}) \qquad (3.23)$$

As indicated by the indices, the total number of rules is Z and the number of LHS statements in each rule is different, representing a general case. The RHS statements include the same consequent fuzzy variable X_p in these Z number of rules. When parts of a fuzzy inference engine (or the entire inference engine) contain Z number of rules with the same consequent variable, the aggregation of contributions from the related rules (all Z of them in this case) is determined by an operation sometimes referred to as ELSE[3] in the literature. For the consequent variable X_p, the aggregation expressed by an operator (denoted by $\Im$) between each implication result is obtained from the consequent predicates or from the corresponding membership functions. The aggregated result O_p for the pth consequent variable (in this case X_p) is given by

$$O_p = P'_{p,q} \Im P'_{p,r} \Im ... \Im P'_{p,s} \equiv \Im_p(P'_q, P'_r, ...P'_s) \qquad (3.24)$$

where P' represents individual implication results. How to obtain the individual implication results is explained in the previous section. Although fuzzy logic theory does not bring any restrictions on the variability of $\Im$, it is often employed in one manner throughout the same fuzzy inference engine and is frequently interpreted as union (max operation) among P's. Note that the choice of $\Im$ depends on the implication operator as explained in Chapter 8. As we mentioned earlier, each consequent fuzzy variable is one independent output of the basic fuzzy inference algorithm, which contains individual implication results as shown in Fig. 3.16. The implication results illustrated in Fig. 3.16 are aggregated in Fig. 3.17 by the union operation. The outcome of the aggregation is a new relation, or a new fuzzy set. The discrete form of the aggregated result shown at the bottom of Fig. 3.17 indicates that the overlapping regions are represented by a single data point that is the maximum between the coinciding two or more data points.

---

[3] ELSE operator is also used as OTHERWISE, as explained in Chapter 8.

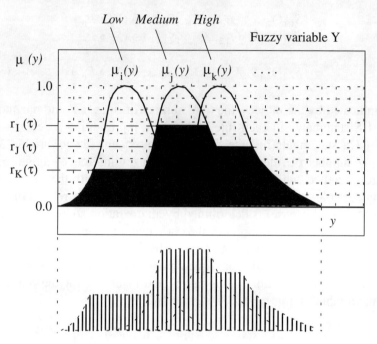

Figure 3.17 Aggregation of individual implication results via union operation.

3.6.3 Implication With Thresholding

Similar to thresholding of the antecedent fuzzy variables, a threshold may be applied to the implication results to eliminate residual truth. However, the consequent threshold affects the results in a different manner than that of the antecedent threshold. Figure 3.18 depicts the threshold operation for a single implication result. The terminal points D′ and E′ indicate that this is a subprocess of the algorithm between terminal points D and E.

The residual truth in the implication process emerges from a very small DOF value that is directly related to its LHS statements. To anticipate cases yielding small DOF is a more difficult task than to anticipate cases yielding a small membership value for a single membership function, because the DOF computation is obviously more involved. Accordingly, it is relatively a more difficult task to design a threshold for the RHS variables than for the LHS variables.

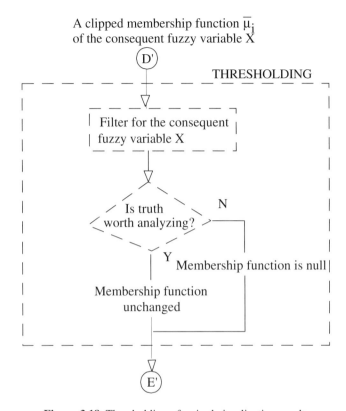

Figure 3.18 Thresholding of a single implication result.

When DOF is low, clipping of the consequent membership function often produces a thin layer of implication result as shown in Fig. 3.19. The α-cut threshold method is illustrated in Fig. 3.19.

Applying a threshold such as α-cuts is referred to as filtering in the block diagram in Fig. 3.18, whereas comparison between the threshold value and the clipped membership function is represented as a decision-making process. The case illustrated in Fig. 3.19 looks like trivial decision making. However, the block diagram in Fig. 3.18 is a generalized representation including other complex thresholding procedures where decision making may become more complicated.

When it is decided that a single clipped membership function is worthless based on a threshold, a null condition is imposed so that the clipped membership function would not contribute to the aggregated re-

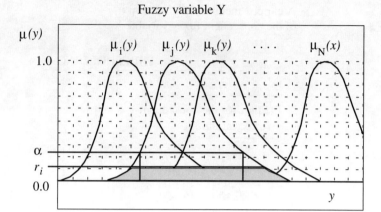

Figure 3.19 A small DOF value producing a thin layer of implication result below the α-cut threshold.

sult. Otherwise, the original clipped membership function will propagate through RHS computations.

A threshold affects the aggregated results when some of the contributing DOF values are small. This is illustrated in Fig. 3.20 for a hypothetical case where the DOF contribution from the Kth rule is considered worthless. In general, the threshold is the property of a consequent variable. However, the threshold may be designed to be the property of each membership function. In such a case, every RHS statement (the predicate and its corresponding membership function) may have a different threshold scheme.

The output vector from the implication computations between the terminal points E and D includes thresholded-aggregated results. Each element in the vector corresponds to the total response from one consequent fuzzy variable, such as the one shown in the bottom of Fig. 3.20. Thus, each element in the vector can be a fuzzy set. Using the discrete data convention, the output of the entire inference engine is an $N \times M$ matrix stored in the memory where N is the number of consequent (action) variables and M is the number of data points representing a fuzzy set.

As one can infer from above, higher threshold levels tend to isolate good rules from bad rules (for the specific input data at step τ) by eliminating contributions from the so-called bad ones. Note that the decision about good and bad, which is employed in a crisp manner using the α-cut threshold in the example above, may be done in a fuzzy manner using different thresholding schemes.

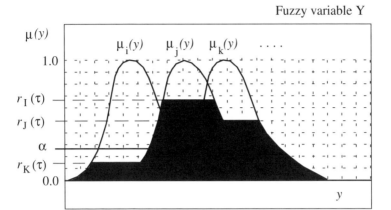
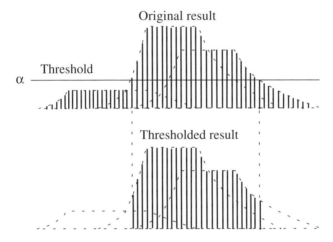

Figure 3.20 The effect of thresholding on the aggregated result.

EXAMPLE 3.10 THRESHOLDING AND AGGREGATION

Assume that an implication computation such as the one described in Example 3.9 was applied to three rules having the same consequent fuzzy variable Y:

IF X IS A THEN Y IS B
IF W IS C THEN Y IS D
IF Z IS E THEN Y IS F

Also assume that the implication results were represented by the following fuzzy sets:

$\mu'_B(y) = 0/1 \cup 0.19/2 \cup 0.38/3 \cup 0.56/4 \cup 0.75/5 \cup 0.75/6 \cup 0.56/7 \cup 0.38/8\ 0.19/9 \cup 0/10$
$\mu'_D(y) = 0/5 \cup 0.25/6 \cup 0.5/7 \cup 0.75/8 \cup 1/9 \cup 0.75/10 \cup 0.5/11 \cup 0.25/12 \cup 0/13$
$\mu'_F(y) = 0/3 \cup 0.1/4 \cup 0.22/5 \cup 0.33/6 \cup 0.22/7 \cup 0.1/8 \cup 0/9$

Suppose this basic fuzzy inference algorithm is designed with 0.4 consequent threshold. The thresholded fuzzy set is the union of singletons with possibility values higher than the threshold.

$$\mu^T(y) = \bigcup_{j=1}^{N} \mu_j(y)/y \quad if \mu_j(y) > \alpha$$
$$\mu_B^T(y) = 0.56/4 \cup 0.75/5 \cup 0.75/6 \cup 0.56/7$$
$$\mu_D^T(y) = 0.5/7 \cup 0.75/8 \cup 1/9 \cup 0.75/10 \cup 0.5/11$$
$$\mu_F^T(y) = 0/0$$

The threshold has completely eliminated the third fuzzy set above, which is considered to be residual truth of no significance. The two fuzzy sets are aggregated by the union operation provided that the implication operator used in this computation was Mamdani, Larsen, or some other operator used with the union aggregation.

$$\mu^A(y) = \mu_B^T \vee \mu_D^T(y) = 0.56/4 \cup 0.75/5 \cup 0.75/6 \cup 0.56/7 \cup 0.75/8 \cup 1/9 \cup 0.75/10 \cup 0.5/11$$

3.6.4 Effect of Weights on Aggregation

The weights are assigned per rule and they have two purposes in general. First, they can be used as control flags during the testing and validation of a fuzzy inference algorithm by simply assigning 0 or 1 to them. A rule with zero weight assignment will produce a null fuzzy set as the implication result and thus will not contribute to the aggregated result. A rule with weight = 1, on the other hand, will preserve the original implication result and fully contribute to the aggregated result. By switching back and forth between 0 and 1, the performance of each rule can be examined selectively during testing.

The second purpose for rule assignment is to incorporate relative importance into each rule. An important point to remember is that the distribution of weights is only effective within the same consequent fuzzy variable due to aggregation process. Thus, only the rules contributing to the same aggregation process will observe relative importance. For example, a rule with weight = 2 contributing to consequent variable Y, and another rule with weight = 4 contributing to consequent variable Z do not mean that the latter will produce twice as strong an effect as the first rule. That is because they contribute to different consequent variables in this example. Also note that importance is not necessarily a possibility or probability measure.

In the basic fuzzy inference algorithm, the most appropriate way to compute the weight effects is to apply them after thresholding. Otherwise, a small (below threshold) implication result would be magnified by

Chap. 3 The Basic Fuzzy Inference Algorithm

a weight to eventually exceed the threshold. This is misleading because the importance of a rule is meaningful only when the original possibility values are above the threshold.

Among the few viable alternatives, weight contribution can be incorporated into results by a simple scaling factor. This is illustrated in Fig. 3.21 where the membership on the right is scaled down by a hypothetical inhibitive weight relative to the membership function on the left. As one may suspect from this example, the inhibited implication result may fall below the original threshold level. In some cases, the same threshold may be applied again to preserve the integrity of computations, but in principle this is against the threshold concept. Scaling can be implemented in the reverse manner; that is, by increasing the height of the implication result on the left at the bottom of Fig. 3.21.

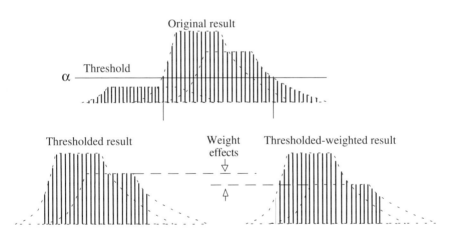

Figure 3.21 Effect of weight assignment on the aggregated result.

EXAMPLE 3.11 RULE IMPORTANCE WEIGHTS

Referring back to Example 3.10, assume that the rule importance weights were assigned during the design as follows:

> Weight = 1   IF X IS A THEN Y IS B
> Weight = 2   IF W IS C THEN Y IS D
> Weight = 3   IF Z IS E THEN Y IS F
> Weight = 2   IF U IS K THEN V IS M
> ....

Now, the thresholded implication results will be adjusted by multiplying the fuzzy sets before they are aggregated.

$$\mu_B^T(y) = W_1 \cdot \mu_B^T(y) = 0.56/4 \cup 0.75/5 \cup 0.75/6 \cup 0.56/7$$
$$\mu_D^T(y) = W_2 \cdot \mu_D^T(y) = 1/7 \cup 1.5/8 \cup 2/9 \cup 1.5/10 \cup 1/11$$

Note that the fourth rule does not contribute to the consequent Y. Therefore, it is not part of the weight adjustment process for Y. In addition, the third fuzzy set will not be included in the weight adjustment process because the threshold has eliminated it. If the weight adjustment were done before the threshold, the third implication result would be a part of the decision, which is conceptually incorrect. Thus, the order must be threshold, weight adjustment, and aggregation. The union aggregation of the two fuzzy sets is given by

$$\mu^A(y) = \mu_B^T \vee \mu_D^T(y) = 0.56/4 \cup 0.75/5 \cup 0.75/6 \cup 1/7 \cup 1.5/8 \cup 2/9 \cup 1.5/10 \cup 1/11$$

Compared to the result in Example 3.9, the aggregated fuzzy set is different, with higher possibility values on the right side of the universe. This set can be normalized if necessary during the defuzzification step. Considering the fourth rule, there will be a similar weight adjustment process for the consequent V among its contributors. For simplicity, this is not shown here.

3.7   OUTPUT PROCESSING

Step 5 between the terminal points E and F in Fig. 3.2 is the generation of outputs in such a way that they can be used in practical applications. Recalling the Step 4 computations, the input-output structure of this process is shown in Fig. 3.22.

There are three forms of information presentation per consequent variable at the output of a fuzzy inference engine.

- Scalar numerical output
- Linguistic output along with its possibility
- A fuzzy set representing a possibility distribution

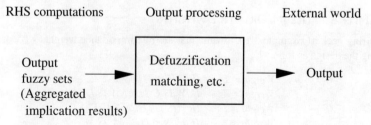

Figure 3.22 Output processing step in the basic fuzzy inference algorithm.

The first form is the most common one, especially in hardware implementation such as fuzzy control. Producing a scalar output from a fuzzy set is called the defuzzification (or decomposition) process and is explained in Chapters 2 and 8. Among the few viable options, the most commonly used method is the center-of-gravity defuzzification. Referring back to the aggregated implication result presented in the previous section, the center-of-gravity defuzzification is shown in Fig. 3.23.

One practical way to calculate the center of gravity is to compute the relative contributions of each singleton to the center-of-gravity point as given below.

$$Y = \frac{\sum_{j=1}^{N} y_j \cdot \mu_j(y)}{\sum_{j=1}^{M} \mu_j(y)} \qquad (3.25)$$

In Eq. (3.25), there are M singletons in the aggregated fuzzy set and each singleton is located at y_j. Because the aggregated implication results are fuzzy sets in two dimensions, the center of gravity is also considered as the center-of-area method. In a computational scheme, this operation is analogous (but not equivalent) to balancing a beam with fuzzy sets exerting gravitational force proportional to their areas, as shown in Fig. 3.24.

The balancing approach is relatively easier to implement because the aggregation step can be omitted. If the rule importance weight adjustments are not implemented before the aggregation process by the scaling technique, then it can also be implemented using this method.

$$Y = \frac{\sum_{j=1}^{N} w_j \overline{C}_j \overline{A}_j}{\sum_{j=1}^{N} w_j \overline{A}_j} \qquad (3.26)$$

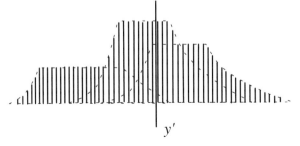

Figure 3.23 Center-of-gravity defuzzification.

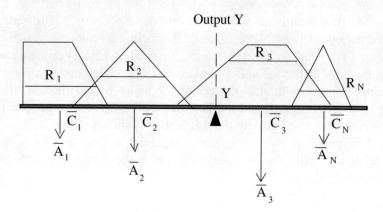

Figure 3.24 Balancing beam analogy for the centroid defuzzification method.

In Eq. (3.26), w, C, and A are weight, center of gravity, and area of each individual implication result, respectively. Note that Eqs. (3.25) and (3.26) produce different results although the differences are often negligible. The difference is due to multiple computations of the same intersected areas in the latter approach.

EXAMPLE 3.12 OUTPUT PROCESSING

Suppose that we have the aggregated implication result defined by the following fuzzy set

$$\mu^A(y) = 0.56/4 \cup 0.75/5 \cup 0.75/6 \cup 1/7 \cup 1.5/8 \cup 2/9 \cup 1.5/10 \cup 1/11$$

which is the result obtained in Example 3.10. Using Eq. (3.25)

$$Y = \frac{\sum_{j=1}^{N} y_j \mu_j(y)}{\sum_{j=1}^{M} \mu_j(y)}$$

$$= \frac{1}{9.06}[4(0.56) + 5(0.75) + 6(0.75) + 7(1) + 8(1.5) + 9(2) + 10(1.5) + 11(1)] = 8.11$$

the scalar output is 8.11. Note that the computation above is algebraic, which does not include any fuzzy set operation.

The second approach given by Eq. (3.26) operates in a different manner. In this case, we use the implication results before they are aggregated and before weight adjustments are made. Referring back to Example 3.10, we have the following thresholded fuzzy sets:

$$\mu_B^T(y) = 0.56/4 \cup 0.75/5 \cup 0.75/6 \cup 0.56/7$$
$$\mu_D^T(y) = 0.5/7 \cup 0.75/8 \cup 1/9 \cup 0.75/10 \cup 0.5/11$$

The center of gravity of each fuzzy set above is known by design; thus, they can be stored in the memory to speed up computations. They are $C_B = 5.5$ and $C_D = 9$ for $\mu_B^T(y)$ and $\mu_D^T(y)$, respectively. In addition, these membership functions, if they are designed to be symmetric, will preserve their center of gravity in the case of thresholding as shown in Example 3.10. The area of each fuzzy set is computed simply by adding the possibilities.

$$A_B = \sum_{i=1}^{M} \Delta y \cdot \mu_{B,i}(y) = 1(0.56) + 1(0.75) + 1(0.75) + 1(0.56) = 2.62$$

$$A_D = \sum_{i=1}^{M} \Delta y \cdot \mu_{D,i}(y) = 1(0.5) + 1(0.75) + 1(1) + 1(0.75) + 1(0.5) = 3.5$$

The rule importance weights were assigned as $W_B = 1$ and $W_D = 2$ in Example 3.10. Using the w, C, and A values, the center of gravity is

$$Y = \frac{\sum_{j=1}^{N} w_j \overline{C_j} \overline{A_j}}{\sum_{j=1}^{N} w_j \overline{A_j}} = \frac{w_B C_B A_B + w_D C_D A_D}{w_B A_B + w_D A_D} = \frac{1 \cdot 5.5 \cdot 2.62 + 2 \cdot 9 \cdot 3.5}{1 \cdot 2.62 + 2 \cdot 3.5} = 8.04$$

In the universe of discourse interval $11 - 4 = 7$ units, the difference between the two results obtained above ($8.11 - 8.04 = 0.07$) corresponds to 1 percent deviation. Neither of the results are right or wrong; they are simply different.

The second form of information presentation at the output is done linguistically by applying a selection criterion to choose an output class. In its simplest form, the selection criterion can be the maximum height (or category matching). In Fig. 3.25, the linguistic output would be *Medium* with the possibility of 0.77. Note that this kind of output is only useful in applications to social sciences, business, or medicine in which an answer stated in natural language has merit. The third form of output presentation, which includes the output fuzzy set itself, is not widely used because the existing hardware or software systems are not suitable to process such information.

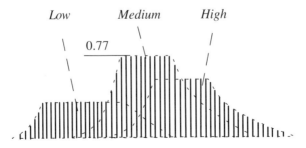

Figure 3.25 Linguistic output selection from an aggregated output fuzzy set.

PROBLEMS

3.1. Consider a fuzzy inference system given by the following rules:

IF Color IS Dark AND Texture IS Soft THEN Fruit_quality IS Rotten

IF Color IS Light OR Texture IS Hard THEN Fruit_quality IS Ripe

The membership functions for *Dark, Light, Soft, Hard, Rotten,* and *Ripe* are expressed by the following fuzzy sets.

$A_{dark}(x) = 0/1 \cup 0.25/2 \cup 0.5/3 \cup 0.75/4 \cup 1/5 \cup 0.75/6 \cup 0.5/7 \cup 0.25/8 \cup 0/9$

$A_{light}(x) = 0/6 \cup 0.25/7 \cup 0.5/8 \cup 0.75/9 \cup 1/10 \cup 0.75/11 \cup 0.5/12 \cup 0.25/13 \cup 0/14$

$A_{soft}(x) = 0/1 \cup 0.25/2 \cup 0.5/3 \cup 0.75/4 \cup 1/5 \cup 0.75/6 \cup 0.5/7 \cup 0.25/8 \cup 0/9$

$A_{hard}(x) = 0/6 \cup 0.25/7 \cup 0.5/8 \cup 0.75/9 \cup 1/10 \cup 0.75/11 \cup 0.5/12 \cup 0.25/13 \cup 0/14$

$A_{rotten}(x) = 0/0 \cup 0.25/0.5 \cup 0.5/1 \cup 0.75/1.5 \cup 1/2 \cup 0.75/2.5 \cup 0.5/3 \cup 0.25/3.5 \cup 0/4$

$A_{ripe}(x) = 0/3 \cup 0.25/3.5 \cup 0.5/4 \cup 0.75/4.5 \cup 1/5 \cup 0.75/5.5 \cup 0.5/6 \cup 0.25/6.5 \cup 0/7$

Evaluate the antecedent membership functions for the inputs Color = 3, Texture = 6 using the max-min composition operator.

3.2. Convert the fuzzy set representation of the membership functions of problem 3.1 into linear functions for both antecedents and consequents.

3.3. Evaluate the membership functions of problem 3.1 for the inputs Color = 3.3, Texture = 6.2 using the continuous function formulation. Compare the results with those obtained in problem 3.1.

3.4. Repeat the membership function evaluation steps of problems 3.1 and 3.3 by assuming the following antecedent thresholds: $\alpha_{color} = 0.55$, $\alpha_{texture} = 0.4$.

3.5. Find the degrees of fulfillment (DOF) for each rule in problem 3.1 with and without the antecedent thresholds for the two input sets: Set 1: Color = 3, Texture = 6, Set 2: Color = 3.3, Texture = 6.2. Use the antecedent thresholds specified in problem 3.4.

3.6. Using the DOF values computed in problem 3.5, create two fuzzy sets corresponding to the DOF value of each rule. Use the same universe of discourse of the related consequents membership functions. Repeat the process for the DOF values obtained from the thresholded operation in problem 3.5. (*Hint: All the elements of the fuzzy sets must have the same singletons.*)

3.7. Find the implication results (fuzzy sets) using the Mamdani implication operator, the fuzzy sets developed in problem 3.6 for clipping, and the consequent membership functions as specified in problem 3.1.

3.8. Apply a consequent threshold of $\alpha_{fruit} = 0.6$ to the implication results obtained in problem 3.7 and reform the implication results.

3.9. Scale each implication result (fuzzy set) obtained in problem 3.7 by the following rule importance weights: Rule 1: 3, Rule 2: 1. Two of the resultant fuzzy sets must be normal while the others must have maximum possibility values less than 1.0. (*Hint: Consider the fuzzy set operation of multiplication by a scalar.*)

3.10. Aggregate the results (fuzzy sets) obtained in problem 3.9 using the union operator. (*Hint: You must obtain two fuzzy sets corresponding to each input set specified in problem 3.5.*)

3.11. Find the defuzzified output using the centroid technique for each aggregated output found in problem 3.10. To accomplish this, apply the balancing beam analogy by computing the areas of each implication result before they are aggregated. Would the results be different if the consequent threshold $\alpha_{fruit} = 0.6$ had been applied?

3.12. Using the fuzzy sets defined in problem 3.1 for the consequent membership functions, apply the centroid defuzzification to each fuzzy set to find the maximum and minimum values that can possibly be obtained for the consequent fuzzy variable *Fruit_quality*. (*Hint: The minimum and maximum values must be different from the boundaries of the universe [0 - 7].*)

REFERENCES

[1] Terano, Toshira, Kiyoji Asai, Michio Sugeno. *Applied Fuzzy Systems*. New York: AP Professional, 1994.

[2] Kruse, R., J. Gebhart, and F. Klawonn. *Foundations of Fuzzy Systems*. Chichester: John Wiley & Sons, 1994

Chapter 4

Conceptual Design

> Design principles start from this chapter. At the conceptual stage, the designer prepares a solution strategy for a given problem by examining the system properties, identifying variables of the problem, and acquiring knowledge that can be used in a fuzzy inference engine in terms of fuzzy IF-THEN rules. The methods and principles stated from this point on are universal, but are non-unique. In other words, they can be replaced by the designer's own methods provided that each new method is justified by the specifics of the problem at hand.

4.1 INTRODUCTION

The conceptual design phase, which can be viewed as "design on paper," involves research and inquiry to acquire an overall design strategy for the problem at hand. In the beginning of this chapter, overall design is defined in the form of a list of elements. When each element is designed, the overall design is considered complete. All other considerations beyond the elements of this list (e.g., hardware or graphical user interface design) are also beyond the scope of the basic fuzzy inference algorithm.

Following the list of design elements, the most widely used design options are outlined. These options are adequate to develop a fuzzy algorithm (in the form of the basic fuzzy inference algorithm) that is as sophisticated as the commercial fuzzy systems listed in Table 1.2, in Chapter 1. The process of design and background requirements can be found in the same section. The design process includes seven steps, most of which are performed sequentially. Nevertheless, there is no reason to follow sequences different from what is suggested if they prove to be functional. Background requirements clarify the domain of occupations involved in fuzzy system development.

Knowledge acquisition is the first conceptual design step. The quality of the basic fuzzy inference algorithm strongly depends on the level and accuracy of the knowledge incorporated into it. If the problem at hand requires expert knowledge, then the knowledge acquisition step involves interaction with experts. Knowledge acquisition may also be performed by extracting information from process data or from analytical expressions. In some problems, commonsense reasoning is adequate without an expert opinion. Several strategies for knowledge acquisition are discussed in this chapter.

Next is the process of inference engine development and the use of linguistic design criteria. The linguistic design criteria presented in this section, which include conservatism, tolerance, optimism, pessimism, robustness, sensitivity, responsiveness, sluggishness, precision, cost, and accuracy, are used in every aspect of design. It is also shown throughout the book that these criteria are the only generalized tool to interpret the design options. Beyond these criteria, either the design choices are quite problem specific or there is not enough information available to complete the design.

Systems and problem types are defined in the last section of this chapter. The purpose of these definitions is to outline the logistics of information handling during design. Following the generalized definitions of systems and information domains, the problem types are identified as forward, inverse, heteroassociative, autoassociative, or complex.

Beginning with this chapter, several design principles are stated that are separated from the main text by boxes. These principles should be interpreted carefully based on the following argument. Designing a fuzzy system has practically infinite degrees of freedom emerging from individualism, style, belief, experience, and creativity. In addition, the objective in fuzzy system design is to provide approximate but "good enough" solutions—solutions that are prone to subjectivity and interpretation. When both goals and performance measures are abstract in this way, it is not possible to apply standard mathematical methods (as in designing an electronic circuit) and to formulate step-by-step instructions for design. Most of the design principles listed in this book describe a behavior in a generalized manner. Thus, the design principles are intended to provide the reader insight and conceptual maturity rather than design prescriptions. As a warm-up exercise, let's state a design principle describing the general philosophy of fuzzy system design.

> Fuzzy system design is an attempt to systematize the natural variations in *human* perception of truth and to imitate rudimentary skills of approximation.

Fuzzy system design is complicated by an unavoidable interrelatedness among its design elements. For example, fuzzy variable design depends on fuzzy rules and decision boundaries, whereas composed rules may require modification of the initial fuzzy variable design. This makes it impossible to lay out a design procedure with concomitant and self-sufficient steps. Therefore, the overall design philosophy is always an iterative one, often requiring going back and forth between the design elements for finer adjustments. Although the design principles stated in this book are dedicated to the calculus of fuzzy IF-THEN rules, they are, to a certain extent, applicable to a broader class of design problems in which solutions are achieved through techniques other than IF-THEN form.

4.2 FUZZY SYSTEM DESIGN AND ITS ELEMENTS

Fuzzy system design in the form of IF-THEN rules consists of deciding what form the design elements should take. The most important design elements scattered around the Appendix and Chapter 3 are collected in Table 4.1 in a condensed form. When these options are determined, the fuzzy system design is considered complete. Depending on the level of complexity, some of the options listed in Table 4.1 may be omitted. The simplest design is marked by asterisks (*) in Table 4.1, indicating the minimum subset to be completed. Translating the design into a computer program or a hardware product is not necessarily an integral part of fuzzy

Table 4.1 The Elements of Design of the Basic Fuzzy Inference Algorithm

CATEGORY	ELEMENTS
ANTECEDENTS	Universe of discourse* Name convention* Membership function* Threshold Linguistic hedge Linguistic input libraries Input fuzzifier
CONSEQUENTS	Universe of discourse* Name convention* Membership function* Threshold Normalization* Output processor
RULES	Logic operators* Rule formation strategy Implication operators* Aggregation operators* Defuzzification* Importance weights

* Minimum subset of design elements to complete a fuzzy inference design

system design, although such a development may be an unavoidable step. The product development phase requires background beyond the scope of this book.

A very important point to remember is that a specific solution to a given problem is not an element of design; rather, it is given or extracted. For example, designing a computerized diagnostics system for the mechanical problems in automobiles requires someone who knows the subject (preferably a mechanic). This knowledge is not to be invented by the fuzzy system designer; instead, it is to be acquired from an expert or extracted from a data set in an appropriate manner. This is one of the design processes discussed in the next section.

The elements listed in Table 4.1 are put under three categories for simplicity. Now we will define briefly what these elements are. Later in the book, their design is explained in more detail. Note that the order shown in Table 4.1 is not the order of the design process.

4.2.1 Antecedent Fuzzy Variable

Universe of Discourse

The numerical range of each antecedent fuzzy variable must be determined by the designer. This is related to membership function development and decision boundary analysis. In many cases, the membership function development stage automatically determines the universe of discourse. The extent of the universe of discourse of an antecedent fuzzy variable indicates the range of inputs in which the fuzzy inference engine is expected to be valid.

Name Convention

The designer must select appropriate names for the fuzzy variable, its predicates (membership functions), and hedges that might possibly appear as external inputs to the system. Because of the transparency property of fuzzy systems, word selection is important (like selecting words when writing an essay).

Membership Function Line Style, Height, Overlap, and Location

One of the most challenging design tasks is to determine the membership functions for each fuzzy variable. Although experiential evidence suggests that membership function shape is often inconsequential, it cannot be designed arbitrarily. The location and the granularity (i.e., the number of membership functions for each variable) are the two relatively more important design attributes. Antecedent membership functions determine how the conditional part of a rule will be satisfied given external inputs. The level of satisfaction (i.e., degrees of fulfillment) determines the strength of action corresponding to these conditions.

Threshold

Another designable parameter is the possibility threshold. The designer can use this parameter to adjust the performance of the fuzzy inference engine. Its design involves the examination of decision boundaries.

Hedge

Linguistic hedges modify membership functions by means of analytical computations established in the literature. The nature of such computations is one of the design elements. In an algorithmic flow, hedges are possible linguistic inputs from the external world.

Linguistic Input Libraries

Because a fuzzy system can process linguistic inputs from the external world as well as numerical inputs, a process in which such inputs are decoded must be designed. One of the standard methods is to build an electronic library of words that includes the predicates and hedges used during the design, then to check external linguistic inputs during input absorption if they match.

Input Fuzzifier

In some cases, a crisp numerical input needs to be fuzzified due to uncertainties implicitly known. A fuzzifier design is similar to hedge design except that there is no linguistic word matching required.

4.2.2 Consequent Fuzzy Variable

Universe of Discourse

Consequent universe of discourse is a part of the action variable design, and its extent determines the range of output values. The design involves considerations such as paralysis (i.e., frozen output behavior) and edge effects.

Name Convention

As with antecedent design, word selection for the consequent fuzzy variable and its predicates is important to preserve the transparency property of fuzzy systems. However, it differs from antecedent design in that no hedge naming will take place for consequents. Linguistic hedges that are encountered during the knowledge acquisition process (e.g., an expert articulates an action by "very much true") are blended into the membership function design (e.g., for "true") or named likewise.

Membership Function Line Style, Height, Overlap, and Location

The appearance of the consequent membership functions determines how strong the decisions will be. Design considerations for consequent membership functions are quite different from those for the antecedents. Some unintentional design attributes may hamper the overall performance.

Threshold

Consequent thresholding is also one of the designable elements. However, its function is more difficult to determine than that of the antecedent thresholding. In most cases, consequent thresholding may be omitted.

Normalization–Denormalization

Consequent universe of discourse may require normalization (0–1) depending on the algorithmic technique selected. For example, the centroid defuzzification technique is often implemented in a normalized domain, and the results are denormalized to the original scale of the universe of discourse.

Output Processor

Often considered within the peripheral computations, output processing includes producing linguistic outputs after the defuzzification is completed. Its design is rather trivial and may include other user interface details.

4.2.3 Rule Composition

Logic Operators

Operators such as AND, OR, ELSE, and others are also subject to design even though their functions are very much standard. In addition to these linguistically recognizable operators, new ones can be designed for specific objectives. Some nonstandard operators are presented in Chapter 7.

Rule Formation Strategy

Strategy such as competitive, cooperative, or hierarchical rule formation allows the designer to develop different rule blocks with different objectives suitable to the problem at hand. Suggested forms presented in Chapter 7 are not prescriptions; rather, they facilitate the design by examples.

Implication Operator

Although there are only 10 implication operators listed in this book, more than 50 implication operators have been proposed in the fuzzy logic literature. Most of them are experimental. The choice of an implication operator is one of the most critical of the design elements. It is also one of the greatest challenges for designers because there is not much help from the theory.

Aggregation Operator

Aggregation means mixing the implication results in an appropriate manner. It must be designed by selecting among few viable options. In most cases, such a choice is dependent on the implication operator.

Defuzzification Method

Also known as decomposition, this process converts implication results from the possibility unit to the unit of the universe of discourse. Defuzzification itself is an approximation by representing the aggregated implication result (which is a fuzzy set) by a single numerical answer (a scalar).

Importance Weights

Not every rule has equal importance in the real world. This fact can be modeled by incorporating importance weights as another element of fuzzy inference design. The appropriate use of weights also allows the designer to perform a sensitivity/robustness analysis during the design stage. Moreover, a dynamic process for allocating weights can turn rules on and off as a useful testing strategy. Importance weights are often incorporated during the aggregation or defuzzification stage.

4.3 DESIGN OPTIONS, PROCESSES, AND BACKGROUND REQUIREMENTS

The list of design options given in Table 4.2 is only a fraction of the methods found in the literature. However, these are the most widely used and tested options in today's industrial applications. The options in brackets in Table 4.2 are relatively less popular.

There is no fixed order for the design of a fuzzy system. Nevertheless, the major design processes outlined in Table 4.3 cannot be shuffled

Table 4.2 Design Options Most Widely Used in Today's Fuzzy Inference Systems

Option	Examples
Knowledge acquisition	Articulated expertise, extraction from data, interpreting data, interpreting formulas
Membership function shape	Trapezoid, triangle, [bell shape, sigmoid]
Logic operators	AND, OR, NOT, ELSE, THEN, [Product, mean, Yager-AND, Yager-OR, consistency]
Implication operators	Mamdani, Larsen, [standard, drastic, Zadeh, Lukasiewicz, Kleen-Dienes, bounded, Gougen, Godelian]
Defuzzification	Center of gravity, maximum possibility, [Center of maximums]
Aggregation	Union, intersection
Threshold	α-cuts, [nonlinear thresholding]

very much from their order of appearance. There are three main steps that are separated by the differences of (1) professions involved, (2) the medium or platform of design, and (3) related theories. The first block of design activities (items 1 and 2 in Table 4.3) includes conceptual design where a specific solution is acquired from experts or extracted from a data set. Also known as the *knowledge acquisition* process, this design step requires interaction between the fuzzy system designer and the expert in the field. The formation of preliminary fuzzy rules includes the identification of fuzzy variables, the determination of the number of categories (membership functions) for each fuzzy variable, consideration of linguistic design criteria, and examination of the properties of the problem at hand.

The second block of design activities (items 3, 4, and 5 in Table 4.3) is performed by the fuzzy system designer, who uses fuzzy logic theory to design the elements of the fuzzy inference system. The third and final block of activities is performed by computer scientists, engineers, and programmers, who translate the fuzzy inference design into a software and/or hardware product.

The background requirements to build a fuzzy inference engine are listed in Table 4.4 for the most general case. This book is aimed at making the reader a fuzzy system designer regardless of the reader's profession and background. The expert in the field, the first line in Table 4.4, can be anyone who has knowledge of the problem at hand. However, an expert in the field of design and a hardware/software specialist are always needed to build a working fuzzy system.

Chap. 4 Conceptual Design

Table 4.3 Major Design Processes

STAGE	DESIGN PROCESS
KNOWLEDGE ACQUISITION	1. Knowledge extraction/acquisition 2. Formation of preliminary fuzzy rules
FUZZY LOGIC THEORY	3. Designing fuzzy variables 4. Designing rule composition 5. Designing rule implication
PRODUCT DEVELOPMENT	6. Software and hardware development 7. Testing and validation

Table 4.4 Background Requirements to Build a Fuzzy Inference Engine

Expert in the field
Fuzzy system designer
Hardware and software specialist

EXAMPLE 4.1 REQUIRED KNOWLEDGE DOMAIN FOR FUZZY SYSTEM DESIGN

To design a fuzzy-logic-based missile control system, the required knowledge domain must include inputs from a software/hardware engineer, a fuzzy system designer, and an expert in missile control. Although such combined expertise may be found in one person for this case, the design of a complete fuzzy system normally requires teamwork to attain all the required expertise in the different areas involved.

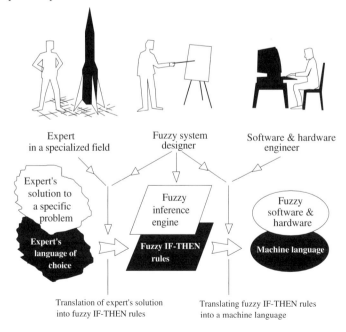

Simpler alternative design processes can be realized by eliminating the need for some of the expertise. For example, a data set can be used to extract the knowledge necessary to form fuzzy rules. This process can be performed under the supervision of an expert in the field, or the expert knowledge can be blended with the rules extracted from data.

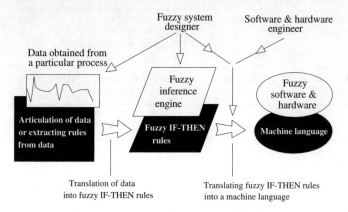

However, simplification of the design process has a price. When knowledge is extracted from a data set in the absence of an expert in the field, the resulting inference engines often do not reflect the full value of the information contained in the data.

4.4 KNOWLEDGE ACQUISITION

Knowledge acquisition extracts useful and reproducible knowledge from (1) experts, (2) data sets, (3) textbook information, and (4) commonsense reasoning applied to a specified objective. Knowledge acquisition has gained an academic definition in the field of artificial intelligence by the emergence of expert systems technology. Some tools, such as TEIRESIAS [1], SEEK [2], and Personal Construct Theory [3] have been developed to automate or to standardize the knowledge acquisition process with limited success. One of the new knowledge investigation paradigms in the medical diagnostics field, for example, suggests that the designer define sets of goals, data, and reasoning and then investigate the knowledge environment, working environment, and expert environment [4].

Knowledge acquisition, in the context of fuzzy system design, can be simple or complex. Simple knowledge acquisition uses commonsense reasoning. For example, if we were developing a fuzzy speech recognizer that maps sound signals to words in English, we would not need a speech pathologist to identify the sound *YES* for us. From the daily experiences of life, anyone can guess between the different tones and accents of the sound *YES*. Complex knowledge acquisition requires commonsense reasoning too, but it also requires experience and expert opinion. Building a medical diagnostics system definitely requires expert physicians. Examples are countless, but to name a few, we have listed simple and complex knowledge acquisition cases in Table 4.5.

As with most of the concepts involved here, the boundary between simple and complex knowledge acquisition is fuzzy. The purpose of such a categorization is to make the reader realize that the term *expert* used in this book does not necessarily refer to rocket scientists, but it does refer to varying levels of accumulated knowledge. The systems that only require simple knowledge acquisition are not simple to develop. In fact, they are at least as (if not more) complex than the systems that require a complex knowledge acquisition process.

Knowledge acquisition from subject matter experts is complicated by unavoidable characteristics of experts. One of the unfortunate facts of life is that two experts almost never have identical views. Recognizing this, the scientific community usually does not rely on one expert's opinion. Expert decisions are usually made through consensus. For example, medical doctors often advise their patients to get a second opinion in cases of serious illness. Multiple consultations are common practice for serious matters in life. Similarly, fuzzy inference systems are often designed to incorporate expertise from multiple sources.

Table 4.5 Simple and Complex Knowledge Acquisition Cases

SIMPLE KNOWLEDGE ACQUISITION CASES		COMPLEX KNOWLEDGE ACQUISITION CASES	
SYSTEM	EXPERT	SYSTEM	EXPERT
Automated car steering system	Any decent driver	Cancer diagnostics	Medical doctors + biologists + ...
Target recognition system in military applications	Any soldier who can recognize the enemy installations	Stock portfolio management	Stock brokers + investors +....
Speech recognition systems	Anyone who knows the language	Ecological modeling	Ecologists + environmental engineers + ...
Automated climate management systems	Anyone who can guess comfortable climate conditions	Air traffic control system	Air traffic controllers + pilots +....

Multiple consultations with several experts inevitably leads to conflict. Between the simple and complex forms of knowledge acquisition as listed in Table 4.5, conflict resolution poses different characteristics. In the simple case, we can easily overcome conflicts by increasing the number of opinions and by approaching the problem in a probabilistic manner. For example, it is easy to form a consensus about "what is parking your car too close to the sidewalk" among 20 people. A conflict that appears among three opinions will most likely diminish among 20 or more opinions.

Consensus approach may not be practical in the complex knowledge acquisition case when expertise is expensive and rare. In such cases, a voting scheme can be applied instead of the consensus approach. To employ voting, different rules composed by different experts are assigned weights in proportion to each expert's level of knowledge and confidence level. Fuzzy voting schemes are used in practice with success (such as in Delphi Systems [5]).

4.4.1 Questioning Approach

Just as different lawyers develop different styles of questioning from years of experience, fuzzy system designers often develop their own style by trial and error. There is no single approach to knowledge acquisition questioning for fuzzy inference systems. Nor will we attempt to claim one in this book. However, we propose a list of important issues given in Table 4.6 that covers practically all the design elements listed earlier. The order of questions should follow the order of the table. This sequence represents one successful approach to systematic knowledge acquisition by questioning experts. Nevertheless, iterations may become necessary in complex problems. Note that this list is also applicable to other modes of knowledge acquisition, including data interpretation and formula interpretation. The issues of Table 4.6 include the following details.

A: In complex problems identifying input and output variables may be a difficult issue. If all the variables cannot be quantified or identified, the fuzzy system designer might have to proceed with only a partial set of variables. For example, economic models are often developed by assuming the existence of a certain set of variables and by accepting that the variables ignored due to lack of data may be influential in real life. Furthermore, there may be unknown input variables.

Table 4.6 A Typical Questioning Sequence for Knowledge Acquisition

A: Identify input and output variables of the problem.
B: Separate crisp and fuzzy variables by best judgment.
C: State the rules using the variables defined and using as many linguistic predicates as possible, provided they represent a variation in logic.
D: Identify the possibilities implied by the predicates quantitatively.
E: Determine a level of threshold for the lowest meaningful possibility value.
F: Differentiate the precision constraints on the consequent variables.
G: Identify how the optimism/pessimism, tolerance/conservatism, responsiveness/sluggishness criteria apply to this problem.
H: Rank the rules according to their importance.
I: Identify what linguistic hedges may be applied in the antecedent domain.
J: Identify operational constraints.
K: Describe the expected performance from the fuzzy inference engine.

B is the first step that encourages the expert towards fuzzy thinking. By differentiating between fuzzy and crisp variables at the beginning, the fuzzy system designer can later alert the expert if he/she mistakenly creates predicates for crisp variables or if he/she never categorizes fuzzy variables during rule generation.

C represents the backbone of design. The expert must be asked to state the rules of the solution using as many predicates (*high, low, medium,* etc.) as possible provided that each category represents a variation in logic.

D prompts the fuzzy system designer to redirect the expert's attention to the characterization of the predicates in terms of possibility distribution (membership functions) after examining the previous steps.

E and D are closely related. Although the human brain appears to think in a fuzzy way, experts are often not able to characterize uncertainty easily. It gets even harder to quantify low possibilities. Thus, low-possibility threshold selection is often determined iteratively.

F is a quality control step by which the design of membership functions is revised. Precision constraints tell the fuzzy system designer which fuzzy variables are to be designed more carefully (e.g., higher granularity or more detailed membership function characterization).

G is also a quality control step. However, the answers are utilized in a different way by the fuzzy system designer. Optimism, tolerance, and responsiveness properties give the designer certain clues as to which design options to use. These properties are discussed later in this book.

H helps the fuzzy system designer to assign weights to different rules or rule banks.

I is a cross check in the presence of the expert to ensure that some of the linguistic hedges (especially those indicating usuality and likelihood) will not be contradictory. For example, *Mary is (always) young* may represent a contradiction (depending on the context) caused by coupling of the hedge (*always*) with the predicate *young*. If the linguistic hedges appear in the original articulation of expertise, they can always be treated as part of membership function development. If they are not articulated in the rules by the expert, then possible hedges that may appear as external inputs must be considered at this stage.

J requires identification of the constraints. Operational constraints may not be articulated as part of the rules. For example, in the design of an automated climate control system, the rules describing the air-conditioning and heating rates for a given climate condition may have not included the cost of operation. If there is a cost constraint, do we choose to tolerate less air-conditioning or less heating? Answers from the expert will determine weight assignment of the rules or rule blocks.

K is obviously very important for the designer to validate the design. This is normally the most time-consuming step because it takes the estimation of the outputs under all possible combinations of inputs. If the expected performance is not known by the expert, then there is a good chance that the rules will be ambiguous. If the expected precision is not known, it can be determined during testing.

4.4.2 Essays

One common approach to knowledge acquisition is either to let the expert articulate his/her expertise freely by writing an essay or to obtain expert knowledge from the literature. Then, the designer translates the essay to a fuzzy IF-THEN form. During translation, the questions shown in Table 4.6 must be answered, otherwise the fuzzy system designer must resolve conflicts with the aid of the expert. Translation from an essay is accomplished by analyzing each sentence separately to identify the variables involved and their predicates and to extract a logical proposition. Sometimes, one sentence yields multiple rules or multiple sentences yield one rule. Sometimes, new variables and predicates are created by the designer. There is no deterministic method to accomplish this process. However, we will suggest an approach in this book by outlining the basic steps of translation from an essay to fuzzy rules.

Identifying Antecedents and Consequents

The very first step is to identify the variables of the problem and to determine if they are fuzzy or crisp. Let's start with simple examples and progress towards more complicated ones.

Temperature is high.

In its simplest form, the sentence above implies that the entity *Temperature* is in a certain state of being, which may or may not change in another instance. Obviously *Temperature* is a variable and *high* is its state or category. Because *high* is a linguistic qualifier (i.e., a predicate), the variable is fuzzy. Note that there is no clue whether this variable is an antecedent or a consequent since there is no reason–action, cause–effect, or input–output relationship in this proposition. Now consider the following sentence.

Crank up the heater to high.

This sentence is similar to the previous one except it implies an action. *Heater* is a variable, possibly a consequent one depending on the context of other sentences associated with this one.

There are cases in which the designer must complete the variable definition by using extra words that are not explicitly stated in a sentence. Considering the previous sentence, the variable definition can be modified to *heater_capacity* or *heater_volume* to enhance its expressive power. Along the same lines, now consider the following sentences.

Mary is young.
Mary is short.

The fuzzy variables here are *Mary's_age* and *Mary's_height* instead of only *Mary*. As illustrated by these examples, the designer must interpret the context and complete the definitions.

As an example to more structured propositions, now consider the next sentence.

Temperature is drastically dropped; turn on the heater to high.

There is a clear reason–action relationship embedded in this sentence. Therefore, *temperature* is an antecedent fuzzy variable with its predicate *drastically_dropped*, and *heater* is a consequent with its predicate *high*.

The determination of whether a variable is fuzzy or not normally depends on the existence of linguistic qualification/quantification. However, not every categorization means fuzziness. For example, consider the next sentence.

> *When traffic light is red, slowly push the brakes.*

The antecedent *traffic_light* has a predicate *red*, which is a category. Obviously, the categories red, green, and yellow are crisp entities in this context; thus, this antecedent is a crisp one. On the other hand, the consequent *braking* is a fuzzy variable since it takes a linguistic value of *slowly* that is subject to design. Crisp variables are often used as Boolean gates for navigation within the rule base as explained in more detail in Chapter 7.

Conventions Used in This Book

Before presenting translation examples, let's review the conventions used in this book for fuzzy IF-THEN rules. Variable names and predicates are denoted by uppercase first letters (e.g., *Temperature*). The same applies to predicates such as *High*. Combined names are denoted by underscore signs, as in *Temperature_profile* or *Very_high*. Linguistic connectors including *is, is not, is less than, must be, was,* and so on, are all denoted by uppercase letters, as in *Temperature IS High*. The same consideration applies to logic operators such as AND, OR, and THEN. When we encounter these conventions in this book, we are encountering statements written in the language of fuzzy IF-THEN rules.

Relational Versus Compositional Propositions

There are two types of structures to look for in a given sentence. The first one is relational propositions indicating the state of being without any conclusion. Consider the following individual sentences.

> *Mary is young.*
> *David is tall.*
> *There is very little mist in the air.*

Sentences in this form correspond to the relational inference in fuzzy logic as described in Chapter 2. Because they only indicate single association, they cannot be translated into a conditional fuzzy rule as they are. However, a collection of them in a given context may be translated into a fuzzy rule. Now consider the following context.

Humidity is quite high. Clouds look very dark. It may rain anytime.

In this context, each relational proposition has a meaning that is also related to other propositions. This collection of propositions can be translated into a conditional fuzzy rule:

IF Humidity IS Quite_high AND Clouds ARE Very_dark
THEN Rain IS Likely

Note that the conditional form IF-THEN is created by the designer's own interpretation. Similarly, the two conditions are assumed to exist simultaneously by using the AND logic operator. This is only implicitly expressed in the context above. Thus, this interpretation is only valid if the translated fuzzy rule serves the design objective. The second type of sentence structure is the compositional proposition. In this type of sentence, the conditions and actions are explicitly stated. Let's consider the following sentence:

If temperature is high and humidity is low then drought risk is high.

The sentence above fits the fuzzy rule structure completely. Thus, we only use capitalization of letters to denote its translation.

IF Temperature IS High AND Humidity IS Low
THEN Drought_risk IS High

Possible mild deviations from the exact form are easily interpretable. For example, the following two examples taken from a daily conversation would yield the same translation as shown above.

Drought is always around the corner when temperature gets high and humidity becomes low.
The two main indicators of drought are high temperature and low humidity.

There are other cases in which the translation becomes quite difficult and sometimes impossible. It is the designer's judgment that plays a major role in translating essays into fuzzy rules. The steps shown below briefly outline this process.

- Look for actions to determine consequents.
- Look for reasons to determine antecedents.
- Identify predicates for fuzzy variables.

- Identify categories (Boolean gates) for crisp variables.
- Determine crisp versus fuzzy by examining the predicates.
- Extract conditional logic from the context that suits the design objective.
- Extract unconditional fuzzy statements from the context that can be used for navigation.

Note that when translation is complete, the design still continues through membership function design, decision boundary analysis, and other design steps explained in this chapter. Also note that the translation method can be used in other forms of knowledge acquisition, such as in questioning the experts.

EXAMPLE 4.2 KNOWLEDGE ACQUISITION FROM AN ESSAY

This example is about the design of a fuzzy decision maker that governs an automatic machine preparing sauce for a spicy dish. We are given the following instructions by the chef.

> Cook 3 cups of chopped onion and garlic in olive oil for 3 or 4 minutes over high heat, stirring frequently. Add parsley and continue cooking a few minutes longer. Add tomatoes, season with salt and pepper, stir once, cover, and simmer over low heat five minutes....

Assuming a sequentially implemented cooking operation, the sentences are translated into a set of fuzzy IF-THEN rules. The translation follows a time line (five time zones) as indicated in the rules below by horizontal lines.

IF Cooking_time IS Zero_minutes THEN Heat IS High
AND Add_onion 3_cups
AND Add_garlic 3_cups
AND Add_oil Some

IF Cooking_time IS Before_3_or_4_minutes THEN Stir_action IS Frequent

IF Cooking_time IS Before_3_or_4_minutes AND Amount_of_oil IS Low
THEN Add_oil More

IF Cooking_time IS 3_or_4_minutes THEN Add_parsley Some

IF Cooking_time IS Few_minutes_longer_than_4_minutes THEN Add_tomatoes Some
AND Add_salt Some AND Add_pepper Some AND Stir_action IS Once AND Flag IS OK
IF Flag IS OK THEN Lid_position IS Closed AND Heat IS Low

IF Cooking_time IS 9_minutes THEN Heat IS Zero

Chap. 4 Conceptual Design

ANTECEDENTS	PREDICATES	CONSEQUENTS	PREDICATES
Cooking_time	Zero_minutes	Stir-action	Frequent
	3_or_4_minutes		Once
	Before_3_or_4_minutes		
	Few_minutes_longer_than_4_minutes		
	9_minutes		
Amount_of_oil	Low	Add_onion	3_cups
Flag	OK	Add_garlic	3_cups
		Add_oil	Some
			More
		Add_parsley	Some
		Flag	OK
		Add_pepper	Some
		Add_salt	Some
		Add_tomatoes	Some
		Lid_position	Closed
		Heat	High
			Low
			Zero

The variables of the problem and the predicates are listed above. As can be seen in the table above, some of the predicates and variables are created by the designer to complete the fuzzy rule set. Now we will go through the questions of Table 4.6 to revise the inference system.

A: Antecedent and consequent variables are listed above.

B: Crisp variables are *Flag* and *Lid_position*. *Flag* is defined by the designer to separate one set of actions from another, (e.g., not to close the lid before the tomatoes are in).

C: Rules are stated by translation. We realize that most of the fuzzy variables have one predicate (one category of state). As long as the external inputs are numerical, one category will serve the purpose. However, the consequents represent single actions that produce scalar defuzzified output. In other words, some actions are crisp. If the actions will be taken by a human (by the definition of *Some Tomatoes,* etc.) then the actions will be approximated or fuzzy. As we can see from the essay, the predicate definitions in the rules yield consistent logic with respect to the chef's directions.

D: This step involves the design of membership functions labeled by the predicates in the rule base. For example, the membership function *3_or_4_minutes* will be designed on the time universe of discourse. If such a categorization requires an expert opinion, then the chef may be asked for help.

E: Lowest possibility determination is again part of the previous step and might require the chef 's opinion. However, a need for threshold design is not very clear in this example. Normally, thresholds are used in rules with the more complex decision boundaries often encountered during knowledge extraction from data.

F: This step requires the chef to point out selectively the consequents that must be more precise than others. For example, the precision of low heat may be much more important that the precision of frequent stirring. This type of information helps the fuzzy system designer to go back, if necessary, to refine the membership function design.

G: Improvement of the fuzzy inference system's performance depends on this step. The chef is needed again to interpret the fuzzy inference rules by considering the optimism/pessimism, tolerance/conservatism, and responsiveness/sluggishness criteria. For example, the chef may specify that the addition of oil must be done conservatively (that is, if the amount of oil is really low) or sluggishly (that is, not to start adding right away). The optimism/pessimism criterion does not apply here because there is no belief mechanism in this particular decision-making task. Once the criteria are defined, the fuzzy system designer may reflect the chef's style by selecting different implication operators or by revising the membership function design.

H: Ranking the rules according to their importances is usually done in diagnostic problems where there are many rules modifying the same consequent. In an automated cooking operation like this one, rule importances are not easily identifiable unless specified by the expert.

I: The linguistic hedges that might be encountered as external input are again mostly applicable to systems that are subject to continuous human interaction. In this example, a set of hedges may be considered for the third rule *IF Amount_of_oil IS (Very, Extremely, Not_quite,* etc.) *Low* with hedges (*Very, Extremely, and Not_quite,* etc.).

J: Assume that tomatoes are extremely expensive and the chef does not want to spoil the sauce by wasting too many tomatoes (other failures are more acceptable). Operational constraints like the one hypothesized here can be characterized like precision constraints. Other modes include assigning rule importances and adjusting thresholds. In this example, the fifth rule becomes important, along with the design of the membership functions adding tomatoes *Some* and cooking time *Few_minutes_longer_than_4_minutes*. Considering the failure scenarios, the rules corresponding to actions before adding tomatoes become more important than the ones after adding tomatoes.

K: The chef describes the taste, color, and smell of the sauce expected as the end product. If this information is not available, the fuzzy inference cannot be validated.

4.4.3 Interpreting Data

Most of the time, the knowledge acquisition process involves analyzing data along with articulating the expert's opinion and personal experience. In this context, data are assumed to contain invaluable knowledge to be used in the design of a fuzzy inference engine. For example, the medical condition of a heart is embedded in electrocardiogram data, or the state of the economy is embedded in financial data, and so on. Such data are only interpretable by experts in those fields. When working with experts, the fuzzy system designer's task is to make sure that the expert's interpretations of a data set yield a fuzzy inference mechanism in an appropriate form. The rule set that is formed by data interpretation may include statements and rules based on commonsense reasoning (or the linguistic articulation of expertise) only. The fuzzy system designer must address the issues listed in Table 4.7 for data interpretation in addition to those in Table 4.6 designed for the linguistic articulation of expertise.

L determines the reliability of decision making through linguistic interpretation. The most important consideration is whether the data set can yield generalized rules describing the actual mechanism rather than specialized rules describing a local event.

For example, should the interpretation of a set of data obtained from the stock market be considered general or local? If the expert is looking at IBM stock prices of the last two months, can rules be stated describing (1) what happened to IBM stocks in the last two months, (2) what will happen to IBM stock prices in the future, (3) what is happening to the IBM corporation, (4) what is happening to the domestic economy? In the absence of a special reason, normally only question (1) can be answered. The validity of the rules may be improved by interpreting more data. Note that for relatively simple systems such as the thermostat control, for example, a data set of certain size may be adequate to state fuzzy rules describing a generalized behavior.

M requires the expert to categorize certain elements of data under separate groups (fuzzy sets) such that the partitioning would serve for the best articulation of the logic. For example, if there are three modes of heart failure that can be detected on a cardiogram, then the designer must identify three categories—not four or two. This is not an easy task when decisions are more complicated than identifying three modes of failure on a single universe.

N helps the fuzzy system designer along with the expert to state mapping rules based on the granularity (i.e., the number of categories characterized by membership functions) selected in the previous step. Decision boundary analysis is explained in Chapter 5.

Table 4.7 A List of Questions for Knowledge Acquisition Via Data Interpretation

L: Assess the reliability of data and determine the extent of its validity.
M: Identify fuzzy categories by employing expert opinion that represents variation in logic, variation in symptoms, and variation in actions.
N: Perform, if possible, a decision boundary analysis using the fuzzy categorizations, and form the rules accordingly.
O: Evaluate the fuzzy inference for a set of pseudo inputs, then adjust fuzzy categorization to improve the performance.

O is a test procedure to validate the consistency between the rules and their granularity. The expert defines hypothetical (or actual if possible) input cases where all possible combinations are covered. If the test runs are not in agreement with the expected performance (see K in Table 4.6), then the steps M, N, and O must be repeated.

4.4.4 Extracting Rules From Data

Knowledge acquisition by extracting rules from data is similar to the previous case in principle. The difference is that data interpretation that requires experts is often needed for complex problems, whereas rule extraction from data is an algorithmic process often based on data that can be interpreted by commonsense reasoning. However, both approaches always converge in industrial applications because commonsense reasoning is usually checked against expert knowledge.

One drawback of this method is that when every data point is used for mapping, all information embedded in data will be translated to fuzzy rules regardless of whether the rules make sense. After the rules are outlined, the ones that do not make sense must be ruled out or modified by the designer. This can be a tedious task for systems with hundreds of rules. There is no deterministic method to accomplish the process of extracting rules from data. However, an approach is suggested in Table 4.8 for the reader's convenience.

To understand the steps in Table 4.8, the cluster concept must be understood first. A *cluster,* also referred to as a *category* in this book, is a collection of data points that represents the same semantically justified property. For example, in the context of the height of a 30-year-old man,

Table 4.8 Steps for Extracting Fuzzy Rules From Data

A: Select an iteration strategy for data clustering, either from few clusters to many (from approximate to precise) or from many clusters to few (from precise to approximate).
B: Create clusters of data both on the input and output spaces for each fuzzy variable by using an objective function such as distance.
C: Identify which input clusters correspond to which output clusters.
D: Articulate fuzzy rules by appropriately naming the clusters and by designing membership functions.
E: Check if there are any conflicting rules. If there are, adjust the corresponding clusters; go to step C.

the data points 5'2", 4'8", and 5'1" can be grouped to represent the category *short*. Similarly, the data points 6'7", and 7'1" can be grouped as the category *tall,* and 6'1" and 5'9" as the category *medium*. We have created three clusters of data called *short, medium,* and *tall* in this example in an arbitrary way. We could have created more clusters or less. The appropriate number of clusters is determined by decision boundaries and by the criterion of precision versus cost. This is explained in Chapter 5. In this example, grouping data points close to each other means we have used *distance* as the objective function for clustering. Cluster boundaries can also be defined by membership function design, which is also explained in Chapter 5.

Once we have the clusters, the next step is to determine the association between the clusters of different fuzzy variables by examining the data. For example, consider a set of data collected from the real world that represents height and weight, such as [6'5", 230 lbs.]. If we have created a cluster called *tall* that entails 6'5", and another cluster called *heavy* that entails 230 lbs., then the fuzzy rule extracted from this data would be *IF Height IS Tall THEN Weight IS Heavy*. Although it looks simple in this example, this process may become complicated when there are complex decision boundaries embedded in the data. Often, conflicts may arise due to improper clustering. If conflicts occur, clustering must be repeated iteratively either from few clusters to many or vice versa.

Creating clusters blindly only by examining the data domain of individual fuzzy variables can be a tedious task when iterations are required due to conflicts. An alternative approach is to create clusters by examining the relationship among data points. This can be done by starting from the first data set such as [6'5", 230 lbs.], then finding the next data set with a similar relationship, such as [6'4", 228 lbs.]. Once similar relationships are identified, clusters can be created, such as [6'4", 6'5"] for height and [228 lbs., 230 lbs.] for weight. This is followed by selecting proper names for the clusters and articulating fuzzy rules. This approach changes the steps B and C in Table 4.8 simply by reversing their order.

EXAMPLE 4.3 EXTRACTING FUZZY RULES FROM A DATA SET

In this example we will model the new car buyer's thinking process based on price and number of problems in the first year (from consumer reports). The sales data reflect how the buyers think but do not state the thought process explicitly. To generate such an explicit representation, we need to articulate it by rule extraction from data. There are 12 car models labeled as A, B, ... L. The sales data are shown below.

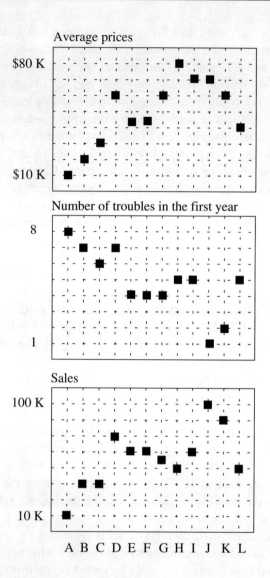

Clustering here is done by deciding on the number of clusters to generate, then by using "distance" as the criterion. The relationships are shown in the next figure by vertical lines. Note that the data points are aligned in a proper order in this example for simplicity. In real life applications, the data points may be more scattered, which causes a more complex appearance. Clusters are formed by horizontal lines and are also shown in the next figure.

Chap. 4 Conceptual Design 157

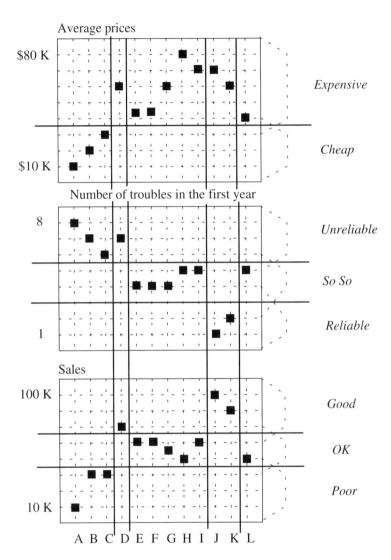

Except for model D, the categorization yields a set of fuzzy rules describing the consumer's thinking when buying a new car. Membership functions are also shown above by dotted lines across from the predicates for each fuzzy set. Fuzzy rules extracted from this data set are stated below.

IF Price IS Cheap AND Car IS Unreliable THEN Sales ARE Poor
IF Price IS Expensive AND Car IS So_So THEN Sales ARE OK
IF Price IS Expensive AND Car IS Reliable THEN Sales ARE Good

Now, let's deal with model D, which conflicts with the rules stated above. Close inspection of the data set indicates that there is no practical way to create different categories to form a consistent logic for all data points including model D.

In such a case, the knowledge acquisition process requires data interpretation by an expert. An expert may say that model D is very fashionable due to its appearance. Thus, the problem cannot be modeled without using this third antecedent variable if we insist on keeping model D in our analysis. Without model D in the analysis, this fuzzy inference engine may be used by a car dealer who makes investment plans. The investor will be warned that the inference engine is not sensitive to fashion.

The reader may have a few questions at this point. First of all, the categorization could be done using a higher number of fuzzy sets. The price range, for example, could have been divided further into three fuzzy sets, including a medium price range. Increasing the number of categories can improve the precision as long as the decision boundaries do not conflict. This decision is often made by the precision constraint (Question F in Table 4.6).

Second, the reader may get confused by what is extracted in the example above because the rules reflect commonsense reasoning, as if anybody could compose them without data. This example was purposely made simple to illustrate rule extraction from data. When the given problem is not as easily understood as the example above, the advantages of this approach will be understood better. For example, extracting a set of rules from financial data might tell us about the decision-making strategies of a firm. Or the same process might tell us about the mechanism of how a certain contagious disease spreads. Examples are only limited by the designer's imagination and by the validity of data.

Third, the reader may wonder about the process of generating categories. This is explained in more detail in Chapter 5, where the decision boundaries are determined on the input product space.

4.4.5 Interpreting Formulas

Another viable source of knowledge is closed form formulas derived by scientists in specific fields. Every analytical expression can be represented as a logical proposition at the expense of losing precision and limited range of applicability. In some cases where expressions are overly complex, such a conversion may not be practical.

To formulate an analytical expression in logic, the key element is to model the behavior of the expression. For example, in a simple expression such as

$$y = x + z \qquad (4.1)$$

the directions in which x and z increase determine the direction of y. Accordingly, if x and z are increasing, then y is also increasing in Eq. (4.1). This relationship can be expressed by a set of fuzzy rules to be valid for certain intervals of x, y, and z (i.e., the universe of discourse). In other words, we can only model this relationship within a specified set of universes.

There are nine basic combinations if we use three categories for the antecedents x and z, and five categories for the consequent y. Increasing the number of categories will increase the number of fuzzy rules as well as increasing the precision level.

IF X IS Positive AND Z IS Positive THEN Y IS Large_positive
IF X IS Positive AND Z IS Zero THEN Y IS Positive
IF X IS Positive AND Z IS Negative THEN Y IS Zero
IF X IS Negative AND Z IS Positive THEN Y IS Zero
IF X IS Negative AND Z IS Zero THEN Y IS Negative
IF X IS Negative AND Z IS Negative THEN Y IS Large_negative
IF X IS Zero AND Z IS Positive THEN Y IS Positive
IF X IS Zero AND Z IS Zero THEN Y IS Zero
IF X IS Zero AND Z IS Negative THEN Y IS Negative

The algebraic operation + is to be modeled by mapping fuzzy rules with appropriate membership function design. Consider the membership functions shown in Fig. 4.1. Using the basic fuzzy inference algorithm with centroid defuzzification, one can verify that all combinations listed below are completely satisfied for crisp numbers: $2 = 1 + 1$, $1 = 1 + 0$, $1 = 0 + 1$, $0 = 0 + 0$, $-1 = 0 - 1$, $-1 = -1 + 0$, $-2 = -1 - 1$. Thus, the inference engine is valid for the antecedent universe $[-1, 1]$ and consequent universe $[-2, 2]$.

By appropriate membership function design, the same set of fuzzy rules can be used to model other algebraic operations such as $y = 2x + 3z$.

The same algebraic formula can be translated into fuzzy rules by directly mapping the input–output data points produced by the expression without modeling the directional behaviors. In this approach, the behavior of the expression is already embedded in data, which eliminates

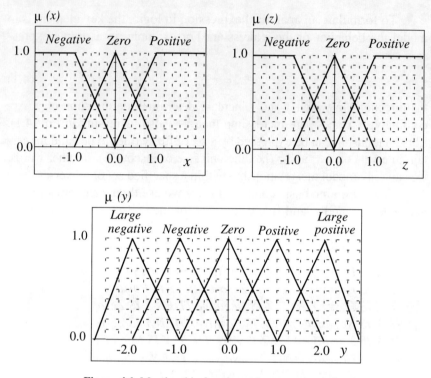

Figure 4.1 Membership functions of $y = x + z$ operation.

heuristic modeling such as in *positive + positive = large_positive* or *zero + positive = positive* as described previously. However, as shown in the following, the same logic model will eventually emerge from data analysis. Consider the following data set obtained from $y = x + z$.

Y	X	Z
0.0	0.0	0.0
1.0	0.0	1.0
1.0	1.0	0.0
2.0	1.0	1.0
−1.0	0.0	−1.0
−1.0	−1.0	0.0
−2.0	−1.0	−1.0

The input product space and the corresponding *y* values are shown in Fig. 4.2. Each distinct *y* value represents one distinct output category.

Chap. 4 Conceptual Design

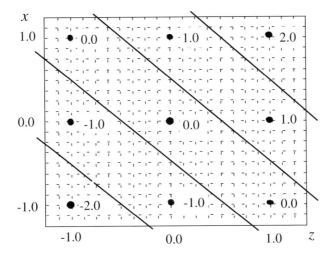

Figure 4.2 Data obtained from y = x + z operation.

This is illustrated by the drawings in Fig. 4.2. We see from this figure that there have to be five output categories. Also, a mapping between the crisp data points yields the same fuzzy rules if we use the same antecedent membership function designs shown in Fig. 4.1.

Considering three crisp data points per antecedent in this example, the design choices shown in Fig. 4.1 are appropriate. Note that the fuzzy regions (i.e., the regions between crisp data points produced by the expression) yield approximate results. By generating more data points and increasing the number of categories, the precision level can be improved at the cost of composing more fuzzy rules.

This approach allows the designer to extract the logic behind an analytical expression and to represent it by a set of fuzzy rules articulated in fuzzy IF-THEN language. Accordingly, a theoretical expression converted into fuzzy IF-THEN language can be combined with the experience of a human articulated in the same language to improve the overall solution. The following example illustrates the use of this technique in application to control.

EXAMPLE 4.4 FUZZY PI CONTROLLER

As an example of the translation of textbook knowledge into fuzzy rules, consider the design of a conventional proportional-integral (PI) controller [6] expressed as

$$u = K_p e + K_i \int e \, dt$$

where u is the control signal, and e is the error between the actual state of operation and desired state of operation. The PI controller is nothing but a simple algebraic expression such as Eq. (4.1) provided that the integral and derivative signals are available. Therefore, the addition of two components of a PI control signal can be performed in a manner similar to that described in this section. The PI expression given above does not explicitly reflect the directional relationship. Therefore, the complete set of PI equations are either

$$u = -K_p e - K_i \int e\, dt \qquad e(t) = y(t) - r$$

or

$$u = K_p e + K_i \int e\, dt \qquad e(t) = r - y(t)$$

where r is the *reference* (i.e., set point). The first form clearly shows the directional dependence. If the error signal is positive (i.e., the output is higher than the reference value) then the control becomes negative to produce a compensation effect (i.e., to lower the output towards the reference value). It is also possible to convert these equations into a differential form in which the control action is defined in terms of du [7]. Using the first form, we can convert the analytical expression into a set of fuzzy rules. Assume we have the following definitions:

CONTROL	ERROR SIGNAL	INTEGRAL ERROR SIGNAL
Huge Negative = HN	Large Negative = NG	Large Negative = NG
Very Large Neg. = VN	Small Negative = SN	Small Negative = SN
Large Negative = NG	Zero = ZR	Zero = ZR
Medium Negative = MN	Small Positive = SP	Small Positive = SP
Small Negative = SN	Large Positive = LP	Large Positive = LP
Zero = ZR		
Small Positive = SP		
Medium Positive = MP		
Large Positive = LP		
Very Large Pos. = VP		
Huge Positive = HP		

The definition of categories entirely depends on the designer. Here, we employed a commonsense logic for the addition of signals as shown below:

Antecedents	Consequent
Zero + Zero	= Zero
Zero + Small	= Small
Zero + Large	= Large
Small + Small	= Medium
Small + Large	= Very Large
Large + Large	= Huge

Using these definitions, fuzzy rules for mapping the PI controller are tabulated as follows.

Chap. 4 Conceptual Design 163

Integral Error

	LN	SN	ZR	SP	LP
LP	ZR	MP	LP	VP	HP
SP	MN	ZR	SP	MP	VP
ZR	LN	SN	ZR	SP	LP
SN	VN	MN	SN	ZR	MP
LN	HN	VN	LN	MN	ZR

Error

There are 25 fuzzy rules above that cover the entire product space. To understand this representation better, we illustrate the rule corresponding to the double-line box.

IF Error IS SN AND Integral_error IS SP THEN Control IS ZR

Every box in the matrix above corresponds to a fuzzy rule similar to the one shown above. The design up to this point included only the general partitioning of the product space for any algebraic addition of two variables.

To convert this table of associations into a specific PI controller, we need to design membership functions; therefore, we need data $y(t)$ from the process under investigation, the reference point r, and the constants of K_p and K_i. Using such data, we can generate the input product space by calculating the error and integral error signals. Consider the following normalized input product space obtained from such computations.

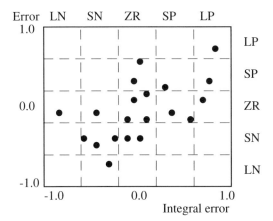

The original partitions shown by dashed lines can be adjusted to define the extent of membership functions more appropriately. An important observation at this point is that some of the categories do not contain any data point; thus they can be eliminated, reducing the number of fuzzy rules. The adjustment and elimination processes are shown below.

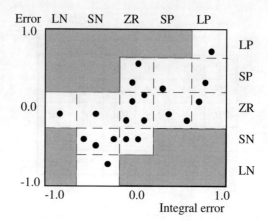

The adjusted product space can be used to design membership functions. This is explained in Chapter 5. Note that the data must represent the range of all possible values of $y(t)$. Changing the values of r, K_p and K_i will also change the distribution of data on the input product space resulting in possible modifications in membership function design [18]. The knowledge acquisition from a PI formula is reduced to the rules indicated by the adjusted product space as shown above.

4.5   THE FIRST PRINCIPLE OF FUZZY INFERENCE DESIGN

Fuzzy systems in the form of IF-THEN rules are different from those developed by introducing fuzziness to the mathematical formulation of generalized solutions. An extreme example is the application of fuzzy set theory to the solution of ordinary differential equations by providing approximate solution to finite mesh building given only the most essential data unique to the problem [4]. In the other extreme, which is this book's primary focus, a typical example comes from daily decision making of a human being, such as selecting the best apple in a grocery stand. Selecting the best apple is based on the evaluation of criteria such as color, size, aroma, hardness, and the memory of taste of an apple, all of which are related in a very complex manner for mathematical modeling.

Fuzzy systems theory does not explain the complexity of how the human brain works when selecting the best apple. Instead, it provides us with a mathematical capability to imitate the process in a very simple way. The methods of imitation, which we refer to as the calculus of fuzzy

IF-THEN rules, are by no means universally proven piecemeal strategies. Furthermore, the freedom of design reflects the very core nature of human intelligence that is the basis of individualistic style, taste, belief, experience, perception, and so forth. The first principle is therefore the expression of the freedom property in designing a fuzzy system.

> The design principles stated in this book or others obtained from the literature, which are based on generalized mathematical interpretations of linguistic terms, can be superseded by a specific context-related criterion.

This principle merely states that uncertainty characterization by means of mathematical modeling is non-unique, and it should not absolutely rely on pre-established norms or methods. Every design problem in practice is potentially unique in one (or more) aspects that may require a special approach not considered in the literature. This principle particularly applies to the membership function design, for which the possibility distribution is primarily determined by the designer. For example, the reader will encounter later in this book discussions on the design of membership function shape in relation to criteria such as tolerance, conservatism, optimism, pessimism, responsiveness, sluggishness, precision, cost, and robustness, most of which are defined based on possibilistic interpretations of uncertainty. The designer will find these discussions useful in evaluating a context-specific reason for forming a particular membership function shape.

The first design principle also applies to the specific solution embedded in a fuzzy inference engine. A fuzzy inference engine does not have to employ a specific context-related solution that obeys common sense. Although this seems like a controversial statement, there is no complex philosophy behind it. For example, if a fuzzy inference engine is developed by embedding laws and regulations, some of which are contradictory, then commonsense reasoning might be violated. It is the job of the fuzzy system designer to build a system that functions as intended (i.e., perhaps legal decision making in the presence of contradictions). A similar consideration applies to those cases in which a particular context-related solution defies statistical or cultural norms.

4.6 LINGUISTIC DESIGN CRITERIA

As stated in the first design principle, mathematical modeling of uncertainty is non-unique and can be developed in a number of ways. This fact, which directly affects fuzzy system design, causes great difficulty in discovering a set of generalized design criteria expressed purely in mathematics. Some of the mathematical criteria defined in the literature such as distance, entropy, and similarity are useful—under certain circumstances—to develop membership functions individually. We discuss these issues in Section 5.2.5. Yet, there is no mathematical criterion that can be applied to the design of an entire fuzzy inference algorithm.

On the other hand, there is a handful of design criteria utilized by designers in today's applications. They are articulated in natural language; therefore, we call them *linguistic design criteria* in this book. These criteria may be considered a checklist of behaviors expected from the system under development. This checklist guides the designer to focus on one (or more) expected behaviors while designing all the elements of the basic fuzzy inference algorithm such as membership functions, rule composition, implication operators, and logic operators. The criteria presented in this section is a collection from practical design experiences, most of which are scattered around the current literature.

4.6.1 Conservatism, Tolerance[2]

We start from the conservatism versus tolerance criterion, which can practically be applied to any problem. An important point that distinguishes this criterion from the others is that the consequences are known, so that the designer can deliberately choose a risk-taking or non-risk-taking path by analyzing the context-related solution obtained from the knowledge acquisition process. This criterion can be applied to both premises and consequences of each rule in the rule base. Consider the following pair of rules.

IF Temperature IS High THEN Cooling_pump MUST Run Full_power
IF Pressure IS Low THEN Cooling_pump MUST Run Low

---

[2] Tolerance in this context is different from the tolerance relations described in the fuzzy logic literature.

The conservatism/tolerance criterion, when applied to a pair of fuzzy rules as shown above, gives us a reason to differentiate the importance of each rule. If we can tolerate the uncertainty in *Temperature* being *High* more than the uncertainty in *Pressure* being *Low,* then we can express this choice in mathematical terms by increasing the weight of the first rule.

In membership function design, tolerance means that the possibility for an element to be the member of a fuzzy set can be exaggerated (i.e., can be high), which yields exaggerated levels of fulfillment of the conditions and therefore leads to rapid formation of an action. Consider the following fuzzy rule.

IF Fire_signal IS High THEN Extinguisher IS On

In other words, we will tolerate the possibility of wrong categorization (of fire signal being high) to act quickly (to extinguish it). Thus, tolerance means that we chose to tolerate exaggerated action. Conservatism is the opposite. The possibility for an element to be the member of a fuzzy set can be suppressed (i.e., can be low), which yields suppressed levels of fulfillment of the conditions and therefore causes the formation of action slowly. Consider the following fuzzy statement.

IF Man's_appearance IS Like_terrorist THEN Investigate_him

By choosing conservatism, we will not tolerate any wrong categorization (mistake in recognition) to act cautiously (when arresting somebody). In Chapters 5 and 6, more elaborate discussions on how to design a membership function are included.

4.6.2 Optimism, Pessimism

This criterion serves the same purpose as the conservatism/tolerance criterion, and it is used when the consequences are not known. Because the consequences are not known, belief factors are involved. For example, consider the following rules.

IF The_number_of_cookies IS Lower_than_expected
                    THEN Mike IS (probably) Guilty
IF The_number_of_candy_bars IS Lower_than_expected
                    THEN Mike IS (most_likely) Guilty

Believing that Mike would prefer candy bars, the second rule might be modified by a linguistic hedge or by an importance weight to express the belief factor. Note that the optimism/pessimism criterion is closely related to the probabilistic assessment of conditions to determine the corresponding actions.

On the membership function level, optimism means that the possibility for an element to be the member of a fuzzy set can be overpredicted (i.e., can be high), which yields overpredicted levels of fulfillment of the conditions and therefore causes the formation of an action rapidly. In other words, if we are optimistic about one fuzzy statement and pessimistic about another then we can employ different design approaches to express this notion.

4.6.3 Responsiveness, Sluggishness

Responsiveness is characterized as the capability of switching from one expert opinion to another rather rapidly or in a competitive manner. Sluggishness is the opposite, such that we keep considering all opinions in decision making in a collective manner, thus the transition from one expert opinion to another is rather smooth and slow.

In evaluating fuzzy rules, there has to be an aggregation process to apply this criterion when more than one rule contributes to one decision (one consequent variable). This is often the case. Responsiveness/sluggishness is expressed by the choice of implication operator, defuzzification method, threshold levels, and membership function overlap design. For example, consider the following rules:

.......THEN Flow IS Low
...... THEN Flow IS Medium
.......THEN Flow IS High

This criterion tells us that the decomposed value of *Flow* will depend on the rate of transition (or the amount of contribution) from one implication result to another. If the system is expected to behave sluggishly, transition from *Medium* to *Low* will be smooth or slow. In the responsive case, this transition will be more abrupt or quicker.

4.6.4 Precision, Accuracy, Cost

This is one of the most widely known criteria. Precision often means high cost. In a mapping process where decision boundaries are given (say 10 decision regions), we need to compose 10 or more rules to describe the mapping inference regardless of the number of data pairs on the input/output spaces. If we approach the problem in a crisp manner, which means one rule per each point on a product space, we will have to compose, say 10,000, rules even though there are only 10 decision regions. Thus, by creating fuzzy sets we reduce the cost of managing 10,000 rules to the cost of managing only 10 by compromising from the most precise (possible) solution of 10,000 rules. When there are a few number of decision regions characterized by vast amounts of data, it is often the case that such a compromise is worthwhile. Furthermore, fuzzy sets make it possible to find a practical solution if there are millions of data and if a crisp solution is impossible.

This criterion is expressed by determining the number of membership functions (i.e., granularity) in an open-end problem in which the decision boundaries are to be determined by the designer. Otherwise, when decision boundaries are extracted from data or from the articulation of expert knowledge, the determination of number of membership functions becomes more deterministic.

The cost criterion has a second function in fuzzy system design. In this case, we are concerned with the cost of a decision, not the cost of solving the problem in practice. Similarly, we are also concerned with the accuracy of a decision in terms of uncertainty characterization. Cost of decision, which is related to the tolerance/conservatism criterion, tells the designer how to distribute weights among different fuzzy rules. A typical example is as follows:

> *IF Weather IS Cold THEN Heater IS On*
> *IF Weather IS Hot THEN Air_Conditioner IS On*

Now, if there is a cost constraint[3] (too high an electric bill), do we choose to suffer from heat or cold? Thus, a weight assignment can express our choice of suffering. The accuracy criterion is somewhat different. Consider the pair of rules below.

---

[3] Cost constraint does not have to be a financial cost; it can be any other measure.

IF Temperature IS Just_about_freezing THEN Phase_change IS Likely
IF Temperature IS Well_below_freezing THEN Phase_change IS Unlikely

Modeling of the phase-change phenomena in this hypothetical example consists of the two membership functions *Just_about_freezing* and *Well_below_freezing*. Assume that the first rule describes an event that occurs in a very narrow band of the universe with delicate transition. The accuracy criterion encourages the designer to characterize the membership function *Just_about_freezing* much more carefully than that of *Well_below_freezing*. In fact the second membership function can be a roughly drawn trapezoid extending towards large negative values. The first one may require some analytical computation, modeling, or statistical analysis. As a result, the accuracy criterion tells the designer which membership functions need a careful uncertainty characterization relative to the others. Normally, an expert is needed for such a judgment.

4.6.5 Robustness, Sensitivity

Robustness is defined as the capability of functioning as intended in the presence of disturbances on inputs. Fuzzy systems are already known to be robust against uncertainties in measurements, because the membership function concept is a tool to characterize uncertainties that are somewhat anticipated. Thus, systems designed by considering fuzziness become immune to uncertainties. However, there may be unanticipated uncertainties in which the robustness of a system is challenged. Partially disturbed inputs and partially missing inputs are two common examples. If a fuzzy system is designed to receive a certain set of external inputs, what do we do when some of them are missing? Consider an example from medical diagnostics as shown below.

IF Reason_for_stopping_the_test IS Severe_chest_pain
                                      THEN Artery IS Perhaps_blocked

If we cannot collect the linguistic input (chest pain) for this rule for some reason (maybe patient cannot articulate pain), do we still go ahead and use this inference engine with partial inputs? If so, which inputs can be tolerated as missing, bad, or noisy? The judgment call again belongs to the expert, and the duty of prompting such questions during knowledge acquisition belongs to the fuzzy system designer.

Sensitivity is the opposite of robustness. The application of the sensitivity criterion to a fuzzy system tells the designer when to declare an "out-of-bounds" situation. Robustness and sensitivity considerations together will help the fuzzy system designer formulate an input acceptance criterion (or requirement) that is practically employed at the input gate within the input data processing stage as shown in Fig. 3.1. In the case of declaring an out-of-bounds situation for a given input data set, the output processor must convey this message to the system's user. Thus, this criterion also helps in deciding the user interface details at the very end of the basic fuzzy inference algorithm.

If the rules embedded in a fuzzy inference engine are known to be valid for a range of values outside the declared universes of discourse, extensions may be considered. Normally, the design of membership functions at both edges will determine the width of the universe of discourse. For example, a set of predicates such as *Medium, High,* and *Very_high* specified by an expert implies that *Very_high* is the upper boundary that may theoretically extend to infinity. Because we usually apply a finite boundary for computational reasons, it may be wise to decide whether input values higher than the upper boundary should be accepted or not. The same considerations apply to the lower boundary of a universe of discourse. As with the previously described criteria, there are a number of ways to incorporate sensitivity/robustness behavior during design, some of which are explicitly illustrated later in this book.

4.6.6 Other Criteria

There are other criteria that the fuzzy system designer may consider while building a fuzzy inference system. Needless to say, some of them overlap conceptually with the criteria already mentioned. Some of them are applied in the same manner in terms of practical design modifications. Unfortunately, none of the design criteria can be applied deterministically, unlike design problems in some other fields such as electrical or aerospace engineering.

Understandability

This criterion ensures that the appearance of the rule base is designed as appropriately as its working mechanism. In other words, the word selection for fuzzy variables, predicates, hedges, and linguistic connectors must be understandable to anyone. Understandability in fuzzy systems is one of the key features because fuzzy logic is a tool for computing with words. Consider the example below

If $(R = H) \vee (W \neq P) \rightarrow W \cong S$
IF It rains hard AND we don't take any precautions
THEN we may get soaked.

The two lines above illustrate two different ways of expressing the same fuzzy rule. The first type of symbolism is sometimes useful in course work or in a textbook in which a certain structural property is discussed. However, the second line represents the essential form that enables designers, users, experts, and students to understand (and modify) a fuzzy inference engine.

Computability

Computability is another important consideration when hardware implementation is concerned. Between a triangular membership function and a bell shape curve occupying the same universe of discourse with the same crisp maximum, the fuzzy system designer may select the triangle because of computational ease within a hardware environment. This is valid if there is no specific reason for selecting the bell shape curve. Similar considerations apply to the choice of implication operators, defuzzification method, input data processing, and logic operator design.

Adaptability

This criterion is one of the most widely researched topics today. As discussed in Chapter 2, adaptive fuzzy systems have the ability to change their internal design structure in accordance with the environmental changes that affect the performance of the original fuzzy system in a negative way. Among many different levels of adaptation, the simplest one is the adaptation of membership function shape using a dedicated set of fuzzy rules. In other words, an adaptive fuzzy system consists of a set of fuzzy rules controlling the properties of another set of fuzzy rules. This hierarchical structure can be repeated as many times as is necessary. When applying this criterion, the fuzzy system designer examines the rules provided by an expert (or rules extracted from data) and questions the invariance property of the design. Consider the following example:

*IF Temperature IS Low AND Humidity IS High AND Atmospheric
_Pressure IS Very_low THEN Rain IS Likely*

This rule, established by a meteorologist, is based on the fundamentals of atmospheric science that is supposed to be valid everywhere on the earth. However, the membership function *Low* temperature, for example, will be different from season to season and also from one geographic location to another. Instead of defining different *Low*s for different circumstances, the fuzzy system designer may approach the problem with an adaptive solution in which extra rules will be composed to control the membership function *Low* (and others). Basically, the designer has two options. The first option is to compose adaptive rules using linguistic hedges that modify *Low*. The second is to compose adaptive rules that numerically modify the shape of *Low*. This type of design is explained in Chapter 7 in more detail.

Learning From Experience

Adaptability can be used to design fuzzy systems by learning from past experience. This topic is already under intense research and development in the form of using learning algorithms and neural networks. Most of the existing research focuses on training a neural network (using a set of numerical data) to determine membership functions [9, 10]. Genetic algorithms that employ evolution principles are also used in some applications to design membership functions [11]. Some of the recent work attacks the problem by finding neural network structures that are functionally equivalent to fuzzy rules or a fuzzy inference mechanism. Another research topic is the discovery of new rules simply by examining a set of input/output data. Finally, the most challenging problem is to design an adaptive fuzzy system that can learn from linguistic data; that is, like reading a book and creating fuzzy rules. The last topic strongly depends on advances in natural language processing and semantic networks using fuzzy logic. The fuzzy system designer must examine the fuzzy solution provided by an expert (or extracted from data) and must decide whether a learning mechanism can be devised to improve the overall performance. This somewhat experimental topic is beyond the limited scope of this book.

4.7 APPLICATION OF THE DESIGN CRITERIA

During the knowledge acquisition process, one of the most important tasks of the designer is to evaluate the rules based on the linguistic criteria described in Section 4.6. The application of the design criteria, including conservatism/tolerance, optimism/pessimism, robustness/sensitivity, precision/accuracy/cost, and responsiveness/sluggishness, makes sense when there are different importances or nuances among the fuzzy statements. For example, there will not be much (if any) point in modifying all of the membership functions using the same criterion. Thus, creating relative differences is essential. The design challenge is to examine the given problem and its fuzzy solution to determine if any difference of nuance exists among the different fuzzy statements or rules. Once such differences are identified based on the specifics of the problem, the next challenge is to modify the design to produce the corresponding effects. On the other hand, the cost criterion, when interpreted as the issue of granularity, applies to the entire fuzzy system. So do the other design criteria, including understandability, computability, adaptability, and learning from experience. Relative differences among the parts of the fuzzy inference system are not relevant for these criteria.

The manner in which linguistic design criteria are utilized is explained in the remainder of this book. However, computability, understandability, and learning from experience are not discussed any further. The most prominent dependencies between the first five pairs of design criteria and the design elements are shown in Fig. 4.3. The adaptability criterion, which is not listed in Fig 4.3, has only been successfully applied to the membership function design in the literature.

Linguistic design criteria are subject to interpretation. The designer may apply a specific criterion during the conceptual design phase that might partially include the considerations stated above, or the criterion might be entirely new and creative. Also note that some fuzzy inference applications can be quite simple, not requiring the application of a linguistic design criterion.

4.8 SYSTEMS ONTOLOGY AND PROBLEM TYPES

To be able to design a fuzzy system for a real-life problem, it is important to know the properties of the problem. In this section, different problem types will be identified in terms of the domain of information a system

Chap. 4 Conceptual Design

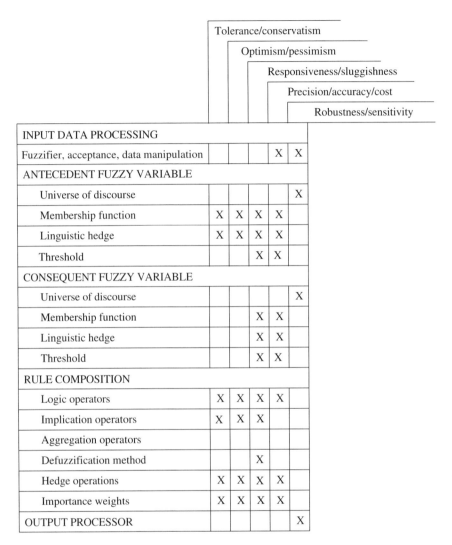

	Tolerance/conservatism	Optimism/pessimism	Responsiveness/sluggishness	Precision/accuracy/cost	Robustness/sensitivity
INPUT DATA PROCESSING					
Fuzzifier, acceptance, data manipulation			X	X	
ANTECEDENT FUZZY VARIABLE					
Universe of discourse					X
Membership function	X	X	X	X	
Linguistic hedge	X	X	X	X	
Threshold			X	X	
CONSEQUENT FUZZY VARIABLE					
Universe of discourse					X
Membership function			X	X	
Linguistic hedge			X	X	
Threshold			X	X	
RULE COMPOSITION					
Logic operators	X	X	X	X	
Implication operators	X	X	X		
Aggregation operators					
Defuzzification method			X		
Hedge operations	X	X	X	X	
Importance weights	X	X	X	X	
OUTPUT PROCESSOR				X	

Figure 4.3 The relationship between the design criteria and the designable elements of the basic fuzzy inference algorithm.

can generate or receive. These discussions are trivial but essential because the information domain used for fuzzy system design changes for each problem type. However, design attributes of the basic fuzzy inference algorithm are invariant with respect to problem types discussed below.

4.8.1 System Definition

A system is defined as a process, event, or mechanism and characterized analytically, algorithmically, symbolically, or linguistically within certain boundaries. A system can be physical, numerical, or conceptual. Each system is defined relative to four basic domains of information as shown in Fig. 4.4

Note that information is the only building block of the system concept. In the absence of information, systems cannot be realized even if they exist. The flow of information is indicated by arrows in Fig. 4.4. In general, the environment affects the input, system, and output domains. If an output affects the environment then there may be potential feedback mechanisms. For the fuzzy system designer, this is a generalized problem layout in which some of the information has to be constructed for different objectives. The construction of different types of information defines different problem types. For simplicity, let's denote input, system, output, and environment by I, S, O, and E, respectively. Table 4.9 lists simple examples to emphasize the fact that the domain definitions are problem dependent, always interpreted by the designer.

4.8.2 Forward Problems

The simplest case is the construction of the output domain information given the system (S), input (I), and environment (E) domain specifica-

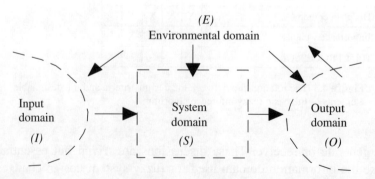

Figure 4.4 A system is characterized by separating four domains of information.

Table 4.9 Simple Examples of Different Information Domains

INPUT	ENVIRONMENT	SYSTEM	OUTPUT
Money supply	Inflation	Economy	Unemployment
Gas	Road condition	Car	Speed
Water	Weather	Tomato	Weight
Electricity	Color of the walls	Lamp	Light intensity
Testimony	Press	Court case	Verdict

tions. This type of problem is known as a forward problem,[4] in the sense that the solution of the problem is in the natural direction of information flow. Typical examples are estimation, prediction, and forecasting problems, as described next.

Estimation, Prediction, Forecasting

Although the terms *estimation, prediction,* and *forecasting* have several connotations in daily language, they all refer to guessing (by computing) an outcome that will happen in the future. Time is implicitly incorporated. In general, time is not necessarily clock time; instead, it is the sequence in which the system's behavior is shaped. Accordingly, the order of inputs and outputs are interrelated. Systems of this nature are known as dynamic systems if the independent variable is time. We will express this type of problem by

$$O(t + 1) = S\{I(t), E(t)\} \tag{4.2}$$

where S represents the relationship between the output and other domains, and t is the sequence. The objective of fuzzy system design is to build a fuzzy inference engine that will imitate the system's sequential or dynamic behavior by producing outputs of the future given existing inputs and environmental effects. In a more generalized form, systems that predict the events of the future are built by the knowledge of the past. Then, the problem is expressed by

---

[4] The term *forward problem* should not be confused with *forward chaining* search in traditional expert systems terminology.

$$O(t+1) = S\{I(t), I(t-1), \ldots I(t-n),$$
$$E(t), E(t-1), \ldots E(t-n), \quad (4.3)$$
$$O(t), O(t-1), \ldots O(t-n)\}$$

where the past and current outputs $O(t)$, $O(t-1)$, ...$O(t-n)$ can also be utilized. When the system behavior is characterized by the articulation of expert knowledge, the resulting fuzzy rules are equivalent to the system characterized by Eq. (4.3). This is because expertise is almost always based on experience in the case of sequential events. If the fuzzy rules are formed by knowledge extraction from data, then the history of each information domain must be available in terms of data. The generalized problem definition along with the domain of information in fuzzy inference design and its implementation are depicted in Fig. 4.5.

There are several conventional estimation techniques in the literature in which the problem is defined as the process of extracting information from data—data that can be used to infer the desired information and might contain errors. Optimal state estimators process measurements to minimize an error estimate (a cost function) of the state of the system by

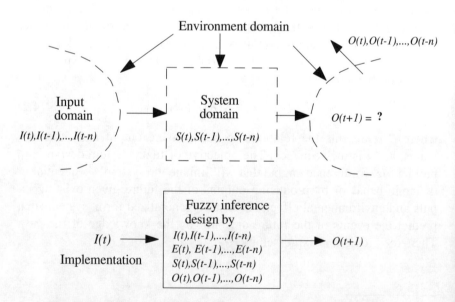

Figure 4.5 Forward sequential problems: estimation, prediction, and forecasting.

utilizing: knowledge of the system domain and input measurement dynamics; assumed statistics of system noises and measurement errors (environment domain); and initial conditions. Conventional estimation problems are divided into different problems (filtering, smoothing, and prediction) based on the span of available data with respect to the reference time point (point of estimation). There are also fuzzy estimation techniques in the literature where conventional crisp mathematics are fuzzified by introducing possibility theory to characterize measurement uncertainties.

When a fuzzy inference algorithm is replaced by the conventional mathematical methods, semantic reasoning (linguistic articulation of expert knowledge or extracting knowledge from data) can be tailored to function like an optimal state estimator by defining a linguistic cost criterion, knowledge of the system, and measurement uncertainties using possibility theory. Both the mathematical approach and fuzzy inference suffer from inexplicable departures from the ideal case. Conventional methods based on pure mathematics rely heavily on data—data that is often contaminated beyond the validity of statistical assumptions. Fuzzy inference relies heavily on semantic reasoning, which is always subjective and relatively inconsistent. Conventional mathematical methods find success in engineering systems of somewhat regular behavior (stationary processes) but they fall short in more complex systems such as economics. Fuzzy inference has potential advantage in complex problems because it can imitate human expertise—a viable (and sometimes the only) option.

Forward problems are not always sequentially correlated. Except forecasting, which strongly emphasizes time dependence, terms such as *cost estimation* or *damage estimation* refer to the assessment of the unknown based on the current and only available measurements. When the time span of available data is highly restricted, the definition of the estimation problem departs from its mathematical origins and turns into a heuristic decision-making problem. When the sequential dependence of measured data no longer exists, the solution of such problems relies on the characterization of lateral dependence, often referred to as *cross section data*. A good example of the difference between estimation by sequentially correlated data and by cross section data is the weather estimation problem. The weather can be estimated by analyzing weather chemistry based on the current temperature, humidity and pressure measurements, or it can be estimated probabilistically by looking at historical data. Obviously, there is no restriction on using both techniques simultaneously.

4.8.3 Inverse Problems

Inverse problems are defined in general as the construction of the input domain information (I), having specified the system (S), output (O), and environment (E) domains.

Control

Control is a special inverse problem where the input domain is split into two domains: inputs that are not controllable, and inputs that can be physically controlled. The control problem as shown in Fig. 4.6 is to reconstruct the control input domain (U) given the exogenous input (I), system (S), output (O), and environment (E) domains.

There are different types and levels of control problems; for example, control systems in engineering (e.g., thermostat, robot, or aircraft control); environmental control (regulating pollutants); population con-

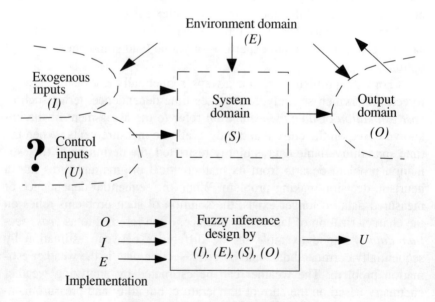

Figure 4.6 Control problem definition.

trol (social measures of control); government control (political measures of control); crime control (human psychology and behavior); monetary control (economics and finance); and military control. One common property among these control concepts is the structure shown in Fig. 4.6. The abundance of linguistic definitions associated with the term *control* suggests an enormous diversity in problem types as well as in their solution.

Despite the diversity, control theory has been mainly developed for engineering systems. It has been based on pure mathematical theories with occasional heuristic justifications. The reason is trivial. Many engineering systems represent the easiest and the most deterministic form of control problem due to the relatively small number of input and output variables in addition to regular (often linear) dynamic behavior. In fact, we normally do not want to build engineering systems with highly nonlinear behavior due to stability concerns. Advances in nonlinear theory and the building of increasingly more complex systems have recently inspired the emergence of the nonlinear control concept in conjunction with adaptive systems theory. In dealing with nonlinear systems, we encounter great difficulties with the formulation of control solutions using crisp mathematics. On the other hand, solutions based on semantic reasoning and heuristic interpretation often suit the nonlinear phenomena. In other words, heuristic solutions start when pure mathematical solutions end on the higher steps of the nonlinearity ladder, sometimes because of their superior performance, and other times because they are the only solution.

Control by means of fuzzy inference, which is landmarked by Mamdani's [12] and Sugeno's [13] pioneering applications to engineering, is the first computational form of semantic reasoning in the quest for handling nonlinearities in a practical way. It is not coincidence that all types of control problems have another common property in addition to what is shown in Fig. 4.6; that is, all the control solutions in those arenas listed previously are provided by human operators in practical life. Therefore, fuzzy control that imitates human reasoning is equally applicable to every control concept that exists in language. This forces the fuzzy system designer to consider the broad picture of control problems as potential targets, unlike the traditional control system designer who specializes in engineering systems only.

Control, as it is formulated in engineering, can be implemented in the manner depicted in Fig. 4.7. This is called cascade control using feedback signals and is one of the most common forms seen in industrial applications. The important point here is that, unlike the forward problems

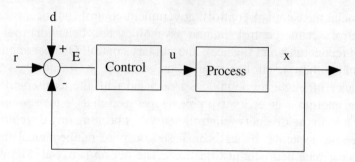

Figure 4.7 A cascade controller using feedback signals (x = output state, r = reference, d = disturbance, E = error, u = control)

discussed earlier, fuzzy inference (or fuzzy controller) does not imitate the process; instead, it imitates a control solution deployed by an expert. The control box in Fig. 4.7 is where fuzzy inference resides.

Another typical layout is shown in Fig. 4.8, where the controller resides on the feedback loop. From the fuzzy inference point of view, different locations of the controller block with respect to other domains mean somewhat altered logic using a different set of input/output variables. For example, the fuzzy inference engine determines control u by looking at the error E (deviation from reference) in Fig. 4.7 and by looking at the output x (feedback signal) in Fig. 4.8. Accordingly, the solution logic (control law) is different for these cases.

Considering the big picture in which a fuzzy control design may be applied to a large number of problems, the number of possible orientations of a controller block may not be as limited as in engineering systems. Therefore, one of the tasks of the fuzzy system designer is to exam-

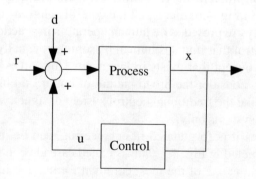

Figure 4.8 A feedback controller.

ine the logistics of a control solution to determine the antecedent and consequent variables along with the solution logic involving these variables.

There are mainly two distinct approaches in developing fuzzy inference for control. First is the imitation of a generic control law such as the proportional-integral-derivative (PID) control that can be applied to any system. Piecemeal solutions such as the PID control always have piecemeal objectives defined in general terms. The two commonly employed objectives are regulation and trajectory control. In both cases, the objective is to keep the process under control so that its behavior matches a desired behavior.

The second approach of fuzzy inference for control is the imitation of a specific control law applicable only to one or more specific systems. A control law of this sort comes from the articulation of expertise (or extraction of knowledge from data). The objectives are never piecemeal, and thus may include modes of control other than just set point control. Typical examples are navigation control and controlling a robotic arm (manipulator). In such cases, the problem is to select a control solution out of many possible ones or to decide a sequence of control actions leading to the goal. Objectives of this kind are very difficult to achieve by generalized mathematical solutions, and in some cases there is no unique solution. A flexible joint robotic arm manipulator with redundant degrees of freedom is a typical example. Control problems fitting this description, which are also called *open-end problems* in this book, constitute the majority of control problems when considering the big picture, including all other linguistic definitions of control. This statement is particularly true for systems other than engineering systems.

Optimal Control

The definition of optimal control is similar to the ones made above except that there are constraints characterized by a cost function. In engineering applications, typical constraints emerge from physical limitations of control actuators such that their motion must be minimized to reduce the rate of wear and their dynamic behavior must not include sudden jumps to avoid harmful stresses on the process. Minimizing the cost of controlling a process while maintaining the desired performance is the challenge to be met.

Within linear systems theory, optimal control solutions are well defined and are generally applicable to a wide range of well-behaved systems. Nonlinear optimal control is a more difficult problem. Among the different solutions suggested in the literature (Pontryagin's maximum principle for example), not many are easily applicable in the practical realm due to computational burdens and limitations on the ability to characterize nonlinearity.

The fuzzy inference approach can be applied to optimal control problems by imitating the actions of a human operator who controls a process under known constraints. This type of optimization is purely heuristic and context specific. Because almost all systems in practical life are subject to some sort of constraint (mostly due to financial cost), the control solution articulated by an experienced human operator can be considered optimal or suboptimal. The fuzzy system designer's main task is to investigate the constraints of the problem and to make sure that the acquired knowledge reflects an optimal control solution for the known constraints. However, this approach is limited by the degree of optimality embedded. Fuzzy optimization by means of minimizing a fuzzy cost function and other similar exotic methods have been the subject of intensive research with few known industrial applications to date.

4.8.4 Heteroassociative Problems

Contrary to forward and inverse problems, some problems involve the challenge of determining the system properties given the inputs, outputs, and environmental effects. They are heteroassociative in the sense that the associations among multiple domains of information must be determined to characterize the unknown system properties. This is illustrated in Fig. 4.9.

There are different reasons for wanting to know the system properties in practice. Each reason defines a distinct problem. We start from modeling.

Modeling

The definition of modeling is to build a replica of an existing process, known phenomenon, or experienced event so that we can perform experiments with reduced cost or risk. For example, a model of an airplane can be made by scaling down its dimensions and building it physically. Alternatively, its model can be developed mathematically by a set of dynamic equations using computers. In both cases, a flight can be simulated without risking an actual flight.

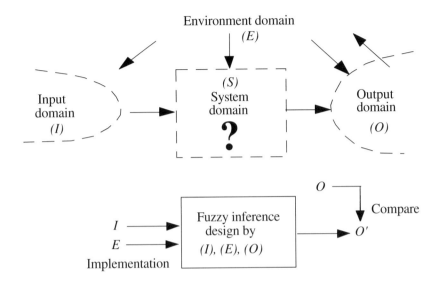

Figure 4.9 Heteroassociative problem where system is the unknown.

Modeling is a widely used concept in engineering as well as in other disciplines. There are basically two alternative and complementary approaches in modeling: deterministic and empirical. In deterministic modeling, the working principles of the process, phenomenon, or event are known and they can be expressed analytically or can be articulated linguistically. Fuzzy inference for modeling uses the articulated knowledge in the form of fuzzy IF-THEN rules. The output domain information is used for validation. In empirical modeling, the working principles of the process are not known; instead, the model is built by examining the input/output data. There are various mathematical techniques, including probabilistic methods and learning algorithms such as neural networks. Fuzzy inference for empirical modeling can be built by rule extraction from the input/output data set. Just like in other empirical modeling techniques, the performance of the fuzzy model depends on the validity of the data set. Note that when the control signals are given (known) in Fig. 4.7, the control problem can be viewed as a modeling problem also.

Classification, Feature Selection

Another form of the heteroassociative system is the classification problem where the input/output data (representing the desired classification) is available and the problem is to build a classifier that satisfies outputs given inputs. The difference between modeling and classification is that we are not interested in the working principles of a system; rather, we need a system that identifies each given input case as one of the classes represented by the output data. Classification and feature selection are different versions of the same general problem, often called *pattern recognition*. Classification is a search for structure in data spaces or samples, whereas feature selection is a search for structure in data items.

A typical example is the classification of army personnel as small, medium, large, and extra-large sizes to give them the right equipment and uniforms. Such a simple classification problem would have height and weight data on the input space and class indicators on the output domain (indicating which input points fall into the small, medium, large, and extra-large categories) as shown in Fig. 4.10.

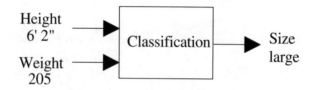

Figure 4.10 Simple classification example.

The problem is heteroassociative in the sense that the class information (output) must be available to design the classifier. We will see later an autoassociative version of this problem. Let's examine Fig. 4.11 now

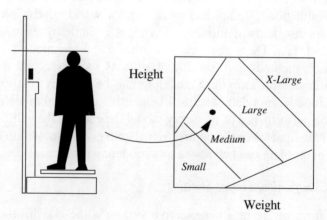

Figure 4.11 Classification of personnel to fit one of the uniform sizes.

where the output knowledge (class boundaries) are drawn on the input space.

For this type of problem, fuzzy system designers perform mapping by fuzzy inference using the rule extraction from data method. Weight and height can be divided into fuzzy subsets so that simple fuzzy rules are composed to represent the mapping. Otherwise, the problem can be solved by crisp rules by defining intervals between crisp boundaries. Once the system design is complete and validated, the fuzzy classifier will be used to evaluate new input data and to classify it as one of the output categories.

Size of a uniform is a crisp concept that requires crisp class boundaries as shown in Fig. 4.11. However, there are classification problems with fuzzy boundaries. Consider classification for fitness level based on the same input space as shown in Fig. 4.12.

Here, the output boundaries are fuzzy, and each fuzzy set is often defined by linguistic terms in accordance with common sense. Also note that fuzzy subsets or fuzzy class boundaries are spatially correlated. In other words, if we move from one fuzzy subset to another with small steps, their location make sense. Decreasing possibility values of *Overweight* corresponds to increasing possibility values of *Athlete* who weighs more than a normal person because of muscle tone.

The feature selection problem is similar to the classification problem logistically. However, there are some differences. First, the output domain information is a target to be matched rather than being a category

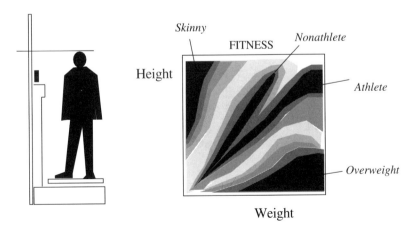

Figure 4.12 Classification with fuzzy boundaries.

to be classified. For example, given the weight and height information we may classify a person to be slim, normal, or overweight. This is casual matching in the sense that there are a fixed number of options and the input case must fall into one. In feature selection, we employ a strict match that is looking for the closest similarity to, say a person with the size of a fashion model, given the weight and height information as depicted in Fig. 4.13.

If the similarity is not adequate, the input case does not have to fall into a category; instead, it is simply not recognized. Second, the output target pattern is always characterized with data (often with a high level of detail) in feature recognition problems, whereas in classification problems, the output domain can simply be labels of categories.

Feature recognition problems normally involve complex targets, as often encountered in image processing. There are purely mathematical methods (conventional and fuzzy) to calculate the similarity between the target and the input. Conventional matching methods (some of which employ dynamic programming techniques) are improved by reducing the computational burden using fuzzified (approximated) relationships. A typical example is fuzzy pattern matching in speech recognition applications [14]. Feature recognition by means of fuzzy inference is often effective in conjunction with peripheral computations converting patterns into approximated equivalents when the patterns are complex.

The fuzzy inference approach is more effective and sometimes the only solution when a feature recognition problem is defined entirely in the linguistic domain. Criminal behavioral pattern recognition—a subject in criminal psychology—is one of the best examples. The fuzzy system

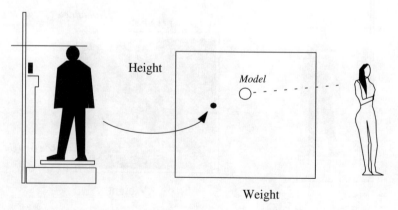

Figure 4.13 Feature recognition problem.

designer may choose several different avenues to follow when the problem at hand is identified as a pattern recognition problem. Design may include expert knowledge, knowledge by interpreting data, rule extraction from data, and peripheral computations for data transformations.

4.8.5 Autoassociative Problems

Autoassociative problems only involve input domain information, and there is no output specification as shown in Fig. 4.14. This type of problem transforms the input domain data into a new form. Most of the autoassociative problems are attacked by purely mathematical methods.

Information Compression

A typical example is the data compression problem, where input domain information is represented by a smaller number of data points. The challenge is to be able to retrieve the original information from the compressed data as accurately as possible. Numerous numerical applications exist in the telecommunications, image processing, and computer hardware fields. Numerical data compression problems today are defined and solved by a pure mathematical approach and there seems to be no room for inference methods based on human reasoning.

However, perhaps the best compression machine is the human brain and its abstraction skill when we redirect our focus on linguistic data. Consider the following fuzzy statement.

The man is fit for the construction job.

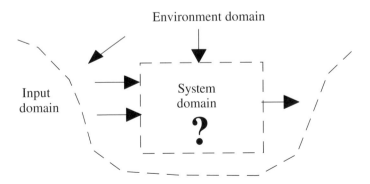

Figure 4.14 Autoassociative problem definition.

To be fit for a construction job, someone has to have many qualities, most of which are implicitly known or known by commonsense reasoning. Thus, the reverse operation—that is, expanding compressed information back to its original size—requires commonsense reasoning and a global dictionary. This is illustrated in Fig. 4.15. The structure shown in Fig. 4.15 is the basic mechanism for summarization (a more complex problem known in the field of artificial intelligence), and it is also related to fuzzy database systems. The biggest challenge is to make a model for commonsense reasoning that will accurately function in both directions as shown in Fig. 4.15. In its most trivial form, a commonsense algorithm is a function that finds a representative phrase connected to the elements of a global dictionary with maximized strength. In this context, the connections a, b, c, . . . , f are fuzzy, each describing a different degree of relevance. Therefore, the global dictionary is a fuzzy dictionary (or a fuzzy thesaurus). From the viewpoint of the basic fuzzy inference algorithm, the problems defined above are far more complex and experimental.

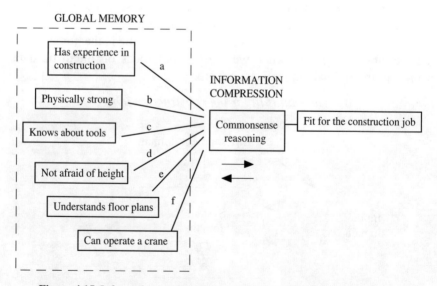

Figure 4.15 Information compression and decompression in a linguistic domain.

Clustering

Clustering is basically a classification problem with an unspecified output domain; that is, either we have output data points without class descriptions or we have to create output class descriptions without output domain data. Let's return to the army personnel example. Instead of having four sizes of uniforms in stock, assume that the army must decide in how many size categories they should make the uniforms. This requires analyzing the input data and creating clusters based on some criteria that justifies every element in the cluster being a member of the cluster. Criteria, also known as the objective function, can be distance, frequency, or some other measure that describes the characteristics of being a member of a cluster. A simpler version of clustering is to determine cluster boundaries having specified the number of clusters. Clustering (crisp clustering and fuzzy clustering) is realized as a mathematical problem in the literature with generalized methods such as c-means clustering [15] that are applicable to a variety of problems. From the fuzzy system design point of view, mathematical clustering methods can be especially useful in determining the number of membership functions (granularity) to be designed in open-end problems.

Clustering by fuzzy inference is not as thoroughly researched, although the human brain performs clustering by inference in practical life. For example, consider the clustering problem in Fig. 4.16.

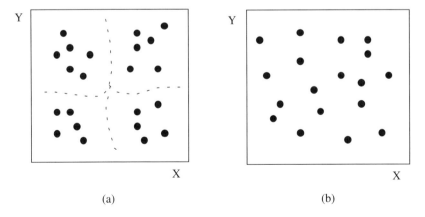

Figure 4.16 Clustering based on distance: (a) Interpretable by human, (b) ambiguous to human.

It is obvious that there are four clusters in Fig. 4.16 (a) when the objective function is based on distance between the elements. We can easily create clusters by simple reasoning, such as *If X and Y are small then it is cluster-1; If X and Y are large than it is cluster-2,* and so forth. On the other hand, the clustering problem in Fig. 4.16 (b) is complex since every element seems to be equidistant from each other. The important question here is whether a mathematical technique can do a better job than the human brain in cases like (b). The reader of this book might naturally disagree with partitioning of the space in (b) based on any mathematical technique, because it would simply be highly debatable. This argument tells us that the fuzzy inference approach employing commonsense reasoning can be as powerful as mathematical techniques in clustering problems.

4.8.6 Complex Problems

Up to this point, we have not discussed environment domain information; we have assumed that its effects on input, system, and output are well defined. In reality, this is never true and there is always some environmental interference. There are different levels of interference that contribute to uncertainty in measurements and in the system's working principles. The simplest form of such effects is formulated as disturbance in control science that raises the robustness issue. It is the fuzzy system designer's judgment call whether environmental effects can be treated as simple disturbances (such as noise on the measurements) for the problem at hand. Fuzzifying crisp inputs at the input gate is one of several techniques to incorporate uncertainties (see Chapter 2) and to handle disturbances. In more complex forms, environmental effects can be so severe that the problem definition becomes uncertain. Consider the complex problem definition shown in Fig. 4.17, in which inputs, system, and outputs are partially unknown.

One of the primary tasks of the fuzzy system designer is to treat every problem as a complex problem at the beginning. Then, the problem can be simplified by eliminating the unknowns one by one during the knowledge acquisition process. Such a refinement is only done in the presence of an expert who can interpret the possibility of unknown elements in a given problem. A qualitative evaluation requires the assessment of each domain separately. The fuzzy system designer employs the following reasoning during knowledge acquisition:

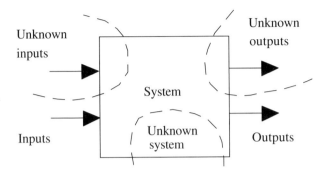

Figure 4.17 Complex problem definition.

- If it is known that some outputs are not measured, then they can be ignored if they are not engaged in any feedback mechanism.
- If it is known that some inputs are unaccounted for, and if these inputs do not contribute to the results of the defined outputs, then they can be ignored.
- If the unknown system characteristics are guaranteed to be unrelated to the results of the defined outputs, then they can be ignored.
- If none of the above is true for inputs and outputs, then the fuzzy system designer asks the expert to evaluate the robustness/sensitivity issue.
- If none of the above is true for a partially unknown system, then the fuzzy inference rules are locally valid and they cannot be used for generalization.

In these evaluation steps, it is implied that we have some approximate knowledge about the unknown elements. A worst-case scenario is that the only knowledge is the existence of unknowns. The worst case is that we do not know of the existence of unknowns. The last two cases significantly decrease the credibility of a fuzzy inference engine, even to the point of terminating the development. Note that the complex problem definition and the qualitative measures for developing reasonable solutions are not special limitations arising out of fuzzy system design. Instead, they are encountered in all kinds of developments (i.e., using neural networks, statistical modeling techniques, etc.) For all practical purposes, the assessment of unknowns in a complex problem is never an exact science, and it solely depends on an expert's interpretation. Let's elaborate more on the qualitative assessment process itemized above by a case study.

Unmeasured Outputs

It is relatively easy to make judgments on unmeasured outputs. In the climate temperature control problem, for example, heat input to a room's atmosphere might result in decrease in humidity. Because humidity regulation is not one of the control objectives in this example, this outcome can be totally ignored. Similarly, all other effects produced by the heat input can be ignored provided that they do not produce feedback effects. For instance, a feedback effect could be that occupants would open the windows in case of too much heat input, thus reducing the temperature of the room. If an unmeasured output is known to produce a feedback effect, then the expert must make a judgment about whether this is a tolerable unknown and whether the solution without this information is a robust one.

Unaccounted Inputs

In the ongoing example, unaccounted inputs to the room's atmosphere could be negligible heat sources such as human body, light bulbs, wall temperature, and oven. We may also consider inputs such as smoke, dust, atmospheric pressure, and humidifier, and treat them as unrelated inputs. However, unaccounted inputs must not affect the process. For instance, heat loss through ventilation sealing would affect the room temperature; thus, it is an input that cannot be ignored. If an unaccounted input is known to affect the process, then the expert must make a judgment about whether this is a tolerable unknown and whether the solution without this information is a robust one.

Partially Ignored System Properties

There may be a number of mechanisms contributing to the atmospheric conditions of the room in the ongoing example. Suppose that there is large furniture in the room creating vortices of airflow over stagnant layers of cold air. Assuming that this mechanism produces negligible effects on the average air temperature, we may ignore this partially unknown system of relationships. Let's now consider another case in which such a commitment cannot be easily made. Suppose that the room has Plexiglas skylights and the heat loss or heat gain (through sunlight) characteristics of the windows are not known, either due to lack of experimental knowledge or due to the high cost of obtaining this knowledge. The expert must make a judgment about this partially unknown system property (perhaps by examining the size of the window in comparison with the size of the room) and decide whether it is tolerable to ignore it. When a partially unknown system property is ignored by reasonable justifications, then the designed system is only valid in a certain operational range or mode (i.e., for rooms with very small skylights).

Qualitative Assessment Summary

The initial problem definition was climate temperature control, possibly using a thermometer and an air heater mounted in a room with a skylight and full of large furniture. Instead of designing a fuzzy inference control right away based on the given information, the fuzzy system designer has treated the problem as a complex one by considering the unknown elements. The considerations included humidity, the possibility of people opening the windows, the human body as a heat source, light bulbs, wall temperature, oven, smoke, dust, atmospheric pressure, humidifier, heat loss from ventilation sealing, vortices of airflow, and heat transfer through the skylight. All these effects exist in reality. When these effects are not known explicitly, conventional mathematical methods are not suitable to incorporate the corresponding uncertainties. Because such an assessment is accomplished only by human judgement in practical life, fuzzy inference imitating human expertise is the only viable solution. The design challenge is to characterize such unknowns by interpretation and incorporate uncertainties using possibility theory for those unknown parameters that are believed to be influential. Note that the discussions derived from the climate temperature control example are equally applicable to any other problem type (i.e., estimation, modeling, etc.).

4.9 USEFUL TOOLS SUPPORTING DESIGN

The contents of this book mainly explain the transition from theory to application for fuzzy inference systems. Due to the abundance of issues related to such a transition (which we call *design*), there is not enough room in one book to cover entirely both fuzzy systems theory and examples of specific industrial applications. However, there are several recently published books on these subjects, most of which will be useful during design. On the theory side, books written by Ross, Tsoukalas, Kruse, Kosko, Yager, Dubois, Driankov, Zimmerman, and Prade can be directly used in conjunction with this book. On the application side, Zadeh, Terano, and Yen's books give plenty of examples from the industry. It is highly recommended that applications similar to the design problem at hand are studied. There are also special purpose books for fuzzy control, such as the ones written by Jamshidi and Driankov, and for fuzzy expert systems written by Hall and Kandel. There are a few books on the market that elaborate on the practical issues of design and programming, such as the one written by Cox, and another by McNeill. Finally, a comprehensive analysis of the engineering design is examined in general in a book written by Bucciarelli. A list of books useful for design is included at the end of this chapter.

Fuzzy system design can be performed entirely on paper. Nevertheless, there are several computer-aided design tools on the market that provide user-friendly interfaces to design the elements of a fuzzy system. These products, listed in an alphabetic order by vendor, are shown in Table 4.10. Most of these products generate source codes in different programming languages by capturing the design specifics employed by the user. Then, these source codes are compiled and linked to specific application environments. Also called embedded programming, this mode of utility as depicted in Fig. 4.18 is the most common one in the field of soft computing. Some software products have their own compilers and GUI templates such that the application software is directly produced within the same development shell.

Table 4.10 Software Tools Available on the Market

Vendor	Product
Aptronix	FIDE™
Fuzzy Systems Engineering	Fuzzy Knowledge Builder™
HyperLogic Corp.	CubiCalc®
INFORM	*fuzzy*TECH®
Integrated Systems, Inc.	RT/Fuzzy™
MathWorks Inc.	MATLAB™
MODiCO, Inc.	*FUZZLE* ™
Togai InfraLogic	TIL Shell®

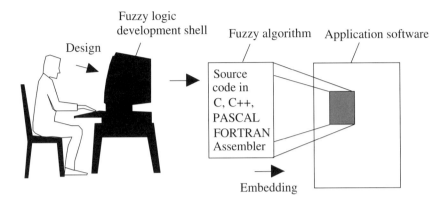

Figure 4.18 Embedded programming.

The software tools [16] listed in Table 4.10 include so-called standard functions as well as unique attributes. Three standard functions are: (1) user interface to design fuzzy variables and membership functions; (2) user interface to edit fuzzy rules; and (3) source code generation capability. Unique attributes differ from one product to another, making each tool somewhat specialized in certain aspects of fuzzy system design. For example, RT/Fuzzy™, TIL Shell®, FIDE™, and MATLAB™ offer simulation capabilities, whereas *fuzzy*TECH® and TIL Shell® offer easy fuzzy microprocessor design. CubiCalc® and *FUZZLE*™ can execute fuzzy inference engines without compiling source codes and linking to application code. *fuzzy*TECH® provides a fuzzy associative memory (FAM) development environment whereas TIL Shell® incorporates neural networks to generate rules. Fuzzy Knowledge Builder™ offers a cellular automata algorithm to interpolate missing rules.

Another common design mode is building fuzzy hardware. Some of the fuzzy logic development shells on the market are equipped with design options specific to hardware implementation (such as ROM, RAM, and E^2 memory allocation properties in microprocessors). In this mode, a fuzzy logic development shell is used to capture a particular design and to produce source codes in machine language so that they can be downloaded (or burnt) on a hardware device. This is depicted in Fig. 4.19.

Downloading of a source code produced in Assembler or more specific languages (such as the Motorola's HC11 language) often requires code optimization due to memory restrictions imposed by the maximum allowable power source on a digital board. However, the existing soft-

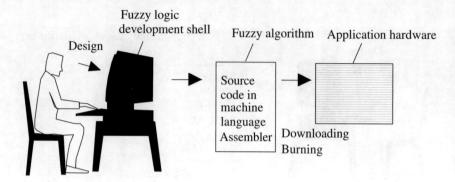

Figure 4.19 Downloading a fuzzy algorithm on digital hardware.

ware tools on the market offer limited capabilities in terms of fuzzy source code optimization. Accordingly, some hardware manufacturers have focused their attention on the traditional analog technology, in which some of the restrictions encountered in digital systems can be avoided. Unfortunately, there is no computerized tool on the market as of today to provide design support for fuzzy analog circuits. Most of the fuzzy hardware realized in the field of control is built by individual manufacturers who develop their own software tools in the process of designing fuzzy hardware.

REFERENCES

[1] Davis, R. and D. B. Lenat. *Knowledge Based Systems in AI.* New York: McGraw-Hill, Inc. 1982.

[2] Politakis, P. and S. M. Weiss. "Using Empirical Analysis to Refine Expert System Knowledge Bases," *Artificial Intelligence* 22, 23 (1984).

[3] Boose, J. H. "Personal Construct Theory and the Transfer of Human Expertise." *Proc. Natl. Conf. Artificial Intelligence,* Austin, Texas, (1984).

[4] Kandel, A., *Fuzzy Expert Systems,* Boca Raton: CRC Press, 1991.

[5] Cox, E. *The Fuzzy Systems Handbook.* New York: AP Professional, 1994.

[6] Kuo, B. *Automatic Control Systems.* Englewood Cliffs, NJ: Prentice-Hall, 1982.

[7] Tang, K. L. and R. J. Mulholland. "Comparing Fuzzy Logic with Classical Controller Design." *IEEE Trans. on Systems, Man, and Cybernetics* 17 (6) (1987): 1085–87.

[8] Smith, S. M. and D. J. Comer. "An Algorithm for Automated Fuzzy Logic Controller Tunning." *Proc. of the IEEE Int. Conf. on Fuzzy Systems* (FUZZ-IEEE92) (1992): 615–21,

[9] Lin, C. T. and Y. C. Lu. "A Neural Fuzzy System with Linguistic Teaching Signal." *IEEE Trans. on Fuzzy Systems* 3 (no. 2) (1995):169, 189.

[10] Lin, Y. and G. A. Cunningham III. "A New Approach to Fuzzy-Neural System Modeling." *IEEE Trans. on Fuzzy Systems,* 3 (no. 2) (1995): 190, 198.

[11] Homaifar, A. and E. McCormick. "Simultaneous Design of Membership Functions and Rule Sets for Fuzzy Controllers Using Genetic Algorithms." *IEEE Trans. on Fuzzy Systems,* 3 (no. 2) (1995): 129, 139.

[12] Mamdani, E. H. "Application of Fuzzy Algorithms for Control of Simple Dynamic Plants," *Proc. of IEEE,* 121(12) (1974): 1585–88.

[13] Sugeno, M. *Industrial Application of Fuzzy Control.* The Netherlands: Elsevier Science Publishers BV, 1985.

[14] Fukimoto, J., T. Nakatani, and M. Yoneyama. "Speaker-Independent Word Recognition Using Fuzzy Pattern Matching." *Fuzzy Sets and Systems* 32 (1989): 181.

[15] Zimmerman, H. J. *Fuzzy Set Theory and Its Applications.* Boston: Kluwer Academic Publishers, 1990.

[16] Personal communication with the vendors.

RECOMMENDED BOOKS FOR DESIGN

Bucciarelli, Louis L. *Designing Engineers.* Cambridge, MA: MIT Press, 1994.

Cox, E. *The Fuzzy Systems Handbook.* New York: AP Professional, 1994.

Driankov, M. *An Introduction to Fuzzy Control.* Aachen, Germany: Springer-Verlag, 1993.

Dubois, D., and Prade, H. *Fuzzy Sets and Systems: Theory and Applications.* Boston: Academic Press, 1980

Hall, L. O. and A. Kandel, *Designing Fuzzy Expert Systems.* Koln, Germany: Verlag TUV Rheinland, 1986.

Jamshidi, M. *Fuzzy Logic & Control.* Englewood Cliffs, NJ: Prentice-Hall, 1994.

Kosko, B. *Neural Networks and Fuzzy Systems.* Englewood Cliffs, NJ: Prentice-Hall, 1992.

Kruse, R., J. Gebhart, F. Klawonn, *Foundations of Fuzzy Systems.* Chichester: John Wiley & Sons, 1994.

McNeill F. M. and E. Thro. *Fuzzy Logic, A Practical Approach.* New York: AP Professional, 1994.

Prade, H., and C. V. Negoita, *Fuzzy Logic in Knowledge Engineering.* Koln, Germany: Verlag TUV Rheinland, 1986.

Ross, T. J. *Fuzzy Logic with Engineering Applications.* New York: McGraw-Hill, Inc. 1995.

Toshira Terano, Kiyoji Asai, and Michio Sugeno, *Applied Fuzzy Systems.* New York: AP Professional, 1994.

Tsoukalas, L. H., and R. E. Uhrig. *Fuzzy and Neural Approaches in Engineering.* New York: John Wiley & Sons, 1997.

Yager, R. R. *Introduction to Fuzzy Logic Applications in Intelligent Systems.* Boston: Kluwer-Nijhoff Publishing, 1992.

Yen, J., R. Langari, and Z. A. Lotfi. *Industrial Applications of Fuzzy Logic and Intelligent Systems.* Piscataway, NJ: IEEE Press, 1994.

Zadeh, L. A., and T. A. Tilly. *Applications of Fuzzy Logic.* Englewood Cliffs, NJ: Prentice-Hall, 1996.

Zimmerman, H. J. *Fuzzy Set Theory and Its Applications.* Boston: Kluwer Academic Publishers, 1990.

CHAPTER 5

Fuzzy Variable Design

> This chapter identifies the important steps in fuzzy variable design and outlines a few general design principles to follow. Topics like membership function design, decision boundaries, and linguistic hedges are examined in detail to give the reader insight and conceptual maturity rather than to offer deterministic design prescriptions, which do not exist. Although the main focus of this book is the calculus of fuzzy IF-THEN rules, the variable design issues presented here are applicable to other types of fuzzy system design that are beyond the scope of this book.

5.1 INTRODUCTION TO FUZZY VARIABLE DESIGN

One of the most important steps in fuzzy system design is the design of its building blocks: the fuzzy variables. Before its design, a fuzzy variable should be identified among the variables of the problem such that there is significant uncertainty involved in characterizing its behavior. The identification requires judgment based on the fuzzy measures listed in Chapters 1 and 2, such as imprecision, inaccuracy, ambiguity, randomness, vagueness, and other measures including the designer's own. It is important to ensure that using a fuzzy variable yields a more realistic representation of the problem than that using its crisp counterpart. It is also necessary that the solution using a fuzzy variable becomes more practical and cost-effective compared to the solution using a crisp variable. Without a clear insight into the benefits of using a fuzzy variable, design attempts can become pointless and cumbersome. Unfortunately, such an insight is not easily gained without some design experience. The basic fuzzy inference mechanism can use both crisp and fuzzy variables; thus, crispness is acceptable and is sometimes necessary.

The overall structure of a fuzzy variable is outlined in Fig. 5.1. A fuzzy variable formed in this manner is also a linguistic variable. The linguistic interaction (i.e., taking values) occurs between the first two layers at the top in Fig. 5.1. Therefore, predicates are also called fuzzy values. For example, if *Age* is a fuzzy variable, the predicate *Young* may be one of the values it can take.

The first design element of a fuzzy variable is the word selection. Because of the semantic importance, word selection has more meaning than just assigning a symbol as is frequently done in calculus. The same consideration applies to the selection of predicates. Semantically appropriate word selection between the first two layers is the essential key that makes a fuzzy engine transparent (i.e., easily understandable) in contrast to all other mathematical methods such as neural networks, which are ex-

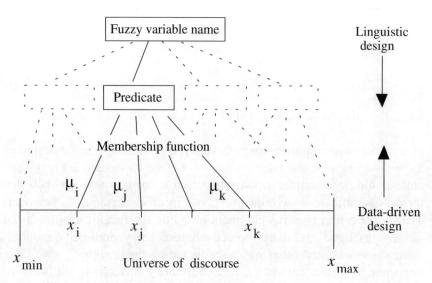

Figure 5.1 Structure of a fuzzy/linguistic variable.

pressed in pure symbolism. The second design element of a fuzzy variable is the membership functions. Designing membership functions is closely related to the overall solution, and considerations beyond the domain depicted in Fig. 5.1 often become necessary. The first challenge includes finding the appropriate number of membership functions (also referred to as the *number of mapping categories* or *granularity* in this book) and their locations on the universe of discourse. The second challenge is to determine the shapes of the membership functions. As we will see in this chapter, determining the number of membership functions and their location corresponds to the mapping of input-output relationships of the basic fuzzy inference mechanism. Determining the shape of a membership function (e.g., triangle, Gaussian curve, etc.) has less impact on the overall solution. The shape effects are the subject of Chapter 6. The third element of variable design is the determination of the universe of discourse. Universe of discourse represents the extent of validity of the fuzzy inference rules on each variable domain. Outside such a domain, the performance of a fuzzy inference engine will be unpredictable or ill defined.

The fuzzy variable design process is outlined in Fig. 5.2. The first goal in fuzzy variable design is to determine the number of membership functions and their location on the universe of discourse. When this goal is achieved, the other properties of a fuzzy variable can be determined easily. There are two main approaches to membership function design. The first approach, which is called *data-driven design* in this book, utilizes a set of numerical data that represents the inference mechanism. For example, a data set from a power plant, including operator's control actions and the variables of the plant, represents decisions made by the operator. A fuzzy inference engine automating the operator's decisions in the form of IF-THEN rules can be developed by examining such a data set. Data-driven membership function development can also be improved by heuristic interpretations during design. The second approach, which is called *linguistic design*, utilizes heuristic interpretation (or expertise articulated in daily language) due to either the lack of appropriate data or because of the nature of the problem. An example is a fuzzy inference engine that automates compliance to environmental regulations. In such a case, membership functions are mainly developed by the rules (or the logic) of decision making for compliance. Referring to Fig. 5.1, data-driven design is from bottom to top, whereas linguistic design is from top to bottom.

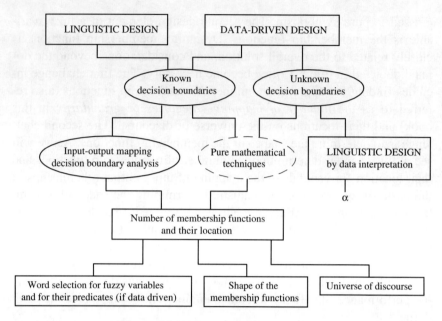

Figure 5.2 Fuzzy variable design process.

In both approaches to membership function development, an important concept called *decision boundary* plays a major role. Decision boundaries indicate which regions in the input space correspond to which regions in the output space in an input-output mapping process. In the linguistic design, decision boundaries are expressed in a qualitative manner. For example, when an expert articulates a solution using daily language such as "moderate exercise is good for the heart but no exercise is bad for the heart," the decision boundary between the two output categories (*good for the heart* and *bad for the heart*) is already determined by the levels of exercise (*moderate* and *none*). The challenge is to divide the input and output spaces by membership function design such that qualitative decision boundaries are represented in appropriate quantities. In the data-driven design, however, decision boundaries might not be as easily recognizable.

Extracting decision boundaries from a data set by membership function design, also known as *partitioning*, can be accomplished in a simple manner when we know which output data points correspond to which input data points. We refer to this case as the case of *known decision boundaries*. The decision boundaries are considered known in the sense that they can be formed by examining the data points and by applying simple geometric principles to define fuzzy regions. This process is labeled as the *decision boundary analysis* in Fig. 5.2 and is fully explained in this chapter. There are also advanced mathematical methods in the literature to accomplish partitioning. Among them, neural networks and genetic algorithms have shown promising results [1, 2]. For monotonically structured simple rules and a small number of variables, these methods can locate membership functions in their best—and sometimes optimal—positions, depending on whether a unique solution exists or not. Although some successful experimental results have been achieved using neural networks and genetic algorithms, there is no evidence that fully justifies their superiority over the simple "rule-of-thumb" methods. Particularly, when there is a handful of variables and the rule structure is nonmonotonic (like most of the fuzzy inference applications in which expert knowledge is embedded), these methods become less effective in finding the best (or optimal) membership functions due to the increased entropy (increased number of possible solutions), multiple objective functions, and increased computational complexity. Advanced numerical techniques for partitioning are beyond the scope of this book.

When the correspondence between the input and output data points is unknown or nonexistent, which we refer to as the case of *unknown decision boundaries*, there are two avenues to take. The first avenue is to have experts interpret the data. Because of the involvement of expert knowledge, this process is considered the same as linguistic design. As an example, consider the development of a fuzzy inference engine to diagnose cancer cells given Pap smear images. In such a classification problem, the expertise gathered from physicians will be needed to interpret the image data such that the decision boundaries are identified. When we have no means of interpreting data, and we do not know the correspondence between input and output data points, the problem becomes quite complicated. In the second avenue, the partitioning problem can be solved mathematically; for example, using the c-means clustering technique. However, purely mathematical methods are limited only to certain classification problems in which the output classes are to be constructed for a given objective function. For example, the c-means clustering technique cannot extract the decision boundaries from the Pap smear images to identify cancer cells in the absence of expert interpretation. Thus, not knowing the decision boundaries via heuristic interpretation makes fuzzy variable design—and the design of the entire fuzzy system for that purpose—a purely mathematical exercise that falls short in solving certain problems.

Another important consideration is related to statistics. Frequency of occurrence can be a useful concept when decision boundaries do not conflict with the statistics of the problem. However, developing membership functions only based on frequency data has limited application, as explained later in this chapter. We will also examine the linguistic modifiers (also called *hedges*) that change the role of a membership function in a controlled manner. For example, a membership function with the predicate *small* can be made *very small* by the hedge *very,* which is defined as a mathematical operation.

5.2 DATA-DRIVEN FUZZY VARIABLE DESIGN

Regardless of how a set of input-output data is obtained (statistical measurements, analytical computations, or articulation of expert knowledge), we will focus on membership function development by examining data and by extracting the decision boundaries. Data-driven design in this context assumes that the decision boundaries are known or they are easily extractable from the data set. This means that we know which point in the input product space belongs to which output class. For the case of unknown decision boundaries, the design becomes a linguistic design as depicted in Fig. 5.2. Linguistic fuzzy variable design is explained in Section 5.3.

5.2.1 Some Useful Concepts

Decision Boundary

The decision boundary concept is an essential element of design for both fuzzy variable development and for rule formation. Variable design and rule formation are closely interrelated; thus, one cannot be performed without considering the other. When we talk about decision boundaries, a mapping process—through which all fuzzy inference problems can be represented—is implied. The decision boundary concept is best understood by a simple mapping example with two input variables and one output variable. Consider a truth table formed using the crisp logic convention as shown in Table 5.1. Assume that the relation of interest is the exclusive-or problem.

Chap. 5 Fuzzy Variable Design

Table 5.1 Exclusive-Or Truth Table

I_1	I_2	O
1	1	0
1	0	1
0	1	1
0	0	0

The input product space of the truth table is shown in Fig. 5.3. As shown in Table 5.1, the points (0, 0) and (1, 1) on the input space produce 0, which means these two points belong to the same output class. Similarly, the points (1, 0) and (0, 1) belong to the other class. To separate the two output classes, we need two linear (or one convex) line as illustrated by dashed lines in Fig. 5.3. Note that these lines could have been drawn in a number of different ways. One of the most paradoxical questions is: Which set of boundaries is correct? There is no answer to this question; they are both equivalent when the only available information is as given in Table 5.1.

From this discussion, we conclude that decision boundaries are not uniquely defined and their function is symbolic rather than analytic in the absence of fuzzy information. When fuzzy information is available (i.e., if we know approximately which output class the point 0.7, 0.7 belongs to), the degrees of freedom in drawing decision boundaries decreases, and sometimes the boundary becomes uniquely defined based on the approximations of the fuzzy space. Note that the fuzziness emerges from the *approximate* knowledge of which output category the point 0.7, 0.7 belongs to. It has nothing to do with the fact that the point 0.7, 0.7 is a fraction, or in between 0 and 1.

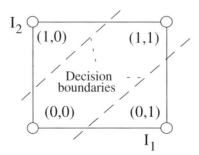

Figure 5.3 Decision boundaries of the exclusive-or truth table.

Decision boundary is a crisp concept. However, decisions can be fuzzy. Thus, once the boundary is crossed in one direction, we can either have a crisp decision or a set of fuzzy decisions. For example, suppose that there is a decision boundary of 65 mph for legal speeding. A crisp decision when the speed is 67 mph or 90 mph would be to issue a speeding ticket with a $100 fine. A fuzzy decision would involve the judgment of how close 67 mph is to the legal limit, and to issue a ticket with a proportional fine of less than $100. A fuzzy decision may also consist of several partial decisions taken at appropriate doses. When we discuss decision boundaries generated by membership functions (fuzzy sets) in this chapter, we will see that fuzzy decisions are determined by the fuzzy surfaces in the vicinity of decision boundaries.

Fuzzy Data

There are two types of data: crisp, and fuzzy. Crisp data have the possibility of 1.0 to represent reality. In practice, most of the measurement systems are considered to produce crisp data even though the data do not always represent reality with the possibility 1.0. When available, statistical measures such as confidence intervals are used to characterize uncertainty of crisp data. With fuzzy data, we consider the uncertainties and assign each data point a possibility value to represent different categories of reality. Fuzzy data can come from a physical measurement system but must be assigned categorical possibility values by human interpretation or by some mathematical criterion. Here, categorical possibility value means that a number obtained from a measurement device is only partial information about the state of the variable. Thus, such a number should only modify our belief in some proportion that the variable is in a particular state of being.

EXAMPLE 5.1 DATA TYPES

Let's examine the three basic forms of data representation. The first form is the crisp data: *Temperature is 50°C*. There is no implication of a category in this representation, and we simply assume that 50° is the reality. Note that we also accept that the temperature is not 49.9999° (or any other close value). The second form incorporates statistical uncertainty (if statistics is applicable): *Temperature is 50°C ± 1°C with 95 percent confidence level*. In this representation, there is an infinite number of values between 49 and 51 degrees that may be true with 95 percent confidence. The third form is the fuzzy data: *Data of 50°C has uncertainty that it could be 49°C with the possibility of 0.92*. Another interpretation is that *temperature of 49°C has the possibility 0.92 to represent the state of temperature 50°C*. Fuzzy data requires a possibilistic measure of being a member of category. Thus, we normally need a possibilistic sampling to be able to characterize the category or truth.

Chap. 5 Fuzzy Variable Design

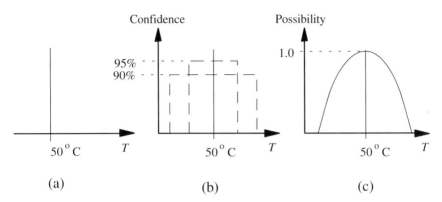

Three different representations of data shown above are: (a) crisp, (b) crisp with confidence intervals, and (c) fuzzy with possibility distribution.

Partitioning

In its simplest form, partitioning is an attempt to find a decision boundary between (at least) two data points on a product space. Because data can either be crisp or fuzzy, there are two cases of partitioning. Let's first consider partitioning based on crisp data. When the regions between crisp points on the input space are not known, one of the most standard methods is to fill the gaps by linear surface extensions [3]. This can be done by determining the most fuzzy points on the input product space that are at equal distance to both crisp points, then intersecting the surfaces at those points as shown in Fig. 5.4. Figure 5.5 shows the same procedure

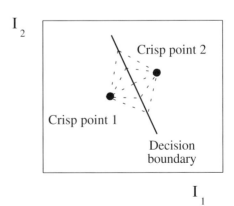

Figure 5.4 An equal distance decision boundary between two crisp points on the product space.

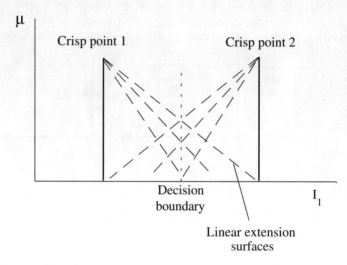

Figure 5.5 The profile of linear extension surfaces yielding the decision boundary in Fig. 5.4.

on one of the universes. Remember that we partition the input space only because two or more crisp points belong to different output classes as specified by a given solution.

Although the decision boundary shown in Fig. 5.4 has some angle with respect to both universes, the symmetric linear surface extensions shown in Fig. 5.5 satisfy the decision boundary in Fig. 5.4. This is explained in more detail in Section 5.2.2.

For the linear extension surfaces to intersect at the same location, it is obvious that extensions must be symmetric. This requirement can be expressed in terms of symmetric shapes for the membership functions. We will summarize this discussion in the form of a design principle.

In the absence of any information between two crisp points on a product space, the equal-distance partitioning can be achieved by membership functions with symmetric shapes to define the decision boundaries.

In the absence of fuzzy information on both input and output spaces, the application of this principle in terms of mapping rules is analogous to the interpolation method in calculus. It is analogous rather than identical because the overall fuzzy inference mechanism is nonlinear due to the implication and defuzzification mechanisms.

In the second case, where some of the regions in between two crisp points are identified by fuzzy data, then the interpolation method becomes a heuristic interpolation. In one sense, fuzzy data are nothing but an interpretation of the decision boundaries. Let's reconsider the equal-distance partitioning employed by symmetric surface extensions. Assume that there is a fuzzy point between the two crisp points on the product space. Assume that this fuzzy data is heuristically categorized to be on one side of the decision boundary. This information pushes the decision boundary away from the fuzzy point as shown in Fig. 5.6.

The new decision boundary can be formed by adjusting the linear (or nonlinear) surface extensions, which often requires asymmetric shapes for membership functions. However, there is no systematic method for such adjustments besides modifying the membership functions point by point while visualizing the changes in the decision boundary on a product space (or spaces).

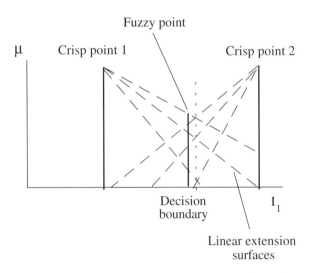

Figure 5.6 A fuzzy data point categorized to be on the left side of the decision boundary requires asymmetric surface extension.

> In the presence of fuzzy data between two crisp points on a product space, equal-distance partitioning can be modified to accommodate the desired decision boundaries by asymmetric membership functions.

The easiest way to employ this design principle is to employ the preceding principle first; that is, taking care of the crisp points first. Then the fuzzy points can be incorporated one by one. In practice, the fuzzy data depicted in Fig. 5.6 could be the stress fraction point computed by various analytical methods in between two measured data points. In such a case, the decision boundary represents resistance to load in a fuzzy inference engine that finds the best loading combinations.

The fuzzy partitioning method mentioned in this context is a straightforward process due to known decision boundaries. When decision boundaries are unknown, fuzzy partitioning becomes more complex. Several generalized comprehensive methods for fuzzy partitioning can be found in the literature in application to fuzzy clustering, pattern recognition, and feature extraction problems [4–6]. Most of these methods are based on the definition of an objective function (e.g., a criterion for classification). However, when decision boundaries are unknown, it is very difficult—if not impossible—to relate a mathematical objective function (such as the Euclidian distance) to the logic of a particular inference solution. In other words, there is no guarantee that an objective function based on purely mathematical techniques will yield mapping categories (i.e., membership functions) suitable for the mapping of the desired inference relationship.

Different Cases of Mapping

A known decision boundary means that all input-output classes are known. This implies that data are dynamic, sequential, or obtained case-by-case. In other words, every point on the product space (every input to the process, system, or event) is distinctly related to a point (or group of points) on the output space in the form of crisp or fuzzy data as depicted in Fig. 5.7.

In such a case, the output classification is easily done by applying the methods described previously. For example, the output classes in Fig. 5.7 can be represented by point membership functions with linear surfaces satisfying all fuzzy data points in each class. This is the most com-

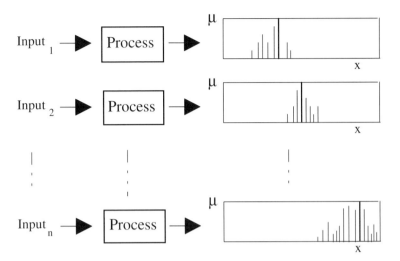

Figure 5.7 An easily identifiable case of output classification.

mon case in practice where fuzzy system applications are employed. Referring to the system descriptions in Chapter 4, this case applies to all problem types.

However, there are other situations where the trivial one-input-to-one-output correspondence does not apply. So-called one-input-to-many-output mapping, for example, is the case where the same input produces different outputs as shown in Fig. 5.8.

A typical example is the hysteresis loop and magnetization curve [7] where one value of magnetizing force produces two distinct values of flux density depending on the direction of approach. One-input-to-many-output mapping cases do not affect the membership function design. De-

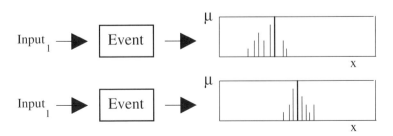

Figure 5.8 The case of one-to-many mapping.

cision boundaries on the product space are to be visualized in the manner described previously. However, one input-to-many-output mapping may significantly affect the fuzzy rule composition step. In systems for which more than one answer is intolerable (such as in diagnostics), this case can become a conflicting case unless the two output classes are identified by new relationships. For example, the two distinct values of flux density for a given magnetizing force can be identified by the direction of approach in hysteresis loops. If this type of auxiliary information is not available, a conflicting situation might arise. This is further discussed in Section 5.4.1.

In the one-input-to-many-output mapping case, a point membership function (see Section 5.2) will be shared by two (or more) output classes as shown in Fig. 5.9. When combining point membership functions, the shared region will be used more than once.

The case of many-input-to-one-output mapping is also trivial from the membership function design point of view. In this case, more than one inputs produces the same output response as shown in Fig. 5.10. In fact, this case is treated in an identical manner to that described in this Section 5.2.

The most complex form is the case where all of the above is mixed up yielding many-input-to-many-output mapping as exemplified in Fig. 5.11. However, as long as the decision boundaries are known (that is, we know which output points belong to which input points), the membership function development can be carried out in the same manner as described in Section 5.2.

The many-input-to-many-output mapping scenario can be a very complex problem if the decision boundaries are not known. In this case,

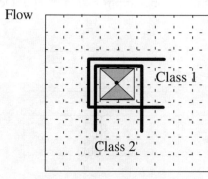

Figure 5.9 A point membership function being shared by two output classes in one-input-to-many-output mapping scenario.

Chap. 5 Fuzzy Variable Design

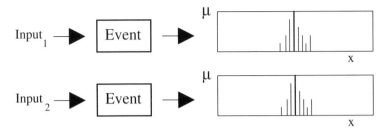

Figure 5.10 The case of many-to-one mapping.

the designer must build them by examining a data set such as that in Fig. 5.11 or by examining the reasoning behind such a relationship.

5.2.2 Point Membership Function

To understand the advanced forms of membership function development, we start from the basic building block which we call *point membership function*. Considering a product space between two input variables for simplicity, let's hypothesize a relation among I_1, I_2, and O defined by a single point. In the presence of a single crisp point but no other informa-

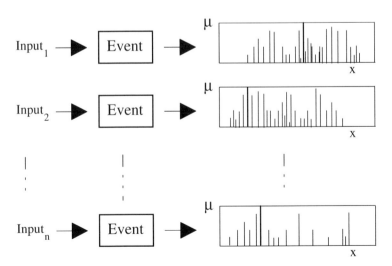

Figure 5.11 The case of many-to-many mapping.

tion, fuzziness around the crisp point may be defined by a pyramid (or a cone) which produces triangular projections on both product spaces. This is illustrated in Fig. 5.12.

We will see what is involved in determining the extent and shape of fuzzy regions later in this section. First, note that when membership functions are to be constructed based on data points of the relationship of interest, the problem setup implies the following mapping rules

$$(I_1 \bullet P_{1,1}) \Theta (I_2 \bullet P_{2,1}) \to O \bullet P_1 \quad (5.1)$$

where Θ = AND. The membership functions corresponding to the first two predicates on the LHS are shown in Fig. 5.12 as the projected triangles. For example, if all the predicates are called *Small*, then Eq. (5.1) would be equivalent to:

IF I_1 IS Small AND I_2 IS Small THEN O IS Small

The formation of the rules is examined in Chapter 7 in more detail. For now, we will only look at the rule formation issue from the membership function development standpoint. The more generalized form of Eq. (5.1) is

$$(I_1 \bullet P_{1,1}) \Theta (I_2 \bullet P_{2,1}) \Theta \ldots \Theta (I_N \bullet P_{N,1}) \to O \bullet P_1 \quad (5.2)$$

which implies $N \times N$ product space with N number of antecedent variables involved. The projection mechanism illustrated in Fig. 5.12 will

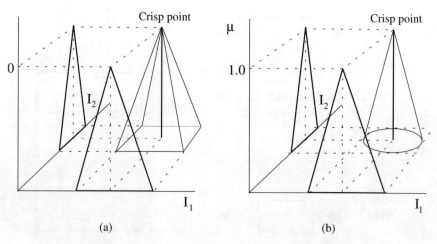

Figure 5.12 Projections of single crisp data surrounded by a pyramid (a) or cone (b) shape fuzziness.

work in the same manner; however, its three-dimensional visualization becomes impossible for obvious reasons.

When there is a fuzzy point in addition to the crisp one with the suggestion that both describe the same input-output relationship (or they both belong to the same output class), the fuzzy extension surfaces can be adjusted accordingly as shown in Fig. 5.13 for a hypothetical fuzzy point. Note that another fuzzy point is shown in the drawing that does not belong to the same output class. Thus, it is not going to affect the shape modification.

The new shapes generated by the projection have different regions with different certainty levels. If we only look at the membership function on the $I_1 \times \mu$ plane, the line connecting the crisp point and the fuzzy point at the top is much better known than the other parts of the membership function. Referring to the markings a, b, and c on the I_1 universe in Fig. 5.13, the interval b is the better known region due to the introduction of the fuzzy data point. The extensions toward the edges (regions a and c), where possibility values go down, are often overlapped by neighboring membership functions. At this point, we can conclude that building of point membership functions should use the best information available that includes a crisp point on the product space and fuzzy points around it that belong to the same output class. The extensions beyond the

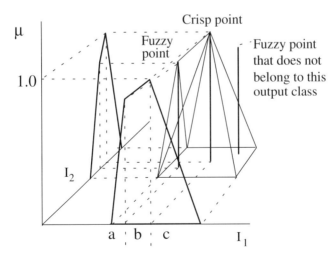

Figure 5.13 Modifying point membership functions by a fuzzy point on the product space.

available data points are to be determined when the decision boundaries are examined.

> A point membership function can be designed by the projection of a surface drawn on the product space that includes a crisp point and fuzzy points around it, all of which belong to the same input-output relation (or one output class).

The projection is from the product space $I_1 \times I_2$ to the planes $I_1 \times \mu$ and $I_2 \times \mu$. When the product space is N-dimensional, this process can be avoided by simply designing membership functions on each plane. However, the product space concept is unavoidable when the decision boundaries are to be examined. One practical way to visualize N-dimensional product space is to view it in 2×2 pairs, while keeping track of the output classes.

5.2.3 Exhaustive, Point-by-Point Construction

When there are N crisp points describing the relationship of interest, point membership function development can be repeated for each point on the product space. This represents an exhaustive solution scheme by which all points will have a mapping rule. We will see later how the number of mapping rules can be reduced to a practical size when the number of points is high.

To illustrate the exhaustive approach, let's first consider a second point defined on the same product space and produce another pair of membership functions for the new relationship. We will use pyramid fuzziness for simplicity. Adding the second pair of membership functions by projections is shown in Fig. 5.14. Two points of relationships imply two mapping rules given by

$$(I_1 \bullet P_{1,1}) \Theta (I_2 \bullet P_{2,1}) \to O \bullet P_1$$
$$(I_1 \bullet P_{1,2}) \Theta (I_2 \bullet P_{2,2}) \to O \bullet P_2 \quad (5.3)$$

where consequent predicates are different, indicating different output classes. When there are two (or more) crisp points on the product space,

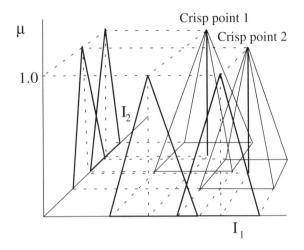

Figure 5.14 Two crisp points on the product space yield two pairs of membership functions via projection.

the shape of the membership functions and their overlap need to be determined. The membership function shape and overlap are discussed in Chapter 6 in detail. However, those discussions assume freedom of design in the absence of a decision boundary constraint. In a mapping process like the one discussed here, decision boundaries play a determining role. To illustrate this phenomena, let's reconsider the two crisp points surrounded by pyramidal fuzzy regions on the product space $I_1 \times I_2$. Using equal-base-area pyramids, the membership functions produced by projection yield an intersection (in the geometric sense) as shown in Fig. 5.15. Note that Fig. 5.15 is the bird's-eye-view of Fig. 5.14 with the exception that the projected membership functions on each axis are leaned backwards for easy visualization. The geometric intersection is a decision boundary in this example because the corresponding outputs belong to different classes. This boundary can be changed from its appearance in Fig. 5.15 by asymmetric adjustment of the fuzzy surfaces.

An important observation can be made at this point. If the fuzzy surfaces selected on the product space were cones instead of pyramids, then the intersecting boundary would be a straight line as shown in Fig. 5.15 (b). This is interesting because we would have the same projected membership functions (triangles). We have already seen in Chapter 2 that

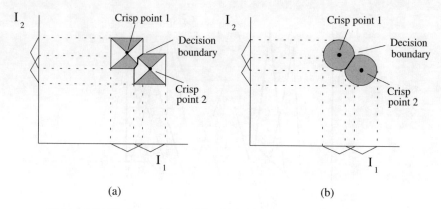

Figure 5.15 Bird's-eye-view of Fig. 5.13 using pyramids (a) and cones (b).

methods like Cartesian product or cylindrical extension cannot reproduce the exact relationship going from the projected membership function domain to the product space domain. Membership functions (fuzzy sets) that are subject to information loss during such transformation are called *interactive*.

Decision boundaries identified on a product space cannot always be exactly reproduced by adjusting the membership function shapes or overlaps.

For all practical purposes, there is no need to reproduce exact decision boundaries determined on the product space by modifying the membership functions. There are two reasons. First, a decision boundary in the absence of any information between the two crisp points is an approximated concept. Thus, there is no unique decision boundary to speak of. Second, approximated decision boundaries often produce the intended results. The design challenge is to form membership functions good enough that the approximated decision boundaries work as intended.

In general, N crisp points on the input product space require N mapping rules with N pairs of point-membership functions (for two antecedent variables). When all output classes are distinct (i.e., all consequent predicates are different), the exhaustive solution becomes the only solution as expressed by Eq. (5.4). Otherwise, the exhaustive solution

Chap. 5 Fuzzy Variable Design

may be reduced to an approximated solution using fewer rules by combining some of the point-membership functions.

$$(I_1 \bullet P_{1,1})\Theta(I_2 \bullet P_{2,1}) \rightarrow O \bullet P_1$$
$$(I_1 \bullet P_{1,2})\Theta(I_2 \bullet P_{2,2}) \rightarrow O \bullet P_2$$
$$\ldots$$
$$(I_1 \bullet P_{1,M})\Theta(I_2 \bullet P_{2,K}) \rightarrow O \bullet P_N$$
(5.4)

To examine combining two or more point-membership functions, we need to discuss more about the decision boundaries. Figure 5.16 illustrates N hypothetical crisp points on the product space, each surrounded by equal-base-area pyramid surface extensions representing fuzzy regions.

Let's first consider the case where each crisp point corresponds to a distinct output class. A bird's-eye-view of Fig. 5.16 is drawn in Fig. 5.17, which shows the intersecting decision boundaries for $N = 7$.

As shown in Fig. 5.17, some regions of the product space are empty that represent undecided cases of input combinations. Without disturbing the decision surfaces between the crisp points, membership functions can be extended to the sides to solve this problem. However, this is not an easy task and it also raises a question of credibility when the extent of the universe is to be classified in the absence of any information. The dotted lines in Fig. 5.17 illustrate the universe of discourse on each antecedent variable when $N = 7$. The single-point membership functions obtained by

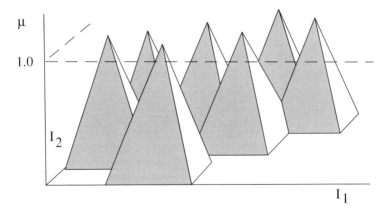

Figure 5.16 N crisp points on the product space surrounded by pyramid surface extensions representing fuzzy regions.

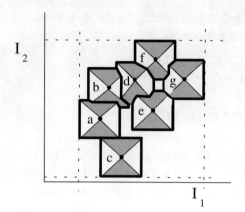

Figure 5.17 Bird's-eye-view of Fig. 5.16 with decision boundaries (N = 7).

projecting the fuzzy surfaces of Fig. 5.16 and by maintaining the decision boundaries of Fig. 5.17 are shown in Fig. 5.18.

The membership functions of Fig. 5.18 can be drawn directly without considering the steps depicted in Figs. 5.16 and 5.17 by placing equal triangles (or other symmetric shapes) on each crisp point on each variable domain. However, when we look backwards from Fig. 5.18 to Fig. 5.17, we see the relationship between such a direct design and decision boundaries. For example, widening the width of each triangle equally would not change the decision boundaries between the crisp points in Fig. 5.17 because the intersecting points remain the same. This leads to the following design principle.

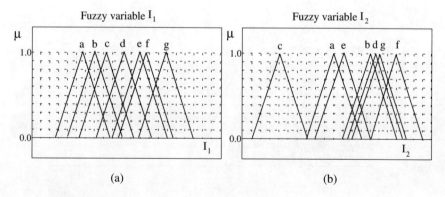

Figure 5.18 Projected point-membership functions corresponding to crisp points a, b, c, d, e, f, g of Fig. 5.16(a) and Fig. 5.17(b).

Chap. 5 Fuzzy Variable Design

> In the absence of fuzzy data points, membership function design for a relationship described by a set of mapping rules can be accomplished by placing equal width, symmetric, convex shapes on each crisp data point.

The principle stated above is based on equal-distance partitioning as described previously. When there are fuzzy points on the product space, asymmetry (or new break points) can be applied to incorporate them.

EXAMPLE 5.2 FUZZY CONTROLLER DESIGN BY EMULATING THE IDEAL PERFORMANCE

Control engineers want to build a fuzzy controller to regulate water level in a pressurized tank by adjusting the valve opening. They want the controller to monitor pressure and flow measurements and to make control decisions accordingly (assuming unreliable level measurement that does not allow a feedback control strategy). The data set of pressure, flow, and valve opening measurements is shown below. In these plots, the horizontal axis is the number of data points taken at different times. The vertical axis is normalized for simplicity. In addition, discrete data points are connected by straight lines for easy visual inspection. The discrete data points are also located on exact grid locations so that the reader can easily track the development steps. Despite all these simplifications, a data set from an actual system would be similar to the one described below.

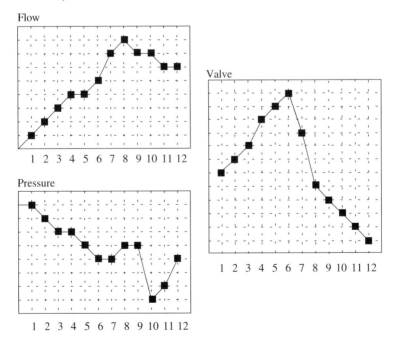

By the problem definition, *Flow* and *Pressure* are the two input variables and *Valve opening* is the output variable. The first step is to form the product space between *Flow* and *Pressure* data. We see that all data points correspond to different control actions in this example. The Flow-Pressure diagram shown below is the bird's-eye-view of a three-dimensional plot where the points indicate the tip of the pyramids and the squares represent their base area.

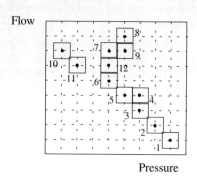

The numbers indicate the data numbers in reference to the previous plots. Note that the first choice of pyramid base areas can be increased to cover more of the product space. We know from the principle in Section 5.2.2 that equal, symmetric extension (of reasonable amount) will not distort the decision boundaries. We now arbitrarily extend the pyramid base areas by equal proportions.

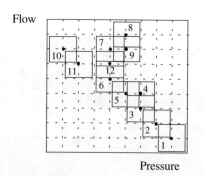

The extensions can be continued until the desired coverage of the product space is accomplished. Point membership functions from this product space are obtained by projection as shown in the following figures. The numbers at the top of each membership function correspond to the labeled data points on the product space shown previously. Notice that some of the projections exactly coincide. Thus, those coinciding membership functions are represented by one membership function (labeled using slash).

Chap. 5   Fuzzy Variable Design   225

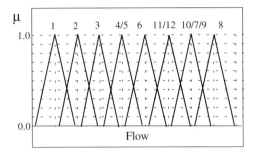

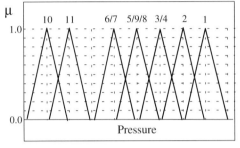

Because every point on the product space belongs to a different output class (which can be seen from the valve opening data plot), the decision boundaries on the product space are already fixed. In the absence of fuzzy data points, which is the case here, the equal-distance partitioning ensures that the decision boundaries embedded in the data are preserved.

Now let's assume that one of the crisp points on the product space has a large uncertainty due to measurement degradation. We assume that the eighth pressure data point, which was 5 units in the pressure plot, is also likely to be 5.3 units with the possibility of 0.9. Such information immediately changes the decision boundary around the corresponding point membership function. Returning to the membership functions defined on the *Pressure* universe of discourse, the previously combined membership function (of labels 5, 9, 8) now splits up to two new membership functions as shown in the following figure.

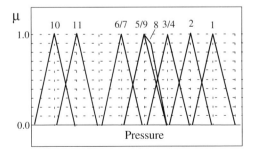

The new membership functions (5,9) and (8) have coinciding surface extensions on the left side of the crisp data point, whereas they have different shapes on the right. The membership function labeled data point (8) has incorporated the fuzzy singleton 0.9/5.3; thus, it is defined by 4 break points now. A fuzzy inference engine replicating the operator's valve adjustment actions can be developed by composing fuzzy rules in the manner expressed by Eq. (5.4). Examining the raw data set tells us that there will be 12 rules, each corresponding to one pair of pressure-flow data points on the input product space.

There may be cases in which all the data points on a product space are fuzzy (i.e., none of them have the possibility of 1.0). In such a case, the maximum possibility (or probability) point in each measurement can be treated as the crisp point during point membership function development. The designer can either normalize the possibility distribution functions to have at least one point equal to 1.0, or keep the original heights. In the second avenue, the designer must consider the height design issues described in Chapter 6.

5.2.4 Reduction of the Number of Point Membership Functions

Given the number of crisp data points N on a product space(s), we applied in the previous section an exhaustive, point-by-point development of membership functions, all of which were used in one mapping rule at least. The number of mapping rules is always determined by the number of output classes M such that each output class requires at least one mapping rule. In this arithmetic, we encounter three distinct cases of design: $N = M$, $N > M$, $N < M$. In all cases, the number of membership functions may be reduced by combining some membership functions that are closely located to each other. Note that this process is an approximation with varying damage on accuracy[1] in case $N = M$. The number of rules can only be reduced in case $N > M$ without significant damage on accuracy. The case of $N < M$ implies that there are at least $M - N$ output categories represented by the same point on the input product space; otherwise, decision boundaries are unknown. $N < M$ case is treated in the same manner as in $N = M$ for known decision boundaries.

---

[1] Distorting decision boundaries damages accuracy whereas having fewer output classes damages precision.

Combining Membership Functions for M = N

Reducing the number of membership functions when $M = N$ is an approximation at the expense of distorting the decision boundaries from their general appearance in Fig. 5.17. There is no measure to tell the designer how much to approximate. For example, the membership functions labeled as b, d, and g in Fig. 5.18 may be combined into one because they are clustered together very closely in comparison to others. This is analogous to finding a cluster with the minimum entropy when statistical methods are used (see Section 5.3). The membership functions of Fig. 5.18 are shown on the left in Fig. 5.19(a), whereas their combined version is shown on the right (b).

The approximation shown in Fig. 5.19 causes the output classification to be somewhat different. To illustrate the deviation, let's examine the effects of approximation. For the input data beyond the crisp point b, the value obtained from the membership function labeled as b in Fig. 5.19(a) would be less than that obtained from the combined membership function (b d g). The discrepancy becomes maximum for a certain value of input data using convex-shaped membership functions. This is illustrated in Fig. 5.20 for the ongoing case study.

This deviation will be translated into a deviation in the degrees of fulfillment of the rule, then will reflect itself in the implication and decomposition processes. There is no point in further iterating the effect produced by the deviation in the ongoing example because it would not yield any conclusive answer as to how much approximation can be expected in general.

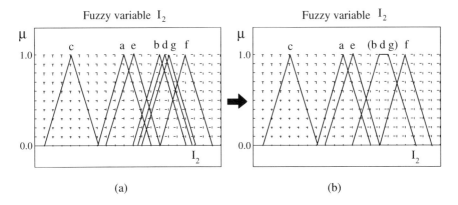

Figure 5.19 Clustering of the membership functions b, d, and g.

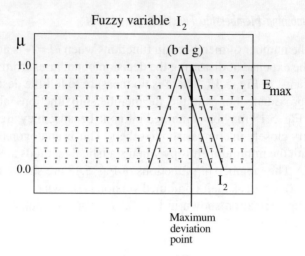

Figure 5.20 Combined membership function (b d g) overestimates the possibility value compared to what would have been obtained from the original membership function b.

This leads to the following design principle.

Some membership functions within close proximity to each other can be represented by one membership function completely entailing all of them. In a case where the number of crisp data points on the product space is equal to the number of output classes, such a simplification yields approximation of the result at a level only determined by the properties of the specific application.

This principle does not suggest that such simplification ought to be avoided. If the decision boundaries are explicitly known, the effect of simplification can be tracked down using a plot such as Fig. 5.20. However, such an attempt can become tedious if the number of crisp points is high or if the product space is of high dimensionality.

Combining Point Membership Functions When $N > M$

In the second case, where the number of output classes M is less than the number of crisp points on the input product space N, the reduction of the

Chap. 5 Fuzzy Variable Design   229

number of membership functions by grouping them as shown above does not yield approximation as in the previous case. In this case, the membership functions to be grouped are determined by the output classes. For example, suppose that the crisp points on the product space belong to only two output classes as shown in Fig. 5.21.

In the ongoing example, the crisp points a, b, c and d, e, f, g belong to output classes 1 and 2, respectively. As we will see later, the shape of the decision boundary determines the number of mapping rules. We start with an approximate linear decision boundary, which is the easiest. Inspection of Fig. 5.21 reveals that the decision boundary can be maintained using only two membership functions that separate the product space vertically. This can be done by grouping a, b, c and d, e, f, g membership functions as shown in Fig. 5.22.

The new decision boundary produced by two membership functions is shown in Fig. 5.23.

The effect of membership function grouping on decision boundaries can be seen by examining Figs. 5.23 and 5.21. The most significant deviation between these two figures occurs between crisp points b and d where the new decision boundary is closer to point d. This means that the product space is given more room for the output class b. If this approximation is undesired, the designer can choose to regroup membership functions by keeping point membership functions b and d separate. Other strategies may be fabricated by considering the specifics of the problem at hand.

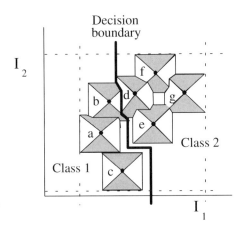

Figure 5.21 Decision boundary separating two output classes.

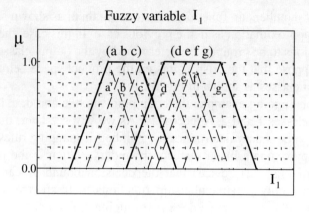

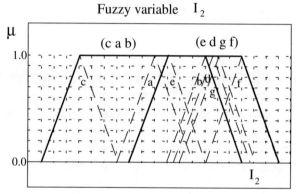

Figure 5.22 Grouping point-membership functions a, b, c and d, e, f, g.

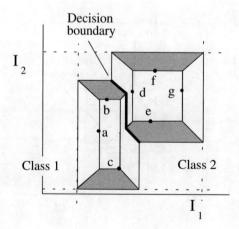

Figure 5.23 The new decision boundary produced by membership function grouping.

Chap. 5 Fuzzy Variable Design

> When an input-output relation (or an output class) is defined by more than one crisp point on the product space, the corresponding point membership functions can be represented using one membership function formed by combining them without significantly disturbing the decision boundaries.

The key challenge in combining point membership functions is the preservation of the decision boundaries embedded in a given solution. To describe a generalized method, we need to consider a generalized decision boundary, which is not possible. However, we will purposely select a complex decision boundary and explain the development such that the method can be applied to all other problems along the same lines.

A typical complex decision boundary is a linearly inseparable one that consists of one decision region completely surrounding another region. Returning to our ongoing discussion, assume now that the two output classes are defined in the manner depicted in Fig. 5.24. The output class 2 is surrounded by the output class 1.

The method consists of placing rectangles on the crisp points that only belong to the same output class, then extending linear surfaces. The rectangles define the possibility 1.0 plane at the top. Extension surfaces define planes that combine the top and the bottom when considering the three-dimensional space with the vertical axis being possibility. This is illustrated in Fig. 5.25.

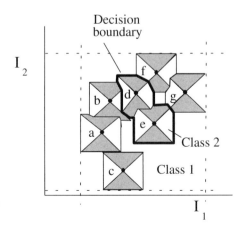

Figure 5.24 A decision boundary separating two nested output classes.

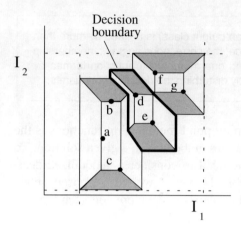

Figure 5.25 Combined point membership functions of Fig. 5.24.

The development steps are outlined as follows:

1. Identify crisp points that belong to each output class. (a, b, c, f, g for output class 1; d, e for output class 2.)

2. Identify possible combinations of rectangles that would include crisp points from the corresponding output class but will not include any crisp points from other classes (d-e is the only combination for output class 2; a-b-c and f-g are the only two possible combinations that comply with the requirement for output class 1).

3. Select from the combinations (if there are more than one) of rectangles that include the greatest number of crisp points (not applicable to our example because there is only one combination for each case).

4. Extend the sides by linear surfaces with equal slope to that of the point membership functions (see Fig. 5.25.)

5. Examine the new decision boundaries and make sure they do not conflict with the original decision boundaries of the problem. Go to step 3 if the current development is not as desired.

6. Develop membership functions by projecting the formed shape on each of the *variable* × *possibility* planes (see Fig. 5.26).

In step 2, placing triangles among the crisp points or forming asymmetric pyramids by combining two crisp points are in fact equivalent to placing

Chap. 5 Fuzzy Variable Design   233

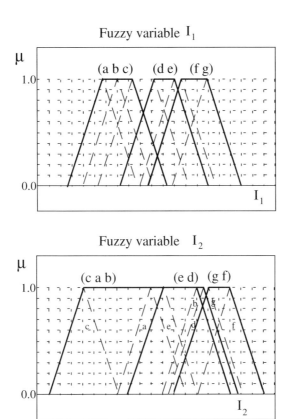

Figure 5.26 Membership functions generated by the projection of the shape in Fig. 5.24.

rectangles as described above. The equivalency is observed when the projected membership functions are examined. For example, instead of a rectangle, we place a triangle among the points a, b, and c in Fig. 5.27. Points d, e, f, and g are connected by lines that define the peaks of the pyramids in Fig. 5.24. The projected membership functions would be the same as in Fig. 5.26. Such discussions have very little (if anything) to do with fuzzy logic, since they come from the basic principles of calculus and geometry.

This method can be extended to product spaces of dimensions higher than two by analyzing all combinations of two dimensional product spaces, then combining the membership functions by the union or intersection operator. Another approach is to take the average among the

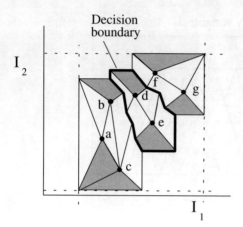

Figure 5.27 Surface formation using shapes other than rectangles.

membership functions formed on different product spaces (i.e., averaging only those corresponding to the same output class). The method can be reduced to directly developing membership functions without product space analysis. This is explained later in this section.

Fuzzy points on the product space can be incorporated by creating new linear or nonlinear surfaces such that possibility values are satisfied at the surfaces. If fuzzy points are inside the rectangles, they can be incorporated by forming a ditch at the top plane. Note that such a ditch would disappear during projection. Thus, the product space analysis is always useful after building membership functions via projection.

EXAMPLE 5.3 MEMBERSHIP FUNCTION DESIGN FOR THE $N > M$ CASE

Let's return to the problem described in Example 5.2. Now, we assume that some of the valve-opening data repeat themselves, creating the $N > M$ case. Our objective is to combine some of the membership functions without distorting the decision boundaries.

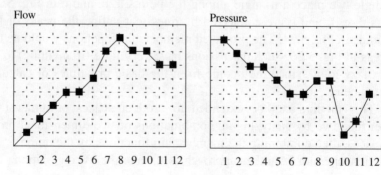

Chap. 5 Fuzzy Variable Design   235

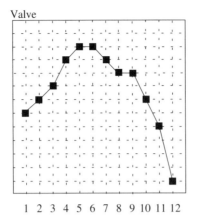

Note that valve data numbers 4-7, 5-6, 8-9, and 2-10 are identical. (In physical measurement systems this might never be the case). Then, the designer must first work on the output classification problem and decide the level of proximity to be considered as the same class. This will have a negligible approximation effect when done properly. Referring back to previously developed product space, the new decision boundaries are now drawn based on the output classes.

Note the difficulty of representing the decision boundaries for the pairs 4-7 and 2-10 in the figure below. There may be cases with more complex decision regions such that they cannot be connected on the product space at all. Such complex structures, including conflicting cases, are explained in Section 5.4. We arbitrarily extended the decision regions for the cases 4-7 and 2-10 in the following left-hand figure so that output classes can be seen. Empty spaces covered with such an extension are in fact not well defined. Thus, keeping the point membership functions for 4, 7, 2, and 10 makes more sense for this particular case, which is shown in the right-hand figure.

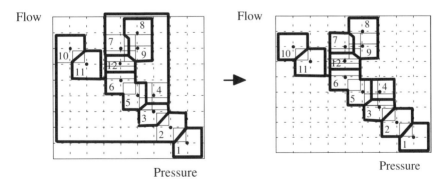

Membership functions from the product space on the right are obtained by projection as shown next. The numbers at the top of each membership function correspond to the labeled data points on the product space shown previously. Coinciding membership functions are denoted by "/" (e.g., 11/12). Combined membership functions are shown in parentheses such as (5 6).

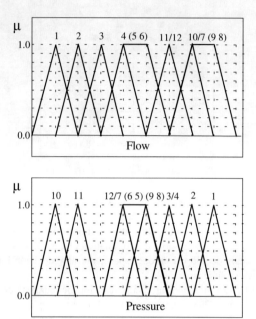

Designing Membership Functions Directly When N > M

Without the visual aid provided by the product space analysis, membership functions can be designed on each universe of discourse directly. This involves identifying the data points that belong to each output class. Assume we have a distribution of data on the universe of variable X as shown in Fig. 5.28.

The method of development includes the following steps:

1. Apply equal-width symmetric point membership functions at each crisp point. The equal-width, symmetric shapes ensure the preservation of decision boundaries. (This is shown in Fig. 5.29 for the output classes a, b, c, d, and e.)

2. Combine those point membership functions that belong to the same output class. (Only the output class a is shown in Fig. 5.30.)

Chap. 5 Fuzzy Variable Design

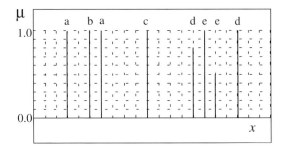

Figure 5.28 Distribution of data points labeled with output classes to which they belong.

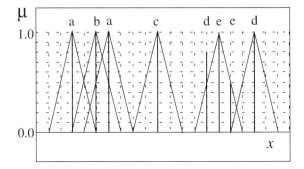

Figure 5.29 Point membership functions.

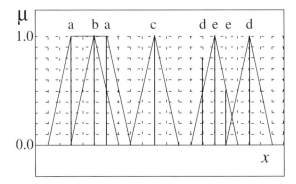

Figure 5.30 Combining point membership functions of the same output class.

3. Incorporate fuzzy points if there are any. Make sure new extensions do not disturb the decision boundaries. (There are two fuzzy points in this example. The first one belongs to group d, whereas the second one belongs to e. Notice the extensions marked with arrows in Fig. 5.31, in which the designer has to make judgment calls as to how far they should go.)

Obviously, such a development can use equal-width symmetric curves (such as bell curves) instead of triangles at step 1. The shape effect on the output response is examined in Chapters 6 and 8.

Designing Membership Functions When N < M

When the number of output categories is greater than the number of input data points, the design considerations of the case $N > M$ will also hold. However, the approximation caused by combination is shifted from the antecedent domain to the consequent domain. In other words, the consequent membership function design will approximate the contribution strength of each fuzzy rule. Compared to the cases $N > M$ and $M = N$, the degree of approximation in the case $N < M$ is more significant because the granularity of the decision variable (precision) is changed by combining membership functions.

> In general, the effect of approximation caused by combining the point membership functions of a consequent fuzzy variable is more drastic than that of an antecedent fuzzy variable.

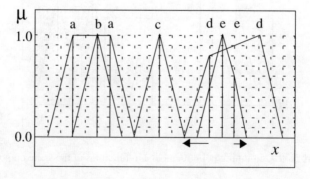

Figure 5.31 Incorporating fuzzy points.

Again, there is no measure to further characterize this principle. However, the designer may consider this principle to focus more attention on the consequent fuzzy variable design to minimize undesired consequences caused by approximations.

5.2.5 Probabilistic Membership Function Design

The analytical derivations of fuzzy measures found in the literature are based on the initial condition that some sort of probability or possibility assignment is provided by experts or by field measurements. Fuzzy measures such as plausibility and belief can be computed for various cases in which there are multiple experts suggesting different estimates of the initial probability assignment about the outcome of an event. In relation to the basic fuzzy inference algorithm described in Chapter 3, we are interested in the mechanism of assigning probabilities that can be applied to membership function design. Regardless of the level of involvement of the probabilistic methods, the first point to remember is that our focus is fuzzy system design and not a probabilistic system design. Therefore, probabilistic theories need not be proven in application to fuzzy inference design.

There is one fundamental issue when assigning probability estimates during membership function design. The designer must identify whether the problem at hand requires *truth characterization* or *truth categorization*. The difference constitutes a delicate criterion through which the nature of design can change drastically. As described previously in Chapter 2, truth characterization applies to problems in which we estimate the outcome of an event through repeated measurements. The question, What is the probability of catching a 12.4 Kg bass in this river? requires probability assessment of this type. Truth categorization applies to problems in which we estimate the category of an event by judgment. The question, What is the probability of catching a huge bass in this river? requires probability assessment of the category huge, which is defined by norms other than probability. *Huge* is subject to interpretation and its boundaries are fuzzy (subjective).

Let's examine truth categorization first. Both human expertise and commonsense reasoning are mostly formed by truth categorization, because it is natural.[2] Most of the fuzzy systems in practice employ truth

---

[2] Natural in the sense that the human brain will form opinions based on the likelihood of truth categories rather than the likelihood of numerical truth values (likelihood of huge bass instead of 12.4 Kg bass).

categorization regardless of whether the embedded rules are formed by expertise or by common sense. In truth categorization, the probability assessment becomes flexible since the probability of data to fall inside the fuzzy boundaries of a category is not well defined. Consequently, probability assignment becomes possibility assignment. To illustrate the complementary nature of possibilistic and probabilistic measures, consider the following fuzzy statement:

Weather_temperature_in_August IS High

The question of how to design a membership function for *High* is answered in the following ways:

- Design by possibility; that is, interpreting what *High* is in August based on the designer's opinion of uncomfortably warm temperature for August.
- Design by subjective probability; that is, interpreting what *High* is in August based on guessed (unrecorded) data, opinion poll, experience, and memory.
- Design by probability; that is, collecting measurement data in August, forming a distribution function, and calling it *High*.

The three alternative ways listed above are all debatable, since the statement *Weather_temperature_in_August IS High* is fuzzy and it contains mathematically unjustifiable linguistic and cultural values. However, notice the restrictions imposed on the third method listed above. That is, no semantics are involved in the development of the statistical distribution called *High*, thus it might not always be in agreement with what we consider *High* to be. Fuzzy variable design for truth categorization is referred to as linguistic design and is explained in Section 5.3.

Truth characterization problems remind us of standard probability problems. Solving such problems by means of the basic fuzzy inference algorithm requires that the fuzzy variables describe likelihood, and the inference mechanism describes a possibilistic mapping between truth characterizations. Next, we will discuss two approaches involving probabilistic fuzzy variable design. Before these discussions, the following design principle will summarize our initial arguments.

> The probabilistic approach is more suitable for truth characterization problems in which the underlying mechanism for fuzzy variable design is *likelihood*. On the other hand, truth categorization problems in which the underlying mechanism for fuzzy variable design is *possibility* are more flexible.

During the knowledge acquisition process as discussed in Chapter 4, the designer must consider the principle stated above if the solution to be implemented includes probabilistic elements. The easiest way to employ this principle is to remember the fish story and compare the variables of the problem accordingly. The designer's main objective is to make sure that the variables of truth categorization were not strictly designed by probabilistic measurement data; instead, they were made flexible by including human judgment and possibilistic measures.

Frequency of Occurrence

Frequency of occurrence is one of the fuzzy measures in identifying uncertainty for particular problems. A membership function can be defined by the frequency of occurrence data that are obtained by repeated measurements of the same event. A hypothetical development is shown in Fig. 5.32.

A frequency plot obtained by repeated measurements of the same event directly yields measurement uncertainty. This corresponds to the definition of point membership function given previously, in which the crisp point was considered to be the tip of a pyramid that represented

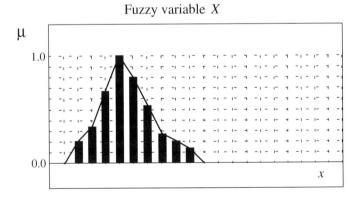

Figure 5.32 Normalized frequency of occurrence data used for probabilistic membership function development based on frequency of occurrence.

fuzzy surfaces. The underlying assumption in such a development is that repeated measurements are obtained from an event that has retained its statistical properties while the measurements were taken. This is almost never true in real life situations, or such a truth is never absolutely known. Thus, frequency of occurrence has an uncertainty associated with the assumptions behind its concept. If the requirement of retention of statistical properties is not satisfied, it would not be known for sure whether the fuzzy points (uncertainty) belong to the same input-output relation (or the same output class). Such an uncertainty means uncertain decision boundaries, which might have further implications depending on the type of development.

> A point membership function development using frequency of occurrence data is based on the requirement that repeated measurements are obtained from the same event that has retained its statistical properties while measurements are taken.

Normalization to 1.0 is necessary to avoid unintentional importance deterioration of the corresponding fuzzy rule. When the rule importances are also to be determined probabilistically, height adjustments can be considered as explained in Chapter 6.

The application of point membership functions to all different events characterized by a fuzzy variable is also possible with the assumption that each event is tracked independently and each data set satisfies the requirements stated previously. An example is shown in Fig. 5.33 for three statistically independent events on the same universe of discourse.

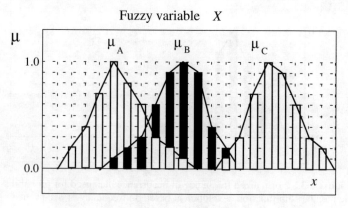

Figure 5.33 Designing a probabilistic fuzzy variable.

Chap. 5 Fuzzy Variable Design   243

Fuzzy variables designed in this manner automatically form decision boundaries. Thus, decision boundaries become purely data dependent, which might not always yield the desired possibilistic mapping between the membership functions of antecedent and consequent variables.

EXAMPLE 5.4 MEMBERSHIP FUNCTION DEVELOPMENT USING FREQUENCY OF OCCURENCE

Assume that we have collected data points for three independent events denoted by A, B, and C on the universe of x [0-21]. The data for event A is:

x_A = [2, 10, 3, 6, 6, 5, 3, 5, 4, 4, 5, 5, 5, 4, 5, 5, 6, 6, 6, 9, 9, 6, 6, 7, 8, 7, 4, 4, 4, 7, 7, 3, 5, 3, 7, 8, 7, 8, 5, 5, 6, 2]

Organizing data for event A yields

Value:	2	3	4	5	6	7	8	9	10
Occurrence:	2	4	7	10	8	6	3	2	1

Normalizing the frequency of occurrence yields the following distribution:

| Distribution: | 0.2 | 0.4 | 0.7 | 1.0 | 0.8 | 0.6 | 0.3 | 0.2 | 0.1 |

Using fuzzy set notation, the membership function is given by:

$$\mu_A(x) = 0.2/2 \cup 0.4/3 \cup 0.7/4 \cup 1.0/5 \cup 0.8/6 \cup 0.6/7 \cup 0.3/8 \cup 0.2/9 \cup 0.1/10$$

For simplicity, we did not include zero singletons that extend beyond the point 10. This distribution is shown in Figure 5.33 as the μ_A. Assume that other events B and C are determined in the same manner and are shown in Figure 5.33. The table below shows three examples of what A, B, C, and x could be in an actual implementation.

As seen in these examples, truth characterization is governed by the principles of the system under investigation rather than by heuristic categorization of events. Considering X is A for the first line in the table, we may have a fuzzy rule of the form

x	A	B	C
Precipitation	Number of times temperatures reached over 100 F° causing drought	Number of storms with winds faster than 80 mph	Number of times the Mississippi River flooded 8' high
Number of hydraulic system failures	Number of failures before the first inflight warning	Number of failures before the first emergency landing	Number of failures before the plane crashed
Number of bees	Bee larvae died due to mulnutrition	Bees resided in the same hive more than 3 months	Bees left the hive due to overpopulation within first 6 weeks

IF Precipitation CAUSES Drought THEN Land IS_NOT For_agriculture.

This fuzzy rule is a part of a fuzzy decision maker estimating the value of land. Note that the connector *CAUSES,* which is used instead of *IS,* is to maintain the linguistic integrity of this phrase. While using this inference engine, the user enters a year-average precipitation value among many other inputs to evaluate the characteristics of the land under investigation. The mapping between the membership functions *Drought* and *For_agriculture* is possibilistic, and it has nothing to do with the statistics of rain. Such problems can be solved by other means, such as using a probability distribution that relates the probability of drought to the probability of farming.

The examples in the table above raise the question of statistically independent events on the same universe of discourse. For example, is the distribution data of event B (number of storms with winds faster than 80 mph) dependent on the distribution data of event C (number of times the Mississippi River flooded 8′ high) on the precipitation universe of discourse? In this particular case, precipitation that occurred during strong storms may have caused some of the floods on the Mississippi River. If this is the case, the number of storms and floods are related. Such a relationship, which is not explicitly characterized in Figure 5.33, can influence the role of the variable *precipitation* depending on the rest of the problem. However, considering the possibilistic mapping mechanism of the basic fuzzy inference algorithm, deviations from the basic axioms of probability are allowed as long as uncertainties are characterized to the best of our knowledge.

Minimum Entropy

In Fig. 5.33, we assumed that each event is tracked independently until a probabilistic distribution is developed. Thus, we created fuzzy data points that are the points with frequency of occurrence less than 1 (1 representing the maximum occurrence). If we have a data set including no fuzzy data (i.e., all data points are assumed to be crisp or are obtained from different single measurements) then we can apply an alternative procedure. There are probabilistic methods in the literature to partition the product space provided that the number of categories is known (i.e., the number of decision regions is known). One of the simplest methods is based on finding a cluster with the minimum entropy and using the cluster boundaries as the intersecting points of membership functions. In Fig. 5.34, we assume that the universe has to be split into two due to a decision boundary constraint. The two output classes are marked with black and white dots for this hypothetical case study.

Entropy estimates can be obtained by first calculating the conditional probabilities of dark and white dots on either x or y domain. Let's assume we want to develop membership functions on the x universe. The

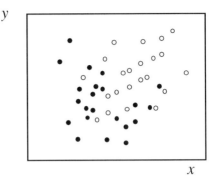

Figure 5.34 A large amount of data on the input product space that belong to two different output classes (black dots versus white dots).

conditional probability of dark dots to reside on one side of the border ξ on the universe x is given by

$$p_d(\xi) = \frac{n_d(\xi)+1}{n(\xi)+1} \qquad (5.5)$$

where n_d is the number of dark (or white) dots that reside on one side of the border ξ and n is the total number of dots in the same region. The entropy [8] is computed by

$$S(\xi) = p^+(\xi)S^+(\xi) + p^-(\xi)S^-(\xi) \qquad (5.6)$$

where + and − superscripts indicate the regions in positive and negative directions from the border ξ and $p(\xi)$ is the probability that all dots reside in that region. Note that $p^+(\xi) = 1 - p^-(\xi)$. The entropies for each region are given by

$$S^+(\xi) = -[p_d^+(\xi)\ln p_d^+(\xi) + p_w^+(\xi)\ln p_w^+(\xi)] \qquad (5.7)$$

$$S^-(\xi) = -[p_d^-(\xi)\ln p_d^-(\xi) + p_w^-(\xi)\ln p_w^-(\xi)] \qquad (5.8)$$

The objective in this computation is to find a border ξ on the universe x such that Eq. (5.6) yields a minimum value. This iterative process is best performed using a computer program. If we had to split the universe into N regions, the procedure is repeated. Note that once an appropriate ξ is found, a membership function shape must be selected such that two of them overlap at point ξ. Thus, this method locates a decision boundary but it does not provide us any information about how to approach the decision boundary (membership function shapes). Figure 5.35 shows the result of this method applied to the data in Fig. 5.34 for both x and y universes using arbitrarily chosen membership function shapes.

Management of Subjective Probabilities

The last two topics illustrated data-driven methods that involve probabilistic fuzzy variable design. When the probability assignments are made subjectively by the opinion of experts, there is no systematical method to formulate this process. It is the expert's own approach to estimate probabilities. However, if there is more than one expert opinion on the same issue, a new problem arises: How to aggregate their opinions?

There are two aspects of this problem: practice and theory. We will first discuss practical considerations. In Chapter 4, we identified the designer as someone other than the expert in the field for problems requiring expert knowledge. This distinction is important to isolate the responsibility of fuzzy system design from the responsibility of forming aggregated expert opinions. It is best to let the experts resolve their conflicting opinions and not to have fuzzy system designers make judgments on aggregation. In most of the practical systems, designs have been created through such a social process. If conflicts are not resolved among

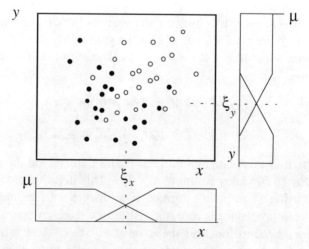

Figure 5.35 Minimum entropy method for finding decision boundaries in both x and y universes.

experts, it is not recommended to apply a mathematical aggregation method and to pursue the design.

In theory, there are many ways to aggregate multiple expert opinions. When the conflicts are mild and manageable, some of these techniques may be applied. The simplest way is the old-fashioned averaging method. A conservative approach is to take intersection, whereas a tolerant approach is to take the union of multiple opinions. Figure 5.36 illustrates two membership functions designed by two experts and their average, intersection, and union.

These operations are performed by treating probability distributions as possibility distributions (fuzzy sets). The aggregation operators found in the literature [9] (i.e., parametrized and nonparametrized t-norms and t-conorms) produce a family of solutions ranging from intersection to union, including the average. The three alternatives presented here are the most practical solutions.

An average, intersection, or union membership function might result in a height less than 1.0, which can be normalized to 1.0 if appropriate. Notice the manner in which the experts have assigned probabilities in Fig. 5.36. There is a single event, and the experts have assigned probability values to x to characterize the likelihood of this event to occur at such locations. Normalization to 1.0 in this context is reasonable. However, when there are multiple events on the same universe of discourse, which

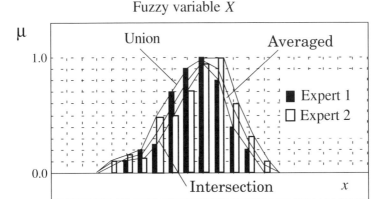

Figure 5.36 Two membership functions designed by two experts based on subjective probability assignment are aggregated by averaging, intersection, and union.

is always the case in the domain of fuzzy inference, experts often assign probabilities to each event such that the likelihood of their relative occurrence is characterized. For example, the probabilities of A, B, and C to occur are 0.2, 0.5, and 0.3, respectively. To convert this one-dimensional information into two-dimensional form suitable for fuzzy variable design, experts also need to specify the most likely location for each event to occur on the universe of discourse. This is summarized in Fig. 5.37 for a hypothetical case.

In Fig. 5.37, the two experts have assigned likelihood of relative occurrence to different events (i.e., A = 0.2, B = 0.7, C = 0.1 expert 1; A = 0.5, B = 0.2, C = 0.3 expert 2) as well as where they could most likely occur on x. If we average, intersect, or union, then normalize the distributions for each event, we may get a reasonable estimate for the most likely locations but not for the most likely relative occurrences. Thus, normalization to 1.0 must be avoided when likelihood of relative occurrence is specified among different events in the form shown in Fig. 5.37. In the realm of the basic fuzzy inference algorithm, a fuzzy variable including membership functions with heights less than 1.0 will affect the output by means of rule importance deterioration as described in Chapters 6 and 7.

There are other methods in the literature, such as Dempster Shaffer's body of evidence theory [10], that offer different ways to estimate combined evidence when the likelihood of relative occurrence is specified by more than one expert. However, there is truly no way to assess how effective such methods can be. The designer should bear in mind

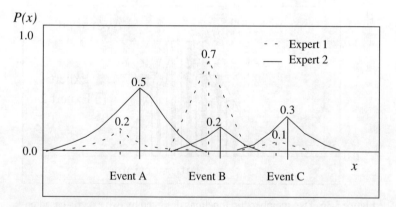

Figure 5.37 Two experts assigning probabilities to three different events on the universe of x.

that theoretical methods for opinion aggregation are rarely used in practice (i.e., in today's commercial products) and these methods do not provide ultimate answers.

5.3 LINGUISTIC FUZZY VARIABLE DESIGN

The fuzzy variable design methods discussed up to this point have treated the predicates as nothing but labels in categorizing different regions of the universe of discourse, because the design procedure was from bottom (data domain) to top (category domain) in reference to Fig. 5.1. Now, we will look into linguistic design issues where the design procedure is from top to bottom. Therefore, semantics will be the driving force in determining the number of membership functions, their locations, and shapes.

Compared to the previous case, this type of design is not based on data either due to the lack of it or because of the nature of the problem at hand. In the previous sections, the decision boundaries were buried in the data set of the input-output mapping relationship. We built membership functions by retaining the original decision boundaries or by creating new ones using data. After categorization, decision boundaries help us to discover the corresponding fuzzy rules. Now, decision boundaries are not buried in a set of data, instead they are buried within the linguistic articulation of expertise. Our goal is to obtain a set of fuzzy rules following a necessary knowledge acquisition procedure, then to interpret the predicates such that their categorization of the numerical universe are appropriate. Here, *appropriate* is a subjective interpretation that often reflects the style of the designer.

> In linguistic fuzzy variable design, the articulation of knowledge in terms of fuzzy rules is the initial requirement by which the number of mapping categories (membership functions) is readily determined. Thus, the design is reduced to determining the location and shape of the membership functions.

To determine the location of the membership functions, the key element is the decision boundaries embedded in the articulation of rules and the type of problem at hand (i.e., truth characterization versus truth categorization). Consider the following fuzzy rules composed for truth characterization:

IF Load IS Too_heavy THEN Bridge WILL Collapse
IF Load IS Moderate THEN Bridge WILL Sustain

The decision boundary between *Too_heavy* and *Moderate* on the universe of *Load*, if not known by analytical methods or by test results, is not something one should guess (or an expert should believe) during design. Therefore, truth characterization problems are not always suitable for linguistic design. Now, consider the following rules:

IF Load IS Too-heavy THEN Number_of_boxes_to_unload IS Many
IF Load IS Moderate THEN Number_of_boxes_to_unload IS None

This is a truth categorization problem with less stringent requirements on the accuracy of the decision boundary between *Too_heavy* and *Moderate*. More importantly, possibilistic mapping between *Load IS Too_heavy* and *Number_of_boxes_to_unload IS Many* does not describe the outcome of a physical event; rather, it forms a flexible relationship between two categories. Linguistic fuzzy variable design is more suitable in such cases.

There might be no clear-cut distinction between truth characterization and truth categorization problems in reality. A general principle is that truth characterization problems require more precise decision boundaries by which truth is directly estimated, whereas truth categorization problems allow flexible decision boundaries by which truth is inferred to fall into a category. It is very much like the difference between estimating your body weight and estimating the similarity of your body to the category *chubby*. Two different objectives are implied.

There are two types of linguistic membership function design when the shapes are concerned. First, the predicates encountered in a set of fuzzy rules are semantically unconstrained, which means infinite degrees of freedom in the design of their corresponding membership functions. Measures of quantity, strength, quality, and association (e.g., *high, small, large, medium, heavy, hot, cool,* etc.) fall into this category of design. Unconstrained predicates are often context specific. For example, the predicate *hot* depends on the context when we need to quantify it.

Second, the predicates are semantically constrained by the language (and culture) that impose commonsense interpretation. Membership functions representing such predicates cannot be designed by infinite degrees of freedom as long as the integrity of their meaning is to be preserved. Measures of comparison (e.g., *less than, quite, almost, very, roughly,* etc.), which are also known as linguistic hedges, fall into this category of design. Constrained predicates are context independent. For example, the role of the predicate *very* is invariant regardless of whether it is used in *very hot* or in *very rich*.

5.3.1 Semantically Unconstrained Predicates

Unconstrained predicates are also in two types. The first type of predicate is not interpretable in a numerical universe of discourse. For example, there is no practical way of defining membership functions for predicates such as *angry* or *red*, although they may be considered as fuzzy concepts. In some cases, such predicates are treated as crisp concepts and are used as Boolean gates (i.e., Boolean gates determine the branch of a decision tree that satisfies given conditions).

The second type of predicate is the measure of quantity, quality, strength, and association that can be numerically represented within a specific context. A short list of such words is shown in Table 5.2. Note that predicates of this sort are too many to list here. Given the case, the designer must interpret them and decide whether they belong to the unconstrained category.

These predicates modify a fuzzy variable and are used in association with the variable itself. Consider the following examples.

Low_temperature	Quantity
High_performance	Quality
Heavy_weight	Strength
Similar_investment	Association
Usually_fast	Likelihood

The associations (denoted by the underscore sign) make it possible to design membership functions on the universe of the corresponding fuzzy variable. Let's consider the first example, *Low_temperature*. As

Table 5.2 Sample of Predicates That Are Semantically Unconstrained

QUANTITY	QUALITY	STRENGTH	ASSOCIATION	LIKELIHOOD
Low	Low	Light	Desirable	Usual
High	High	Heavy	Similar	General
Small	Acceptable	Weak	Acceptable	Often
Large	Poor	Strong	Risky	Rare
Medium	Good	Silent	Dangerous	Never

characterized in Fig. 5.38, the design of the corresponding membership function depends on individualism, experience, style, and beliefs of each designer. Between different designers attacking the same problem, there is theoretically an infinite number of design possibilities for *Low_temperature*. Note that in many applications in the literature, the association is dropped from the notation for simplicity (e.g., *Low* instead of *Low_temperature*), thus making it difficult to identify semantically unconstrained predicates.

An important question the reader may have at this point is the validity of such a design. Although some experiential evidence might suggest that differences between individual designs will produce negligible effects, there is no proof that such an assumption will be valid for all cases. The term *negligible effects* here implies the cost criterion. The design of membership functions for semantically unconstrained predicates, which is encountered in truth categorization problems, requires an iterative approach until the desired cost behavior is achieved. The membership function shape analysis presented in Chapter 6 is directly applicable to this type of design.

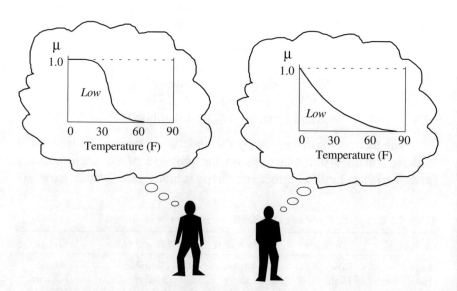

Figure 5.38 It is quite unlikely that two designers will have the same interpretation of a semantically unconstrained predicate such as *Low_temperature*.

5.3.2 Linguistic Hedges

We mentioned earlier that a group of predicates are semantically constrained in the sense that their interpretation is somewhat standard when translated into the calculus of fuzzy systems. For example, the hedge *very* is not as difficult to interpret as the unconstrained predicate *Low_temperature*, which is context specific. A short list of linguistic hedges and their somewhat standard role in fuzzy logic are listed in Table 5.3. Note that this list may be extended by examining the English language (or any other language), which is beyond the scope of this book.

An important point to remember is that there is a certain hierarchy between the semantically unconstrained predicates and linguistic hedges. Hedges modify predicates, whereas predicates modify fuzzy variables. For example,

$$Low\_temperature$$
$$very\ (Low\_temperature)$$
$$very\ very\ (Low\_temperature)$$

Therefore, the function of hedges is to modify an already defined membership function. The easiest way to identify a linguistic hedge is to combine it with a fuzzy variable, which should not make sense (e.g., *very_Temperature*).

The functions listed in Table 5.3 have been defined in the literature [11]. Although the general trend they represent can be formulated analytically, there is no single correct way of designing them. In addition, the selection of words for the intended function might vary among designers.

Table 5.3 Linguistic Hedges and Their Function

HEDGES	FUNCTION
Very, extremely	Concentration
Somewhat	Dilution
Definitely, nearly	Intensification
More or less	Relaxation
Not	Negation
Below, above	Restriction

Concentration

To concentrate a fuzzy set, just like concentrating a liquid, is to shrink its spread over the universe of discourse while maintaining the original possibility distribution profile. Because fuzzy sets have possibility values between 0 and 1, taking powers (greater than 1.0) of a membership function produces a concentration effect, as shown in Fig. 5.39.

Note that there is no justification for representing the word *Very* by the second power of the original fuzzy set. Thus, such conventions used in the literature are subjective in general. The level of concentration can be adjusted by changing the power and can be given any appropriate word of this type.

Dilution

Similar to its meaning in daily language, dilution works opposite to concentration by spreading the membership function over a larger area on the universe of discourse while maintaining the original characteristics of the membership function. Different levels of dilution can be obtained by raising the power of fuzzy sets less than 1.0. A typical application is the square root. The effect of dilution can be seen in Fig. 5.40.

Again, there is no justification for the word *Somewhat* to be a more fuzzy concept than *Not-so*. Such word selections are in fact not very important as long as they are consistently applied and their relative roles are known.

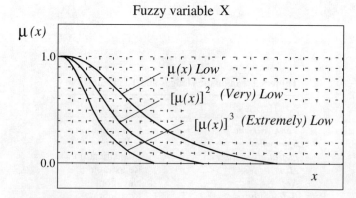

Figure 5.39 Concentration of the membership function *Low* by linguistic hedges.

Chap. 5 Fuzzy Variable Design

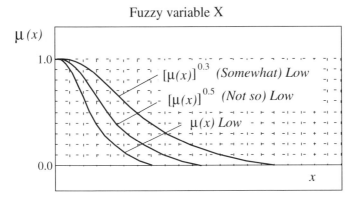

Figure 5.40 Dilution of the membership function *Low* by linguistic hedges.

Contrast Intensification

This is a similar concept to concentration. However, the profile of the possibility distribution is now under modification. In contrast intensification, the membership function is widened for possibility values higher than 0.5 and narrowed for possibility values less than 0.5. The effect is displayed in Fig. 5.41. As illustrated in this figure, the hedge to be used in this context must increase the confidence in identifying the corresponding fuzzy region. Such an operation can be performed by

$$\mu_i(x) = 2[\mu(x)]^2 \qquad 0 \leq \mu(x) \leq 0.5$$
$$\mu_i(x) = 1 - 2[1 - \mu(x)]^2 \qquad 0.5 \leq \mu(x) \leq 1.0 \qquad (5.9)$$

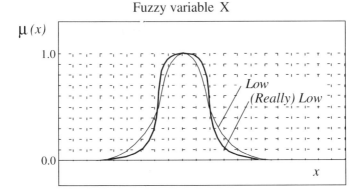

Figure 5.41 The effect of contrast intensification via linguistic hedge such as *Really*.

Contrast Relaxation

Contrast relaxation is the reverse of contrast intensification. Because intensification increases confidence, relaxation decreases confidence. This is similar to a standard fuzzification operation where a fuzzy set is made more fuzzy, as shown in Chapter 2. When an appropriate linguistic hedge such as *More-or-less* is applied to an unconstrained predicate such as *Low*, this effect must be modeled by one of the fuzzification techniques described in Chapter 2.

Negation

Negation denoted by *NOT* was also described in Chapter 2 to be the mirror image of the membership function with respect to the possibility = 0.5 line. However, we have defined *NOT* to be a linguistic connector just like *IS, GREATER THAN,* and *LESS THAN*. The difference between a linguistic connector and a linguistic hedge is that a connector defines an *association* whereas a hedge defines a *modification*. Thus, the hedge *NOT* is different than the connector *NOT*. The following examples illustrate this difference.

NOT as a linguistic connector: *NOT* as a linguistic hedge:
Temperature IS-NOT (VERY) Low *Temperature IS (NOT VERY) Low*

On the left, *IS-NOT* is a linguistic connector. It forms an association between the fuzzy variable *Temperature* and hedged predicate *(VERY) Low*. Thus, the resulting effect must be modeled by $1 - \mu$. On the right, *NOT VERY* is a linguistic hedge with a dilution (opposite of concentration implied by *VERY*) effect. Therefore, *NOT VERY* must be modeled by a dilution method rather than by $1 - \mu$. The confusion is apparent due to the inseparability among the words *IS, NOT,* and *VERY*. However, the resolution of this confusion is trivial and is expressed by the following design principle.

Linguistic hedges, which are often external inputs to a fuzzy inference engine, describe the conditions of the antecedents. Linguistic connectors are determined by the rules during design that describe the solution logic.

Chap. 5 Fuzzy Variable Design

This principle applies in general (not only to the resolution of *NOT* hedge/connector). Therefore, returning to the previous example, the following distinction should be made.

Temperature IS-NOT (VERY) Low → *IS-NOT* is part of the rule base,
VERY is an external linguistic input.

Temperature IS (NOT VERY) Low → *IS* is part of the rule base,
NOT VERY is an external linguistic input.

Restriction

Restriction implied by linguistic hedges such as *Above* and *Below* are modeled by the half-negation operation. Half is determined with respect to the line passing through the center of maximum possibility values as shown in Fig. 5.42. *Below* retains the left part and *Above* retains the right part for obvious reasons.

Restriction can be analytically modeled by other approaches. This function can also be represented by other hedges such as *More, Less, Smaller, Greater*, and so on. The implementation of the restriction function can be confusing in the same way. Let's consider two cases:

Air-Temperature IS LESS THAN Water-Temperature

Air-Temperature IS (LESS THAN) Warm

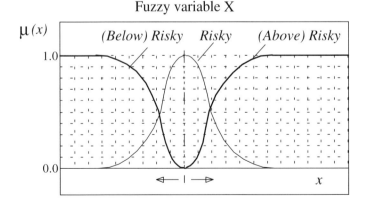

Figure 5.42 Linguistic hedges *Below* and *Above* modeled by half-negation.

The first case is a binary comparison between two variables (or two fuzzy variables), and the statement is not fuzzy. In this context, *LESS THAN* is not a linguistic hedge, instead it is a linguistic connector. In the second case, *LESS THAN* is a linguistic hedge and can be modeled by the restriction function. Handling of linguistic connectors such as *LESS THAN* is discussed in Section 5.4.2.

5.4 PRACTICAL DESIGN CONSIDERATIONS

We have seen in the previous sections the calculus of fuzzy variable design and the involvement of heuristic reasoning. All the issues addressed up to this point can be derived from the theoretical concepts frequently encountered in the fuzzy logic literature. Now, we will examine more practical issues that are not often covered in the literature.

5.4.1 Auxiliary Variable Design for Conflicts

In some circumstances, the definitions of fuzzy input and output variables identified for a given problem might not be adequate to solve the problem. A typical example is the one-to-many mapping case discussed in Section 5.2.1. The problem is that there are two (or more) different output classes (actions) for the same input and we don't know which one to compose a rule for. Assume we have composed a rule for both of them as shown below.

If X IS A AND Y IS B THEN Z IS C
If X IS A AND Y IS B THEN Z IS D

Conflict is apparent because the left-hand sides of the rules are the same while the consequents are different. For example, *C* can be *Hot* and *D* can be *Cold*. If we keep both rules in the rule base, the output will neither be *C* or *D* when *X* is *A* and *Y* is *B*. The output will be formed by the aggregation of *C* and *D*, and the result of the defuzzification process will yield an answer between *C* and *D*. The conflict is depicted in Fig. 5.43 in the input product space.

Normally this kind of problem is encountered when the process is driven by some unknown input variables. One practical solution is to generate auxiliary variables to distinguish the output classes. Given a data

Chap. 5 Fuzzy Variable Design

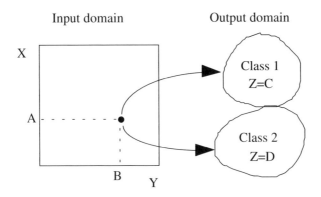

Figure 5.43 One-to-many mapping in which a unique solution cannot be reached.

set from a sequentially correlated process, an auxiliary variable can be generated by the first, second, or higher derivatives of one of the input variables. This is done to distinguish the output classes by examining the past behavior of the system and by finding differences in approaching to the common point. Now suppose we generate an auxiliary variable:

$$X_a = \frac{dX}{dt} \equiv X(t) - X(t-1) \qquad (5.10)$$

By incorporating the auxiliary variable, the product space becomes three-dimensional, as shown in Fig. 5.44. The only requirement is that the auxiliary variable must have distinct values for those common points on the two-dimensional product space.

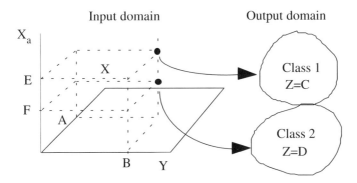

Figure 5.44 Using an auxiliary variable to distinguish output classes.

The treatment described above allows expanding the rules with the new differentiating elements as shown below. The underlined parts illustrate the resolution of the conflict by an auxiliary fuzzy variable.

If X IS A AND Y IS B AND X_a IS E   THEN Z IS C
If X IS A AND Y IS B AND X_a IS F   THEN Z IS D

Converting an auxiliary variable into an auxiliary fuzzy variable does not require any special treatment. Note that such solutions are local in the sense that there is no guarantee of their validity if there are other unknown modes of operations where the approach to common points exhibits different behavior.

If the membership functions are developed linguistically rather than using a set of data, then an auxiliary variable creation must justify a certain logical premise. Otherwise, fuzzy rules will simply be illogical or ambiguous. Consider the following one-to-many mapping clauses:

If Weather IS Nice AND I HAVE FREE_time THEN I WILL Swim
If Weather IS Nice AND I HAVE FREE_time THEN I WILL Read

Let's generate an auxiliary fuzzy variable so that we will know what to do when the weather is nice and we have some free time.

If Weather WAS Nice (yesterday) AND Weather IS Nice (now)
                    AND I HAVE Free_time THEN I WILL Swim
If Weather WAS Bad (yesterday) AND Weather IS Nice (now)
                    AND I HAVE Free_time THEN I WILL Read

The status of the *weather yesterday* becomes an auxiliary fuzzy variable by justifying a logical premise that the water would be too cold to swim if the weather was bad yesterday. In this example, we paid special attention to generating an auxiliary variable using time difference in a sequentially correlated event so that the reader would make an easy association between this and the previous example. In general, variable design in a purely linguistic domain offers much more flexibility in finding an auxiliary logic. In a large rule base in which experts articulate hundreds of rules, conflicting rules are often encountered due to an oversight and their resolution always requires further interpretations of the expert's articulation.

5.4.2 Auxiliary Variable Design for Comparison

In Chapter 2, we mentioned that the two basic linguistic connectors in fuzzy logic are *IS* and *IS_NOT*. However, there are two other basic linguistic connectors in standard computer terminology, which are *GREATER THAN* and *LESS THAN*. These basic comparators of binary logic are widely used in articulating expertise even though the variables of the problem are fuzzy. The question is how to represent a fuzzy comparison appropriately. Let's consider the following statement.

If Weather_temperature IS LESS THAN Water_temperature THEN....

This is a crisp statement whether the variables are fuzzy or not. To make a fuzzy comparison, that is, if it is *little less than* or *quite less than*, an auxiliary variable can be formed by taking the difference (numerical subtraction) between the two fuzzy variables. Accordingly, a new universe of discourse can be defined between the maximum and minimum differences of two fuzzy variables.

In Fig. 5.45, the membership functions *Positive* and *Negative* correspond to the *GREATER THAN* and *LESS THAN* operations. Naming the auxiliary variable as *Weather_Temperature_Difference*, the previous statement can be restated as follows.

If Weather_temperature_difference IS Large_negative THEN.....

This approach allows a fuzzified comparison mechanism by providing a new universe of discourse where different levels of comparison can

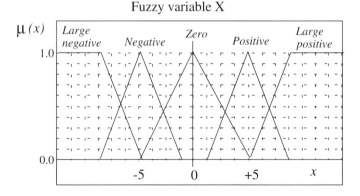

Figure 5.45 Fuzzy comparison using an auxiliary variable.

be represented by membership functions such as *Medium_negative*, or *Huge_positive*.

5.4.3 Threshold Design

Threshold is a mechanism to control the level of uncertainty in a fuzzy system. The main idea is to discard very low possibilities in the antecedent domain and to eliminate the corresponding implication results in the consequent domain. It is easier to estimate the effects of thresholding on antecedents than that on consequents, because the consequent membership functions are subject to aggregation and defuzzification before the effects can be observed.

The simplest form of a threshold is the α-cut that limits (along with the resolution principle) the spread of a membership function on the universe of discourse. A threshold changes the membership function shapes as illustrated in Fig. 5.46. If the level of threshold is above the intersection points (a, b, and c in Fig. 5.46), then the membership functions do not overlap. This is also called *dead zones* and is explained next. Note that the α-cut threshold as expressed by Eq. (5.11) creates a sudden drop in the evaluation of membership functions that is sometimes not desired.

$$\mu(x) = 0 \quad \quad if \quad \mu(x)\big|_{x_i}^{x_j} \leq \alpha \quad \quad (5.11)$$

$$\mu(x) = f\{\mu(x)\} \quad if \quad \mu(x)\big|_{x_i}^{x_j} \leq \alpha \quad \quad (5.12)$$

To avoid sudden drop, a nonlinear threshold can be employed where the possibility value goes down to zero in a slower pace. As shown in Eq. (5.12), a nonlinear threshold can be devised by applying the contrast intensification treatment to a membership function for its values lower than the specified α-cut. This results in a shape modification in a manner similar to what is shown in Fig. 5.47.

The linguistic equivalence of the α-cut threshold clearly illustrates that this method loses fuzziness at certain locations on the universe of discourse by defining crisp boundaries. For example, assume that the membership function *High* around a temperature value of 80° C is being subjected to an α-cut yielding 60° C and 100° C boundaries. Then, the linguistic equivalence is as follows.

There is no possibility of Temperature values under 60° C and over 100° C to be considered High, whereas there is a significant possibility for 61° C and 99° C.

Chap. 5 Fuzzy Variable Design 263

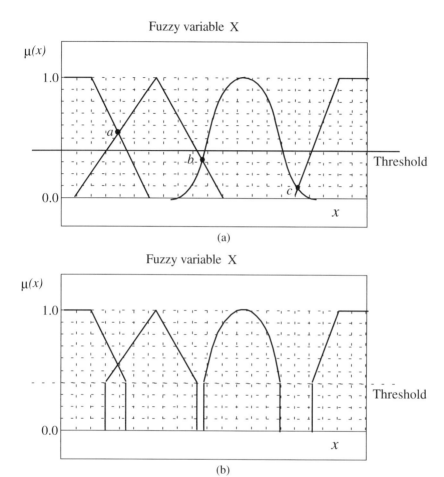

Figure 5.46 The effect of an α-cut threshold (a) on the membership function shape (b).

Thus, nonlinear thresholding makes more sense in general—especially for high threshold levels—by not introducing crispness without a specific reason. Consequently, nonlinear threshold suits human reasoning better at the expense of increasing computational complexity of the basic fuzzy inference algorithm.

Threshold design for consequent fuzzy variables is often unnecessary or too cumbersome to deal with. The reason is related to the implication-defuzzification process. Applying a threshold to a consequent membership function results in narrowing the width of the membership function, as shown in Fig. 5.46, which reduces the contribution strength

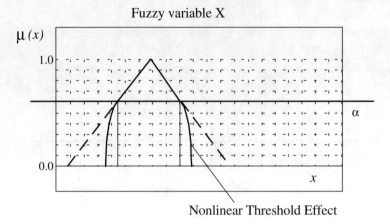

Figure 5.47 Nonlinear thresholding to avoid sudden drop in possibility evaluation.

of the membership function to the aggregated result. The effect is untraceable and sometimes unpredictable because the consequent membership functions are also subject to continuous shape modification due to the implication process. In addition, a similar but more predictable effect can be imposed by applying importance weights to each rule that are easily implemented without computational complexity.

The reader may wonder why an extra shape modification scenario via thresholds is necessary and why the original membership functions are not designed in these shapes. The purpose of thresholds is to tune the fuzzy system by a controllable parameter so that the system's sensitivity to ambiguous, uncertain, and fuzzy inputs is determined on a per variable basis. There is no recipe for threshold design. The only practical method is to set the antecedent thresholds to zero during the initial design, then adjust them during testing, just like tuning a controller.

5.4.4 Dead Zones

During the development of membership functions or during rule composition, some parts of the antecedent universe of discourse might be left uncovered due to an oversight or because of some specific reason. We will illustrate three different ways of creating dead zones either intentionally or unintentionally. Later, we will examine the effect of a dead zone in an antecedent fuzzy variable. The first way of creating dead zones is to

Chap. 5 Fuzzy Variable Design   265

design membership functions such that some parts of the universe of discourse are not covered by any of the membership functions, as shown in Fig. 5.48. If an input occurs between those intervals, the fuzzy variable evaluation will produce 0 possibility.

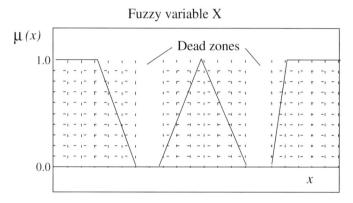

Figure 5.48 Dead zones in the antecedent universe of discourse.

The second way of creating dead zones is to design thresholds higher than intersecting points. This is illustrated in Fig. 5.49.

The third way of creating dead zones is when one (or more) of the membership functions defined are not used in any of the rules. This is illustrated in Fig. 5.50.

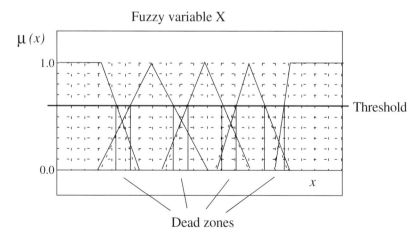

Figure 5.49 Dead zones in the antecedent universe of discourse due to high thresholding.

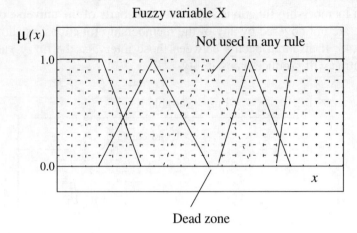

Figure 5.50 A dead zone in the antecedent universe of discourse due to an unused membership function.

Considering the entire fuzzy inference mechanism, we know that zero possibility generates zero degrees of fulfillment on the left side conditions. However, given the zero degrees of fulfillment, the implication result may be 0 or 1 depending on the implication operator selected, as will be shown in Chapter 8. Furthermore, defuzzified (or decomposed) results may produce an answer such as 0. Note that 0 may be a viable answer (e.g., 0% change in the valve position) if the consequent universe of discourse extends to negative values. The problem is that the fuzzy inference engine generates an answer for an undefined region in the antecedent space. This is conceptually unacceptable and should be considered an error.

The solution to problems emerging from unintentional dead zones comes from the peripheral computations, where the input absorption process is monitored and occurrence of undefined possibility evaluations is detected. The fuzzy inference engine may be equipped with such a diagnostics function to tackle this problem. We conclude these discussions with a design principle as stated below.

Unintentionally created dead zones on the antecedent universe of discourse may result in misleading outputs from a fuzzy inference engine when inputs hit the undefined regions of the antecedent domain.

5.4.5 Active Range of Consequents

One important design issue is to be able to predict the output behavior of the fuzzy inference engine. Thus, designing a consequent fuzzy variable and its membership functions requires special attention. Problems can arise if the membership function design does not include considerations of the defuzzification mechanism. Let's start from the simplest case and progress towards more complex forms. The active range of action is defined by the following expression.

$$X^a = \frac{x_R - x_L}{x_{max} - x_{min}} \qquad (5.13)$$

Here the subscripts R and L represent the rightmost and leftmost locations of the defuzzified output. The denominator is the universe of discourse. Theoretically, the maximum value of the active range is 1.0, which occurs when the defuzzified output stretches to both ends of the universe of discourse. The active range of a consequent fuzzy variable that consists of a single symmetric membership function is

$$X^a = \frac{0}{x_{max} - x_{min}} = 0 \qquad (5.14)$$

because a symmetric membership function will produce the same defuzzified output whether it is being computed by the center-of-gravity method or by any other method. Figure 5.51 illustrates that different values of the degrees of fulfillment produce the same defuzzified output. In

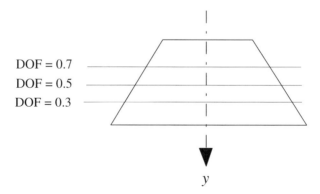

Figure 5.51 The active range of a single symmetric consequent membership function is zero due to the implication and defuzzification processes.

this example, a Mamdani implication operator is used that clips the original membership function by the degrees of fulfillment to compute the implication result (shape under the line); then the defuzzified output is found by the center-of-gravity method.

This is hardly the case in practice because consequent fuzzy variables are always designed with more than one membership function. When there is more than one membership function, the aggregation process becomes an important factor. Consider the following case, where all the membership functions are symmetrical.

The active range in Fig. 5.52 can be computed by Eq. (5.13). Note that the dark bands on each side of the most stretched locations are not achievable. Therefore, the designer should be aware of the active range concept so that the consequent fuzzy variable can be designed to include the desired outputs within the active range.

> The designer must consider the active range concept while designing a consequent fuzzy variable so that the desired outputs become achievable.

It is apparent that membership functions with asymmetric shapes can easily be incorporated within this principle by simply computing their rightmost and leftmost components and selecting the extreme case depending on which side of the universe they are located. For example, the asymmetric membership function shown in Fig. 5.53 produces de-

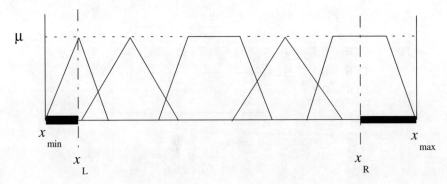

Figure 5.52 A high number of consequent membership functions increases the active range.

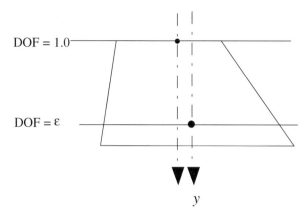

Figure 5.53 A single asymmetric membership function produces defuzzified output between two extreme cases, DOF = 1.0 and DOF very small.

fuzzified outputs between two extremes (i.e. when DOF = 1.0 and when DOF is very small). If this membership function is located at the left corner of the universe of discourse, the leftmost defuzzified output is the one with DOF = 1.0. Otherwise, the right-most defuzzified output is the one with DOF very small.

5.4.6 Paralysis

Paralysis is defined as a situation in which the outputs of a fuzzy inference engine remain the same (or change very little) while its inputs are changing. Of course, not every stationary output response will be considered paralysis. The paralysis definition also includes the requirement that the stationary output behavior is not intended or designed. A few reasons for paralysis are discussed in different sections of this book. Here, we will examine the case in which a consequent membership function design causes unintentional paralysis.

Unintentional paralysis emerges when the defuzzification method is center-of-gravity and the universe of discourse comprises large numbers. Because the maximum height of the membership functions is 1.0, a numerically large universe of discourse causes the shapes of the consequent membership functions to resemble a thin strip. This geometrical trick is illustrated in Fig. 5.54, where a hypothetical consequent fuzzy variable is shown to scale.

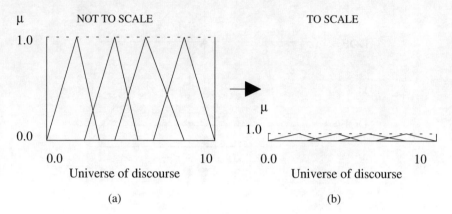

Figure 5.54 A consequent fuzzy variable with a universe interval of 10 units shown not to scale (a) and shown to scale (b).

The problem is that a geometrical shape reaching an infinitesimally thin line shape as the universe of discourse gets larger has the center of gravity around the center. The changes in the shape at different initiation steps via aggregation-implication processes would tend to generate a very small active range due to the thin-strip geometry. Let's go back to Fig. 5.54 (a) and redraw the figure by normalizing the universe of discourse with a hypothetical aggregation.

The center-of-gravity line in Fig. 5.55 (b) is closer to the center than that of the Fig. 5.55 (a). When we renormalize the defuzzified output in Fig. 5.55 (a), the result does not match that in Fig. 5.55 (b). This suggests the following design principle.

Consequent fuzzy variable design is sensitive to normalization; thus, the design must employ a consistent normalization strategy to avoid inadvertant geometrical effects on the output behavior.

It is observed that most of the commercially available software products for fuzzy system development employ normalization on the consequent universes. Thus, the consequent universe is automatically normalized to 1.0 and the results are renormalized given the minimum and maximum universe declarations by the designer. However, this prin-

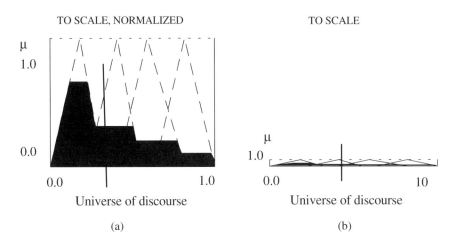

Figure 5.55 A consequent fuzzy variable with a universe interval of 10 units shown to scale with normalized universe (a) and shown to scale without normalization (b).

ciple must be considered by those who write their own fuzzy inference algorithms.

5.5 SUMMARY

Fuzzy variable design as discussed in this chapter consists of determining the number of membership functions, their location and shape, words to represent predicates, and the universe of discourse. There are two ways to design a fuzzy variable, as shown in Fig. 5.2. The linguistic design is based on interpretations of expert knowledge in the absence of data. The data-driven design is based on the availability of data in addition to expert's knowledge and interpretation of data. The latter is the most common case in fuzzy system design. In both cases, knowledge of decision boundaries plays a major role. In an input-output mapping problem for which IF-THEN rules are composed, design gets complicated if the decision boundaries are unknown despite the availability of data. If the decision boundaries are unknown and if the output data are not specified (a typical classification problem), then the designer may have the flexibility to construct the decision boundaries for a given objective function depending on the problem type. When decision boundaries are known, the design becomes straightforward and simple. Thus, the first challenge for the designer is to examine the decision boundaries of the problem at hand and to identify a strategy to complete the design.

The simplest form of membership function design is to apply equal-distance partitioning using pryamidal surfaces on the input product space. This can be achieved by treating every crisp point on the input product space by a point membership function, then combining those points which are in close proximity to each other. Most of the time, a simple judgment such as close proximity results in a successful design and reasonable performance. The fundamental criterion in such an approximation is the desired (or tolerable) level of granularity to be embedded in the solution. More advanced forms of partitioning can be achieved by statistical and artificial intelligence methods when data is reliable. However, there is no generalized theory or experimental proof that guarantees improved performance by using number-crunching exotic methods.

We also examined the basic principles of linguistic design, which includes the design of semantically constrained and unconstrained predicates. The formulation found in the literature along with the designer's intuition often yields reasonable design performance even in the absence of design data. The linguistic design requires that the solution is articulated (by experts when available) in natural languages, preferably in the form of IF-THEN rules. It is always possible to modify a linguistically designed variable by examining relevant data and by using analytical methods. The last part of this chapter is dedicated to more practical design considerations such as resolving conflicts, fuzzy comparison, dead zones, active range of consequents, and paralysis.

PROBLEMS

5.1. Determining decision boundaries from data when the number of data points on the input product space is greater than the number of output classes ($N > M$):
Find the decision boundaries embedded in the sequentially correlated data given below by placing equal-width pryamids on the input product space $x_1 \times x_2$. and by combining membership functions of the same output class y. How many rules are required? What is the effect of combining the membership functions corresponding to each output class?

y	1	1	2	2	1	0	1	0	0
x_1	0.10	0.25	0.25	0.35	0.33	0.65	0.65	0.77	0.55
x_2	0.12	0.20	0.55	0.65	0.40	0.24	0.71	0.44	0.15

5.2. Determining decision boundaries from data when the number of data points on the input product space is equal to the number of output classes ($N = M$):
Find the decision boundaries embedded in the sequentially correlated data given below by placing equal-width pryamids on the input product space $x_1 \times x_2$. What is the total number of membership functions including all universes that yield the most costly solution?

y	1.00	0.97	1.88	2.10	0.85	0.21	1.05	0.30	0.14
x_1	0.10	0.25	0.25	0.35	0.33	0.65	0.65	0.77	0.55
x_2	0.12	0.20	0.55	0.65	0.40	0.24	0.71	0.44	0.15

5.3. Reducing the Number of Membership Functions:
In Fig. 5.20, the maximum deviation point E_{max} caused by combining two membership functions was illustrated. What would be the maximum E_{max} if the total number of membership functions in Problem 5.2 is reduced to 8 on x_1 universe? How many different solutions can be found? What would be the least costly solution among them? (Hint: Apply equal distance partitioning when designing initial point membership functions.)

5.4. Known Versus Unknown Decision Boundaries:
What does the following problem description imply about whether the decision boundaries are known (embedded in data) or unknown (to be constructed per given objective): *Given a data set on the weight and height of each bottle, find the number of size classes for boxes to be manufactured with minimum cost.*

5.5. Determining number of membership functions and their location:
Given the sequentially uncorrelated data set below, develop the antecedent membership functions of a fuzzy inference engine that estimates the status of a tropical storm. Evaluate the suitability of the following methods for this problem:
(a) Equal-distance partitioning
(b) Frequency of occurrence (by selecting an appropriate grid thickness in each histogram)
(c) Minimum entropy method
(d) Linguistic design
(e) Neural networks

The rules are given as:

IF Wind IS High AND Temperature IS Medium AND (Precipitation IS High OR Medium) AND Pressure IS Low THEN Storm IS Hurricane
IF Wind IS Medium AND Pressure IS Low THEN Storm IS Developing
IF Temperature IS Low AND Pressure IS High THEN Storm IS Moving
IF Wind IS Low AND Precipitation IS Small AND Pressure IS Medium THEN Storm IS Over.

Wind	Temperature	Precipitation	Pressure
35	55	2.2	1.1
38	54	2.5	1.12
33	67	2.5	1.21
51	72	4.4	0.94
55	57	3.6	0.96
57	55	5.5	0.88
66	44	5.5	0.91
65	55	4.9	1.0
88	43	11	1.0
97	45	12	0.88
57	66	1.1	0.98
66	57	3.3	0.99
87	44	9	0.88
77	46	6.7	0.86
87	40	6.9	0.99
66	51	4.4	1.11
112	35	21	0.81
105	38	14	0.79
120	36	18.8	0.8
125	34	15.2	0.77

(Hint: Output categories are unknown in this problem. The most costly solution will require 20 membership functions for each variable. By examining the rules, four output categories can be formed using a clustering criteria such as distance. Then, the number of membership functions can be reduced.)

REFERENCES

[1] Tagaki, H., and I. Hayashi. "Neural Network Driven Fuzzy Reasoning." *Int'l. Jour. Approximate Reasoning* 5 (1991): 191–211.

[2] Lee, M., and H. Tagaki. "Integrating Design Stages of Fuzzy Systems Using Genetic Algorithms." *IEEE Trans.* Paper 0-7803-0614, 1993.

[3] Vadiee, N. "Fuzzy Rule-Based Expert Systems." In: *Fuzzy Logic and Control: Software and Hardware Applications.* Englewood Cliffs, NJ: Prentice-Hall, 1993.

[4] Bezdek, J. C. and J. D. Harris. "Fuzzy Partitions and Relations." *FSS* 1, (1978): 111–127.

[5] Ruspini, E. "New Experimental Results in Fuzzy Clustering." *Information Science* 6 (1973): 273–284.

[6] Gupta, M. M., and T. Yamakawa. *Fuzzy Computing Theory, Hardware, and Applications.* New York: Amsterdam, 1988.

[7] Fitzgerald, A. E., D. E. Higginbotham, and A. Grabel. *Basic Electrical Engineering.* Kogagusha: McGraw-Hill, 1975.

[8] Christensen, R. *Fundamentals of Inductive Reasoning.* Lincoln, MA: Entropy Ltd., 1980.

[9] Zimmerman, H. J. *Fuzzy Set Theory and Its Applications.* Boston: Kluwer Academic Publishers, 1990, p. 40.

[10] Shaffer, G. *A Mathematical Theory of Evidence.* Princeton, NJ: Princeton University Press, 1976.

[11] Zadeh, L. A. "The Concept of Linguistic Variable and Its Application to Approximate Reasoning." *Information Sciences* 8 (1975): 199–249.

Chapter 6

Membership Function Shape Analysis

This chapter describes the basic geometric forms used in the design of membership functions, including their interpretation both quantitatively and qualitatively. Frequently asked questions involving the height, line style, and overlapping of membership functions are discussed with examples. Although membership function shape is known to be inconsequential in many practical applications, its design cannot be arbitrary, as illustrated in this chapter. Compared to the topics covered in the last chapter, topics presented here may be interpreted as fine tuning of the design elements.

6.1 INTRODUCTION TO SHAPE ANALYSIS

Uncertainty characterization is a broad theoretical subject studied in many diverse fields ranging from physics to economics. In the context of fuzzy system design, uncertainties are represented by membership functions and by their individual properties. In the last chapter, we discussed membership function design from the point of view of fuzzy variable design. Those discussions mainly included the determination of the number (granularity) and location of membership functions along with some unavoidable discussions on their shape. In this chapter, we will examine their shape without being concerned about their granularity and location on the universe of discourse. To use an analogy, in Chapter 5 we have decided how many chairs we would buy to decorate an empty room, and where we would put them. Now, we will discuss how they should look.

Our discussions will be centered around height, line style, and overlap designs. In general, shape effects emerge when membership functions are designed differently with respect to each other. Otherwise (that is, when all the membership functions have the same shape), the effects are less profound and sometimes inconsequential. The location and granularity are the two relatively more important issues compared to the shape of each membership function. When the membership functions are to be modified due to some undesired performance, the designer must first consider the location and granularity issues before considering the shape effects. This leads to the following design principle, which holds for moderate shape variations.

> In general, the number of membership functions and their locations on a universe of discourse affect the performance of the basic fuzzy inference algorithm relatively more severely than the effects caused by the shape variations among the membership functions.

The geometrical shape of a membership function is the characterization of uncertainty in the corresponding fuzzy variable. Therefore, a high level of detail (e.g., high resolution) in shape design must be considered as a conceptual error (i.e., uncertainty must not be defined in detail). One exception to this rule is the probabilistic design in which uncertainty is purely represented by probability distributions in the presence of reliable data, as we have seen in Section 5.2.5. Note that this is a very special case and is rarely encountered in the existing industrial or commercial applications of fuzzy logic. Nevertheless, the shape of a membership function cannot be formed arbitrarily because arbitrary design can produce unpredictable results in the basic fuzzy inference algorithm. The design challenge is to employ a reasonable level of detail when forming membership functions so that the basic fuzzy inference algorithm behaves as expected. The linear surface extensions based on the equal-distance partitioning described in Section 5.2.1 satisfy the requirement of a "reasonable level of detail" in most cases. However, we will examine the shape effects in more detail to identify some important points useful for design.

There are two ways to represent the shape of a membership function mathematically. The first one is the fuzzy set representation, as illustrated in Section 2.1. For example, a triangular membership function can be represented by the union of singletons $A(x) = \bigcup \mu(x)/x = 0.0/1 \cup 0.5/2 \cup 1.0/3 \cup 0.5/4 \cup 0.0/5$. The second way to represent the same membership function is to use the functional form in calculus. In the ongoing example, the membership function $A(x)$ can be represented by $A(x) = 0.5(x - 1)$ for $1 \leq x \leq 3$ and $A(x) = 1 - 0.5(x - 3)$ for $3 \leq x \leq 5$. These representations are equivalent. The functional representation, since it automatically interpolates the points in between singletons, is more practical. Thus, when the functional form is used, it is assumed that the possibility distribution between two (or more) singletons can be represented by a fitted curve. This assumption is reasonable for all practical purposes. Continuous membership functions—particularly in the antecedent domain—are also very practical when the antecedent variables are evaluated by continuous inputs (i.e., inputs that can have any value on the universe of discourse rather than only values that correspond to the singletons of a fuzzy set). The membership function shapes described in the next sections are defined in the functional, continuous form. They can easily be converted into fuzzy set form by universe sampling. If we presented them in fuzzy set notation, it would be much more difficult to convert them into functional form. Nevertheless, both representations are used in some cases to complete certain arguments.

Piece-Wise-Linear

These shapes consist of linear lines connected at different break points. The triangle (Fig. 6.1(a)) and the trapezoid (Fig. 6.1(b)) are the two geometric shapes commonly used to represent uncertainties. Multi-breakpoint shapes are used when the uncertainty is known in an approximate manner, either via detailed articulation of expertise or by means of available data. This is illustrated in Fig. 6.1(c).

Curves

Curves are also commonly used, yet they are not as practical as the piecewise-linear shapes due to computational complexity and difficulty in hardware implementation. The two most common shapes are the bell and Π-shaped curves. An extended version of the Π-shaped curve is shown in Fig. 6.2(b). When uncertainty can be characterized in a relatively more

Chap. 6 Membership Function Shape Analysis

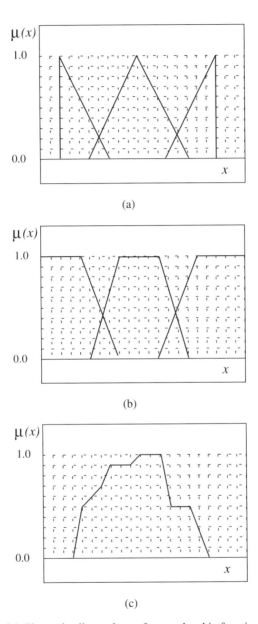

Figure 6.1 Piece-wise-linear shapes for membership function design.

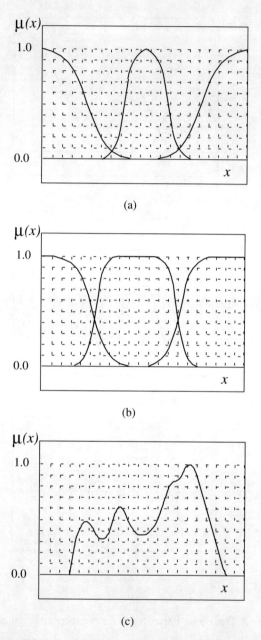

Figure 6.2 Curved shapes for membership function design.

detailed manner, a membership function can be designed by a curve-fitting method among identified points as shown in Fig. 6.2(c). This will be a rare event in practice.

Analytical forms of these functions can be found in elementary calculus books. However, we will list some of them in the form encountered in the fuzzy logic literature.

S-Shaped Function

A typical sigmoidal (S) shape can be expressed in a number of ways in calculus. The following formulation is expressed in terms of parameters defined on the universe of discourse. Three parameters are α (point on the universe where possibility is 0), γ (point on the universe where possibility is 1), and β (point on the universe where possibility is 0.5).

$$\begin{aligned}
S(x;\alpha,\beta,\gamma) &= 0 & x &\leq a \\
S(x;\alpha,\beta,\gamma) &= 2\left(\frac{x-\alpha}{\gamma-\alpha}\right)^2 & \alpha &\leq x \leq \beta \\
S(x;\alpha,\beta,\gamma) &= 1 - 2\left(\frac{x-\gamma}{\gamma-\alpha}\right)^2 & \beta &\leq x \leq \gamma \\
S(x;\alpha,\beta,\gamma) &= 1 & x &\geq \gamma
\end{aligned} \quad (6.1)$$

Π-Shaped Function

This function resembles a curve fitted to a triangle. It is analytically expressed by combining an S-shaped function with its mirror image.

$$\begin{aligned}
\Pi(x;\delta,\gamma) &= S(x;\gamma-\beta,\gamma-\beta/2,\gamma) & x &\leq \gamma \\
\Pi(x;\delta,\gamma) &= 1 - S(x;\gamma,\gamma+\beta/2,\gamma+\beta) & x &\geq \gamma
\end{aligned} \quad (6.2)$$

The extended version of the Π-shaped function has a plateau resembling a trapezoid. The extended Π-Shaped function is expressed by

$$\begin{aligned}
\Pi(x;\delta,\gamma_1) &= S(x;\gamma_1-\beta,\gamma_1-\beta/2,\gamma_1) & x &\leq \gamma_1 \\
\Pi(x;\delta,\gamma_1,\gamma_2) &= 1.0 & \gamma_1 &< x < \gamma_2 \\
\Pi(x;\delta,\gamma_2) &= 1 - S(x;\gamma_2,\gamma_2+\beta/2,\gamma_2+\beta) & x &\geq \gamma_2
\end{aligned} \quad (6.3)$$

Other Functions

Other functions that can be used in building membership functions are listed below.

Hyperbolic curve $\quad f_s(x) = \dfrac{1}{1+\left(\dfrac{x}{\beta}\right)^3} \quad f_s(x) = 1 - \dfrac{1}{1+\left(\dfrac{x}{\beta}\right)^3}$ (6.4)

Bell-shaped curve $\quad B(x;\gamma,\beta) = \dfrac{1}{1+\left(\dfrac{x-\gamma}{\beta}\right)^2}$ (6.5)

Gaussian curve $\quad G(x;k,\gamma) = e^{-k(\gamma-x)^2}$ (6.6)

Convex Shapes

As illustrated in Fig. 6.3, nonconvex membership functions include regions of low possibilities scattered around the regions of high possibili-

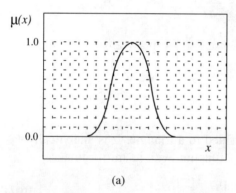

(a)

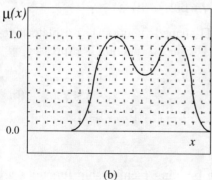

(b)

Figure 6.3 Convex (a) and nonconvex (b) membership functions.

ties. A membership function design of this kind is a rare event because there are not many cases in the real world in which one can define an uncertainty profile in such a deterministic (or precise) manner. Even when probabilistic measures are used, nonconvex distributions are either rarely seen or they represent unfavorable statistics.

In some other cases, a nonconvex membership function is a combination of two or more convex membership functions, and there is no decision boundary requirement to split them. Most of the applications today use simple convex membership functions.

6.2 MEMBERSHIP FUNCTION HEIGHT

Regardless of its shape, the height of a membership function carries vital information and its design can be very important. Let's consider an arbitrarily constructed fuzzy variable Y (Fig. 6.4) with membership functions of different shapes and heights.

The height of a membership function determines the maximum possibility value. For antecedent fuzzy variables, the height determines the maximum possibility values produced during the evaluation (input absorption). If the fuzzy variable Y in Fig. 6.4 is an antecedent, then its de-

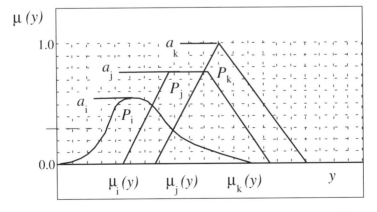

Figure 6.4 An arbitrarily constructed fuzzy variable Y with membership functions of different heights.

sign as shown in this figure suggests the following context-specific knowledge:

The possibility of fuzzy variable Y to be (P_i) is a_i at maximum.
The possibility of fuzzy variable Y to be (P_j) is a_j at maximum.
The possibility of fuzzy variable Y to be (P_k) is a_k at maximum.

The first design principle says that the maximum possibility distribution (height) is determined by a specific solution, and there is no further analysis required. Other principles will be stated for cases when the first principle does not apply and there is no specific solution requiring different heights.

Universe (Truth) Categorization

The universe categorization task is to identify and label different regions of the universe of discourse often required for many-to-one (or many) mapping problems. Consider the case where Y is *Temperature* and the predicates are *Low, Medium*, and *High*. When predicates are fuzzy quantifiers, designing membership functions with different heights does not make sense, because quantifiers function as labels categorizing the different regions of the universe, and there should be at least one point in the universe with ideal categorization. To illustrate this point better, consider the hypothetical case shown in Fig. 6.5.

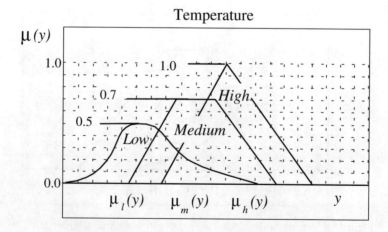

Figure 6.5 Fuzzy variable *Temperature* defined with membership functions of different height.

As shown in Fig. 6.5, this type of categorization yields the following reasoning:

The possibility of *Temperature* to be *Low* is 0.5 at maximum (at all times).
The possibility of *Temperature* to be *Medium* is 0.7 at maximum (at all times).
The possibility of *Temperature* to be *High* is 1.0 at maximum (at all times).

As one would immediately infer from above, there is no single value of *Temperature* to identify the categories *Low* and *Medium* completely (1.0). Therefore, a membership function height other than 1.0 yields an incomplete categorization. Incomplete categorization results in incomplete degrees of fulfillment on the LHS conditions of a rule even though the LHS statements are linguistically satisfied at a maximum level (e.g., when a *Temperature* value matches *Low* 100 percent.) This, in turn, results in undermining the importance of the rule among other rules and reduces its contribution to the result. For these reasons, a new principle is stated as follows:

> To avoid unintentional importance deterioration of fuzzy rules, heights of the membership functions of an antecedent fuzzy variable must be all equal to 1.0 if the predicates are fuzzy quantifiers used for nonprobabilistic categorization of the universe of discourse.

Based on this principle, the previous example is redrawn in Fig. 6.6 so that all categories have at least one point of complete categorization (1.0). Thus, a rule using this fuzzy variable will not suffer from unintentional importance deterioration due to height differences.

The arguments above can be turned around in a sense-making way if there is probability involved, as explained next.

Universe (Truth) Characterization

The evaluation of an antecedent fuzzy variable, when it is designed for universe characterization yields a measure of compliance with the semantics of the predicates used. In such a case, the designer is not concerned about unintentional importance deterioration of the rules. On the contrary, the effect of membership function height on rule importance is one

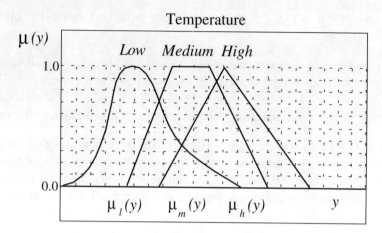

Figure 6.6 Nonprobabilistic categorization of the universe using equal height membership functions.

of the outcomes of interest. Consider the example shown in Fig. 6.7. The fuzzy variable is called *Risky_speed* with predicates *Bad_vision*, *Teenager*, and *Race_car_driver*. Apparently, the fuzzy qualifiers indicate the level of driving skill or competency at different speeds. Thus, they categorize the universe as well as characterize driving skills via height or incomplete categorization (i.e., maximum possibility values less than 1.0). Let's examine the linguistic equivalence of Fig. 6.7 for the extreme cases of maximum possibilities:

The possibility of *Risky_speed* to be that of a *Race_car_driver* is 0.48 at maximum (at all times).
The possibility of *Risky_Speed* to be that of a *Teenager* is 0.87 at maximum (at all times).
The possibility of *Risky_Speed* to be that of a driver with *Bad_vision* is 1.0 at maximum.

An important design question is how to determine the heights. There are basically two mechanisms at work here. The first one is to apply heuristics to the characterization of each qualifier, which is an interpretation of the first design principle. In other words, the specific solution must have required fuzzy qualifiers to be distributed in this way. The second mechanism, which may still be considered as the interpretation of the first principle, involves a probabilistic approach. For this purpose, Fig. 6.7 can be viewed as a modified probability distribution function obtained from the records of accidents that occurred at different speeds with

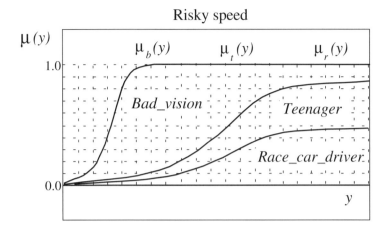

Figure 6.7 Membership functions labeled with predicates that characterize relative driving skills as well as categorizing the entire universe.

different drivers. Note that the area under the curve does not have to be equal to 1.0 in membership function design even though probability measures are used in its construction. Based on the discussions above, an interpretation of the first principle is stated as follows:

> Heights of the membership functions of an antecedent fuzzy variable can be designed differently based on heuristic or probabilistic measures if the predicates are fuzzy qualifiers and if the corresponding rule importance deterioration is one of the outcomes of interest.

Designing membership functions based on the principle stated above assumes that the rule importance deterioration is one of the outcomes of interest rather than using height design as a control mechanism for importance distribution among rules. An important point to remember is that rule importances can be directly controlled via different parameters (such as weights), which are explained in more detail in Section 7.5.3.

Composition With AND Operators

We have examined membership function height design within the framework of designing individual fuzzy variables in the last section. Height design has one other important side effect that shows up when fuzzy rules are composed using the AND logic operator. Let's consider the following LHS portion of a hypothetical rule:

$$[X \bullet P_{x,i}] \ominus [Y \bullet P_{y,j}] \ominus [Z \bullet P_{z,k}] \rightarrow$$

where the logic operator ($\ominus$) is AND. Suppose that the three predicates of the fuzzy variables X, Y, and Z are designed as shown in Fig. 6.8. Because composition using the AND operator yields the degree of fulfillment through min operator, a membership function with the smallest height produces a bottleneck effect.

The shaded areas in Fig. 6.8 will never contribute to the degrees of fulfillment, and the information contained within them is discarded in the LHS statements. This is expressed in fuzzy set terminology by

$$r = \vee[\mu_{x,i}(x), \mu_{y,j}(y), \mu_{z,k}(z)] \leq \sup[\mu^*_{x,i}(x)] \quad \forall x,y,z \ x \in X, \ y \in Y, \ z \in Z \quad (6.7)$$

where superscript * indicates the membership function with the smallest height. Equation (6.7) suggests that the degrees of fulfillment r will never exceed the maximum value of the membership function with the smallest height. Note that the bottleneck effect would not emerge using the OR logic operator. This leads the following design principle:

> When composing rules with the AND (min) operator, the heights of membership functions of an antecedent fuzzy variable must be designed all equal to 1.0 to avoid a bottleneck effect that deteriorates rule importance by removing a portion of information embedded in membership functions.

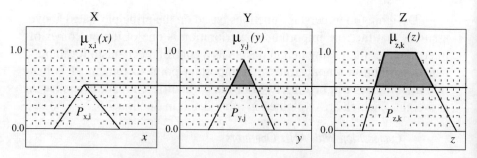

Figure 6.8 The membership function with the smallest height produces a bottleneck effect in composition using the AND operator.

Consequent Fuzzy Variables

In the context of height design, the discussions up to this point were in reference to membership functions of antecedent fuzzy variables that belong to the LHS of a rule. We have seen that height design affects the degrees of fulfillment of a rule and the rule importance in a direct way. On the consequent side, the membership function height design requires different considerations.

The height of a membership function of a consequent fuzzy variable determines the degrees of commensurateness or the strength of conclusion based on the degrees of fulfillment of the LHS conditions. Designing the height smaller than 1.0 is a direct attempt (via the implication process) to undermine the rule's importance and decrease its contribution to the final conclusion. The effect of height design on the final output depends on the aggregation and defuzzification processes. We will first consider the basic effect via the implication process.

The implication process operates on the consequent membership functions by changing their shape according to the degree of fulfillment of each rule. There are two cases to analyze when the membership function height is less than 1.0: (1) LHS degrees of fulfillment value is higher than the height of the membership function, and (2) vice versa. Let's start by analyzing the first case. For simplicity, we will consider a rule of the form IF X THEN Y, with membership functions shown in Fig. 6.9, using the Mamdani implication operator.

As depicted in Fig. 6.9, when the consequent membership function height is less than 1.0, a range of input values between x_1 and x_2 will produce a range of DOF values that will be higher than the membership function (referred to as the frozen region in Fig. 6.9). For these values of DOF, the implication result will remain the same using the Mamdani implication operator, thus resembling a paralyzed situation. If the first design principle of fuzzy systems does not apply (i.e., there is no special reason for such a design), then the following principle should be considered at all times.

Designing a consequent membership function of height less than 1.0 can cause paralysis (depending on the LHS design of each rule) where the contribution of the implication result to the final output remains the same (maximum) for a range of input values.

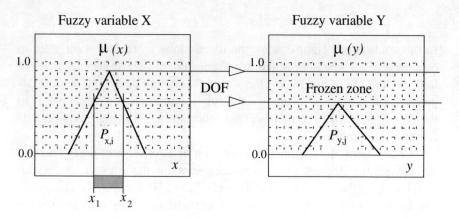

Figure 6.9 The implication result becomes paralyzed for inputs between x_1 and x_2 when the consequent membership function height is less than the maximum possible value of DOF.

Note that the term *paralysis* has a negative connotation as if it is something to be avoided. This is not true, because paralysis may be desired depending on the design objective (back to the first principle). For example, the decision implied by the rule *If it is more than 5 dollars, I will not buy* is true for 6, 7, or 8 dollars in which the action remains the same. The interval of 5–8 dollars could be the interval x_1–x_2 shown in Fig. 6.9. As long as there is another reason for defining dollar values above 5 dollars (perhaps other rules), there will be no conceptual design error in using heights smaller than 1.0 in this manner. Therefore, the principle stated above should not be taken as an avoidance criterion. Another important point is that some implication operators explained in Chapter 8, which are based on negation, operate in a reverse manner that defies Fig. 6.9. However, the only effect is a shift in the frozen zone; thus, the paralysis phenomenon prevails. One last note about the paralysis behavior is that it might not be detectable during the implementation of the basic fuzzy inference algorithm due to contributions from other rules to the same consequent variable.

In the second case, decreasing the height of a membership function (while keeping its spread fixed) produces a deteriorating or exaggerating effect on the implication result regardless of the type of the implication operator. This is an easily reached conclusion because we have already seen in Chapter 2 that the implication result is proportional to the area of the consequent membership function.

> The height of a consequent membership function is a direct measure of the strength of decision (or action) of the corresponding rule in reference to the other rules.

Notice the emphasis of relativity in the principle stated above. Height of a consequent membership function is only meaningful in comparison to the heights of other membership functions. Because each consequent membership function represents one rule, such a comparison is equivalent to comparing the rules. Accordingly, height determines the strength of contribution from each rule, whereas location determines the actual decision value. To illustrate this, consider Fig. 6.10, where three rules have the same consequent variable with different predicates. Using convex shapes, the height of each membership function defines the strength of decision.

There are many real-life scenarios fitting this representation. For example, assume that there are three partners in a corporation each holding different shares, therefore entitled to different votes on making a new

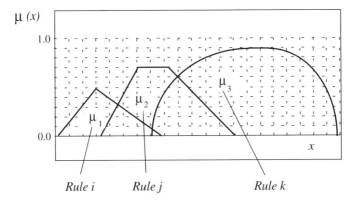

Figure 6.10 Three rules defining the action variable X by three membership functions of different heights.

corporate policy. Assume that each partner has a different rule about next year's operation:

(Partner-i) Existing stock pile is huge therefore let's reduce productivity by 20 percent.
(Partner-j) Last year's debts were high therefore let's keep productivity around nominal levels.
(Partner-k) Last year's sales were good therefore let's increase productivity 100 percent.

A quick glance at Fig. 6.10 tells us roughly that productivity will be increased. That is because the partner-k's rule has a consequent membership function (defined by the predicate *increase by 100 percent*) with the tallest height (biggest area) determined by the number of shares. If this corporation employed fuzzy implication, productivity would be increased less than 100 percent but more than the nominal value. How such a final decision is made is the matter of aggregation-defuzzification process. Although this example seems quite specific, many decision-making tasks in real-life situations can be modeled this way quite realistically. Of course, the main design challenge is to identify measures of strength (like the number of shares) in determining the heights.

Now let's consider the aggregation-defuzzification process and its effect in combination with height design. A consequent fuzzy variable having membership functions of different heights is shown in Fig. 6.11, where center-of-gravity defuzzification method is utilized for decomposition.

It is clear that smaller heights will produce smaller areas when the spread of the membership function on the universe is kept fixed. The bal-

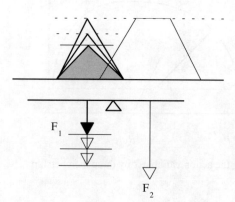

Figure 6.11 Designing one of the membership functions of a consequent variable at different heights varies its contribution (or its weight in a balancing beam analogy).

ancing beam analogy shown in Fig. 6.11 depicts the center-of-gravity defuzzification for each shape. The weight produced by the membership function on the left decreases as the height decreases, exerting less gravitational force on the left, and thus causing the balance point (decision) to shift to the right, moving away from the area that shrank. Most of the defuzzification methods behave (in response to different membership function heights) in a manner similar to the one described above.

Designing membership functions with different heights while keeping the area fixed (instead of the spread fixed) changes the nature of the ongoing discussion. Note that this is also discussed under the topic of overlapping in this chapter. As one would expect, the balance point will be located at the same point if the designer changes the height by keeping the area constant. This is shown in Fig. 6.12 for the ideal (crisp) case using a centroid defuzzification scheme.[1]

The question of why a designer would consider different heights yielding equal areas can be answered by the concept of *sensitivity*. Let's consider two consequent membership functions of equal area but different heights, as shown in Fig. 6.13. Assume that the designer will either select A or B as the membership function describing an action (decision) at the RHS of a rule.

As we have mentioned before, B causes paralysis for a range of DOF values (1–0.7 in Fig. 6.13). Besides the paralysis criterion, there is another important property called sensitivity. Remember that DOF = 1.0

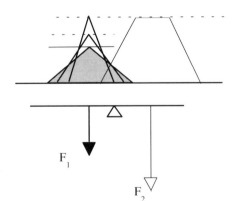

Figure 6.12 Designing a membership function with different heights but equal areas does not affect the strength of decision F_1 at the ideal case (crisp solution).

---

[1] Not quite true for centroid methods in which intersected areas are counted only once.

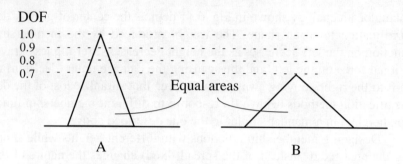

Figure 6.13 Designing a consequent membership function as A or as B requires the consideration of sensitivity to the inputs.

means full compliance between the inputs and the LHS conditions of the rule. The membership function A is cut by the DOF values in the vicinity of 1.0, and is affected by a small deviation from the ideal case. Therefore, A is more sensitive to the input absorption process than B because B is not affected by DOF above the level 0.7. This leads to the following design principle:

> Between two equal-area consequent membership functions located on the same point, the taller one is more sensitive and more responsive to the fuzzy regions of the input space located in the immediate neighborhood of a crisp solution.

Notice the term *responsive* above. It emphasizes the fact that more responsive behavior results in quicker formation of the implication result. Therefore, the strength of decision portrayed in Fig. 6.13 will not be the same away from the ideal case. Let's consider an example of how this principle may be used. We will use partner-k's rule with an aesthetic modification:

IF Last_year's_sales WERE Good THEN Productivity_increase
$$MUST\_BE\ 100\_percent$$

The predicate *100_percent* is represented by two equal-area membership functions in Fig. 6.14, each of which the designer is considering. The taller one, μ_{3A}, will be immediately affected by any DOF value less than

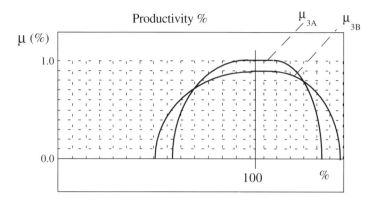

Figure 6.14 Two equal-area membership functions under consideration.

1.0. Considering the rule above, this means that if last year's sales did not match the definition of *good* completely, then even the smallest deviation from *good* will hinder the decision strength by the implication process. On the other hand, the selection of the shorter membership function μ_{3B} will produce a full-strength decision until the DOF falls down to 0.9 in this example. Remember that decisions mentioned here are not the final (aggregated) decisions, but the individual ones.

The design challenge is to study the properties of the problem at hand and to extract useful knowledge such that a consideration like the one above becomes possible. Note that these discussions are valid within the framework of the implication operators and defuzzification methods selected in this book. Implication operators other than Mamdani, and defuzzification methods other than centroid, will exhibit different behavior in response to variations in membership function height. However, the principle stated above holds unless the basic inference mechanism is changed to a form not mentioned in this book.

6.3 MEMBERSHIP FUNCTION LINE STYLE

A membership function can be defined by break points connected with straight lines (point-wise-linear) or it can be defined by a curve. To examine the effects produced by the line style, we need to turn back to the basic concepts. We know from the basic definition of membership function that the width of occupancy is a measure of fuzziness. For example, between the two membership functions labeled as A and B in Fig. 6.15, A is more fuzzy than B.

More fuzzy means there is a larger number of possible points on the universe of discourse to be categorized as A. Thus, by designing the base-width of a membership function, we are directly determining the level of fuzziness. Then, the main question is the effect produced by shape differences between two membership functions with equal base-widths, or with the same level of fuzziness.

Let's examine the shape differences between two convex membership functions with equal base-width. The most frequently asked question among the designers is the difference between a piece-wise-linear shape and its curvature equivalents (i.e., curves satisfying three points of a triangle). As shown in Fig. 6.16, there is an infinite number of ways of fitting a curve around a triangle membership function A. Shapes C and D exhibit approach to the crisp point with two different slopes, whereas shape B exhibits change of slope while approaching the crisp point.

It is very difficult, and also pointless, to search for an explicit relationship or formula to characterize the effect produced by shape differences on the output response of a fuzzy inference engine. However, we can examine the general behavior by considering the degrees of fulfillment of the left-hand-side conditions of a simple inference rule of the form IF X is A THEN . . . . where A is the antecedent membership function under investigation. Let's consider the input range x_1–x_2. When the inputs are changing from x_1 to x_2 during the input absorption process, the

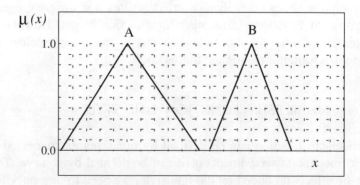

Figure 6.15 Membership function A is more fuzzy than B.

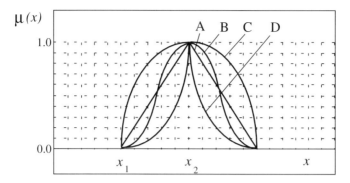

Figure 6.16 Curved and triangular membership functions with equal base-widths.

degrees of fulfillment change in accordance with the slopes of A, B, C, or D, whichever shape is used. This tells us that the degree of fulfillment will increase more rapidly in case C than in cases A, B, and D, yielding a quicker formation of the implication results. At points x_1 and x_2, all degrees of fulfillments are the same. In the fuzzy region, the membership function C will produce more action (or stronger decision) in comparison to others, which means that we tolerate fuzziness (or we accept risks). On the opposite side, by designing a membership function like the shape D, we are not as tolerant to fuzziness (or we don't like to take risks). Thus, curves on both sides of the linear surface line of A give us a chance to express our willingness to take the risk of suffering from approximation when fuzziness is encountered.

> The shape of the line of an antecedent membership function that connects possibility = 0 to possibility = 1 is a tool to express the level of tolerance employed by the corresponding fuzzy statement in reference to all other fuzzy statements.

A very important point to remember here is the emphasis on relativity. In other words, if we design all antecedent membership functions using the same curve, we are stuck with one level of tolerance employed in all fuzzy statements. Otherwise, we will be selectively distributing tolerances among the fuzzy statements. The tolerance assignment by means of membership function shape also determines rule importances and dominance characteristics that are discussed in Section 7.7.

EXAMPLE 6.1 MEMBERSHIP FUNCTION LINE STYLE

Suppose that we have collected several fuzzy facts articulated by an expert and have translated them into fuzzy statements as shown below.

Lion's_roar SOUNDS LIKE In_Heat
Lion's_roar SOUNDS LIKE Stressed
Lion's_roar SOUNDS LIKE Yawning

In this example, we have a digital audio system to measure the sound levels recorded from a lion's cage. Through repeated measurements, the universe of discourse of the fuzzy variable *Lion's_roar* is well defined in the units of dB. The development of an automated care-taking system requires that the system must make decisions at different speeds among the three cases stated above. One of the design requirements is to be quite conservative when the lion is stressed (by making sure that the roar really means the animal is in stress) so that the animals can be shifted to a larger cage. Another requirement is to act quickly when the lion's roar sounds like males are in heat (by detecting quickly) to make mating arrangements or to isolate different sexes. Designing three membership functions that correspond to three cases in the manner shown below satisfies these requirements in an approximate way.

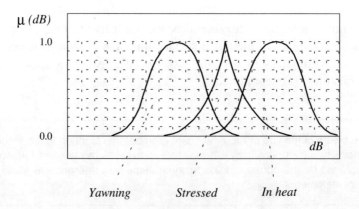

The membership function for the predicate *Stressed* is a conservative (nontolerant) design where the corresponding action (decision) will not be taken as fast as that determined by the tolerant membership function of the predicate *In_Heat*. Conversely speaking, the tolerant membership function *In_Heat* will take the risk of wrong detection for the benefit of quick action. Note that the base-width of each membership function is equal, and the tolerance effects are only produced by the shape of membership functions. Also note that the terms *fast* and *quickly* used in this context do not necessarily imply a dynamic behavior; instead, they imply high possibility, strong likelihood, or optimism.

Designing Curves

Although difficult in practice, it is possible to design membership function curves based on a criterion called the fulfillment table. Consider the two antecedent membership functions (C, D) shown in Fig. 6.17. We know from the previous discussions that C is tolerant and D is conservative towards fuzzy information. Now, the question is how to design them to satisfy a given objective.

When evaluated by the point values on the universe of discourse, a membership function yields degrees of fulfillment, which tells us how much the corresponding fuzzy statements will contribute to the fuzzy actions. A fulfillment table is a systematic implementation of this idea. Let's consider Table 6.1. Fuzzy statements are listed in the leftmost column (such as *Temperature IS Low*) and fuzzy actions are listed in the rightmost column (such as *Heater IS On*).

We assumed canonical fuzzy rules for simplicity. The columns in between indicate the degree of contribution (fulfillment) at different grid locations on the universe of discourse that corresponds to a range between $\mu = 0$ to $\mu = 1$; for example, $f_{1,1} = 0.4$ and $f_{2,1} = 0.85$ for $F_1 = u_k$ in Fig. 6.17. Note that such a table of information is often constructed by experts or it is extracted from a data set. Once such a table is available, the membership function can be formed by a curve-fitting method that would satisfy all the points in the table.

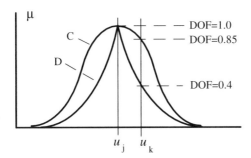

Figure 6.17 Tolerant (C) and conservative (D) shapes of a membership function.

Table 6.1 Fulfillment Table of Monotonically Composed Fuzzy Rules

STATEMENTS	F_1	F_2	..	F_m	ACTIONS
$X_1 \bullet P_i$	$f_{1,1}$	$f_{1,2}$	..	$f_{1,m}$	$Y_1 \bullet P_i$
$X_2 \bullet P_j$	$f_{2,1}$	$f_{2,2}$	..	$f_{2,m}$	$Y_2 \bullet P_j$
...	..	..	..	..	...
$X_n \bullet P_k$	$f_{n,1}$	$f_{n,1}$	..	$f_{n,m}$	$Y_n \bullet P_k$

EXAMPLE 6.2 DESIGN BY FULFILLMENT TABLE

Assume that the following fulfillment table is available to the designer.

STATEMENTS	1/4	1/2	3/4	1	ACTIONS
Stock_price IS Low	0.1	0.4	0.9	1	*Buy_many*
Stock_price IS Medium	0.05	0.2	0.45	1	*Invest_few*
Stock_price IS High	0.25	0.5	0.75	1	*Sell_some*

The stock market expert also tells us that *Low* is $2.00, *Medium* is $5.00, and *High* is $7.00 for this particular stock with base-widths (uncertainties) $4.00, $8.00, and $4.00 respectively. Then the fuzzy system designer builds the following membership functions:

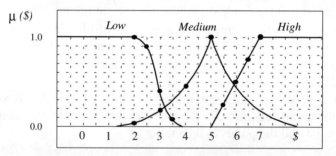

Let's take the development of the first membership function called *Low*. It is situated around the crisp point $2.00 with a base-width $4.00 that extends between $0 and $4.00. The 1/4, 1/2, and 3/4 percentiles ($3.50, $3.00, $2.50) are 0.1, 0.4, and 0.9 in the table. The corresponding points are marked on the drawing above as the DOF values, then a curve is fitted. Note that this is only a mechanism to design the membership functions of an antecedent fuzzy variable and it is not a direct mapping between the input and output spaces.

The problem with this method is that the solution given by a fulfillment table is not how the experts articulate a solution. Thus, it should be obtained through the knowledge acquisition process by asking more detailed questions such as, how much weight should be given to a decision when its trigger mechanism (antecedent membership function) is partially satisfied? Although human thinking is mostly fuzzy, experts often have no answers when it comes to articulation in this manner. The next example shows development of a fulfillment table by examining a data set.

EXAMPLE 6.3 DEVELOPMENT OF A FULFILLMENT TABLE FROM DATA

Suppose the problem in the previous example is given without the fulfillment table. Instead, experts give us the following data set.

BUYS	STOCK PRICE	INVESTS	STOCK PRICE	SELLS	STOCK PRICE
99	$1.00	1	$2.00	37	$5.50
100	$1.50	10	$3.00	25	$6.00
100	$2.00	23	$4.00	12	$6.50
90	$2.50	50	$5.00	2	$7.00
40	$3.00				
10	$3.50				

In addition, the experts define the crisp points for *Low*, *Medium*, and *High* stock prices as $2.00, $5.00, and $7.00. The fuzzy system designer creates three categories of actions. For the dollar values of 1.00, 1.50, 2.00, 2.50, 3.00, and 3.50, the designer picks the buy figures 99, 100, 100, 90, 40, and 10, then forms the fuzzy set *Low*. Taking the values 100, 90, 40, and 10, the fulfillment table can be formed by normalizing the buys (1.0, 0.9, 0.4, 0.1) for each percentile as shown below.

STATEMENTS	1/4	1/2	3/4	1	ACTIONS
Stock_price IS Low	0.1	0.4	0.9	1	Buy_many
Stock_price IS Medium	0.02	0.2	0.45	1	Invest_few
Stock_price IS High	0.25	0.5	0.75	1	Sell_some

Next, the designer selects the dollar values 3.00, 4.00, 5.00, and 6.00 and the corresponding investments 1, 10, 23, 50 to form the second category *Medium*. Investments 50, 23, 10, and 1 are normalized (1.0, 0.45, 0.2, 0.02) for three percentiles. This is the second line of the fulfillment table. The last line is developed in the similar manner. Remember that the fulfillment table only indicates the relative speed of action and it is not a direct mapping between inputs and outputs.

The fulfillment table approach may not always yield a single solution. Consider a case where a fuzzy statement is used in more than one rule contributing to different actions as shown in Table 6.2.

In such a case, the designer either chooses to develop separate antecedent membership functions for each action or designs one antecedent membership function that is the closest fit to all points.

Table 6.2 Fulfillment Table With Complex Relationships

STATEMENT	F_1	F_2	..	F_m	ACTIONS
$X_1 \bullet P_i$	$f_{1,1}$	$f_{1,2}$	..	$f_{1,m}$	$Y_1 \bullet P_i$
	$f_{2,1}$	$f_{2,2}$	..	$f_{2,m}$	$Y_2 \bullet P_j$
	..	..		..	...
	$f_{n,1}$	$f_{n,1}$	..	$f_{n,m}$	$Y_n \bullet P_k$

Triangle Versus Trapezoid

Designing membership functions without a comprehensive linguistic knowledge of fuzziness (no fulfillment tables available) requires a piece-wise-linear approach as the most straightforward form of uncertainty representation. In the family of piece-wise-linear membership functions, the triangle and trapezoid are the most frequently used shapes. Accordingly, the difference between them is one of the most frequently asked questions.

A triangle and a trapezoid with equal base-widths on the same universe of discourse are, again, an expression of optimism versus pessimism or tolerance versus conservatism criteria. Let's consider Fig. 6.18.

The trapezoidal membership function A includes several crisp points in the definition of the fuzzy category whereas the triangle membership function B only includes one. For obvious reasons, A is much more tolerant (optimistic) than B. Note also that the trapezoidal shape produces DOF = 1 in large regions of the universe of discourse, which causes the corresponding rule to fire at full strength. This is very much like applying

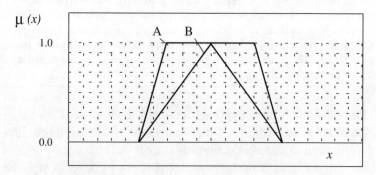

Figure 6.18 Triangular versus trapezoidal membership function with equal base-widths.

low standards to be qualified in an exam. The triangle, on the other hand, is analogous to applying high standards such that only one performance will qualify fully. There is a frequently overlooked statistical relationship in using each of them that affects the outcome of an inference process sometimes in an undesirable way. This is explained in Section 7.7.

> A trapezoidal membership function, by including many crisp points, is equivalent to applying low standards to the antecedents to be qualified as full strength contributors to the overall result. A triangular membership function is equivalent to applying high standards to the antecedents such that only one performance will qualify as a full strength contributor.

There is nothing wrong with applying low standards, nor is there an advantage in applying high standards, to an antecedent condition.

EXAMPLE 6.4 APPLYING LOW AND HIGH STANDARDS IN MEMBERSHIP FUNCTION DESIGN

Suppose there are two fuzzy statements under consideration and there is no knowledge of how fuzzy they can be.

IF Candidate IS Medium_height THEN He/She CAN BE On_our_rowing_team.
IF Candidate IS Size_A THEN He/She CAN BE On_Olympic_team.

Now consider the logic that height differences cause an imbalance in a rowing boat, therefore rowers must have similar heights. However, a window of acceptable heights may include a range of values, all of which qualify. On the other hand, consider forming an Olympic team where sizes are identified more strictly. Therefore, there is only one Size_A athlete; all others approximately qualify. In this example, *Medium_height* will be a trapezoidal shape whereas *Size_A* will be a triangle.

The trapezoid may be viewed as a cooperative condition seeker, whereas the triangle is a competitive one. In applications to many real-life problems, extra room for qualification is a more frequently encountered case. In control design for example, rules are often composed to fire at full strength for a range of input values along with some partial firings that correspond to the fuzzy surfaces. Fuzziness in such cases is interpreted as many "possibility = 1" singletons rather than one "possibility = 1" singleton. Nevertheless, the differences are not as profound when all the membership functions are designed either as triangles or trapezoids.

Consequent Membership Functions

In the context of membership function line style, the discussions up to this point focused on antecedent membership functions, which determine the degrees of fulfillment and the rate of contribution to the action. The line style design of consequent membership functions requires a completely different perspective. We know from the basic definition of membership function that the width (or the area under the curve) of membership function is a measure of fuzziness. Besides the fuzziness, we have the following property for consequent membership functions.

> The size of a consequent membership function reflects the strength of decision when some of the most frequently used defuzzification techniques are employed.

This is a similar design principle to that described under the topic of height analysis. Thus, all arguments about the height design—those causing area variations—are applicable to the line style design if the line shape causes changes in the area of the consequent membership functions. Next, we will examine the RHS operations with consequent membership functions subject to line style design.

A fuzzier (wider) consequent membership function has more decision strength in comparison to thinner and crisper ones. Let's iterate this property by a new case study. Consider three membership functions as shown in Fig. 6.19 (a). The same membership functions are shown in

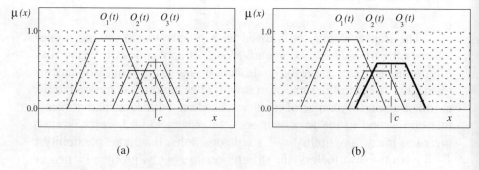

Figure 6.19 Three hypothetical implication results that belong to consequent X.

Fig. 6.19 (b) except the third one which is widened around its crisp center point.

Applying union aggregation and the center-of-gravity defuzzification method, we see that the defuzzified output shifts to the right towards the widened membership function. This is illustrated in Figs. 6.20 (a) and (b).

When the center of maximum possibilities method is employed as the defuzzification technique, the effect of widening a consequent membership function is similar to that of the center-of-gravity method as shown in Fig. 6.21 (a) and (b).

The center-of-mass defuzzification technique is affected by a widened membership function in the opposite manner, as depicted in Fig. 6.22 (a) and (b).

As a result, we can conclude that the size of a consequent membership function will cause the production of different defuzzified outputs from the same implication process. However, not all the defuzzification methods are sensitive to membership function width such as the center of maximum possibility method (e.g., center of the tallest membership function in Fig. 6.22).

The general method for a consequent membership function design is to place a symmetrical geometrical shape (convex) around the crisp point on the consequent universe of discourse with equal width. In the presence of evidence, widths can be adjusted according to the value of each decision (each individual implication result); however, such adjust-

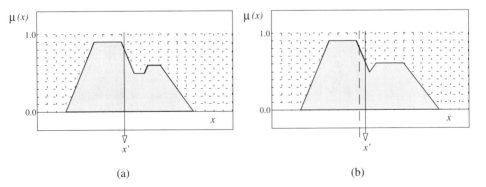

Figure 6.20 Union aggregation and the center-of-gravity method for defuzzification.

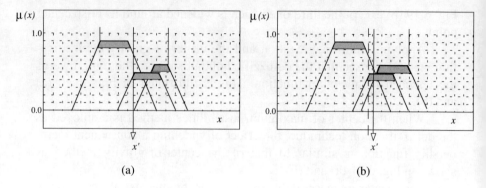

Figure 6.21 Center of maximum possibilities for defuzzification.

ments are difficult to formulate deterministically and are often done in a heuristic manner.

Consequent membership functions are rarely designed point by point to yield a desired mapping. This requires a commensurateness table where the degree of commensurateness is listed for the corresponding degrees of fulfillments from a fulfillment table. Now, consider Table 6.3.

Degree of commensurateness indicates strength of decision. The linguistic equivalence of the relationship between the degree of fulfillment and degree of commensurateness is given below by an example.

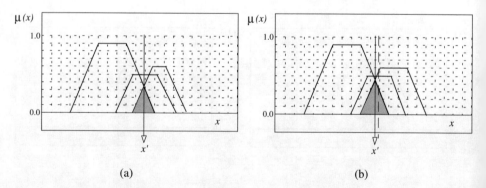

Figure 6.22 Center of mass with the highest density of intersection for defuzzification.

Table 6.3 Commensurateness Table of Monotonically Composed Fuzzy Rules

STATEMENTS	F_1	F_2	..	F_M	ACTIONS
$X_1 \bullet P_i$	$C_{1,1}$	$C_{1,2}$	..	$C_{1,m}$	$Y_1 \bullet P_i$
$X_2 \bullet P_j$	$C_{2,1}$	$C_{2,2}$	..	$C_{2,m}$	$Y_2 \bullet P_j$
...	..	..		..	...
$X_n \bullet P_k$	$C_{n,1}$	$C_{n,1}$	..	$C_{n,m}$	$Y_n \bullet P_k$

If the condition *Temperature IS Low* is fulfilled by 0% then
the action *Valve IS Wide_open* is true by 0%.

If the condition *Temperature IS Low* is fulfilled by 30% then
the action *Valve IS Wide_open* is true by 50%.

If the condition *Temperature IS Low* is fulfilled by 60% then
the action *Valve IS Wide_open* is true by 80%.

If the condition *Temperature IS Low* is fulfilled by 100% then
the action *Valve IS Wide_open* is true by 100%.

Rules are normally written to satisfy the crisp relationship shown last. The first relationship, which represents the outer boundaries, may also be specified. However, the fuzzy regions in between are approximations, which are often not articulated in this detail.

In most practical problems, degree of commensurateness is not explicitly known; instead decomposed output points are known. Thus, degree of commensurateness values $C_{i,j}$ can be replaced by the output values $Y_{i,j}$ provided that the mapping between $f_{i,j}$ and $Y_{i,j}$ exists via defuzzification. For example, selecting the center of maximum possibility defuzzification method, a consequent membership function can be designed in such a way that the desired mapping exists as shown in Fig. 6.23.

Note that when the rules are more complex or when there is aggregation due to multiple contributions, this process becomes too cumbersome to handle. For such reasons, the general principle is to pursue the standard design steps (i.e., locate membership functions on the desired output points and design them to comply with the articulated logic) instead of constructing fulfillment and commensurateness tables.

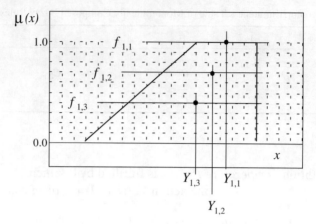

Figure 6.23 Designing a consequent membership function shape to achieve a desired mapping between the degrees of fulfillment and the outputs.

6.4 OVERLAPPING

One of the important design considerations is the amount of overlap between the membership functions defined over the same universe of discourse. We will examine the effect of overlap on the output behavior, then identify some useful principles to remember during design. Antecedent and consequent fuzzy variable design, when overlap is considered, have different characteristics; therefore we will examine them separately. Let's first define overlap.

Using piece-wise-linear membership functions (such as triangles and trapezoids) we define overlap by the ratio between the area under the curve and the area above the curve between two crisp points on the universe of discourse. Considering two membership functions shown in Fig. 6.24, intersecting points a, b, c, and d indicate four possible overlaps among an infinite number of options.

Let's define the rectangle (width x_j-x_k and height equal to 1.0) by a crisp set W having n singletons

$$W(x) = \bigcup_{i=1,n} 1.0/x_i \quad x_j \leq x_i \leq x_k \qquad (6.8)$$

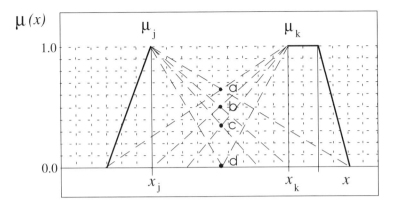

Figure 6.24 Overlapping between two membership functions.

We will consider this rectangle to be the control rectangle for overlap formulation. The total area under the membership function curves is a new fuzzy set defined by the union of two fuzzy sets:

$$\mu_T(x) = \mu_j(x) \vee \mu_k(x) \quad x_j \leq x \leq x_k \tag{6.9}$$

For an arbitrary selection of overlap indicated by point c in Fig. 6.24, the fuzzy set represented by the area under the curve within the control rectangle is shown in Fig. 6.25.

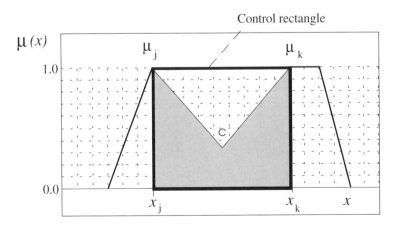

Figure 6.25 Fuzzy set representing the area under the membership function curves within the control rectangle.

The total area (geometrically) can be computed by adding (integrating for the continuous case) all the elements of the new set

$$\overline{A}_T = \sum_{i=1}^{n} \overline{\mu}_A(x_i)\Delta x \qquad (6.10)$$

where Δx is the distance between two singletons. The area of the control rectangle is given by

$$\overline{A}_C = |x_j - x_k| \cdot 1 \qquad (6.11)$$

where 1.0 is the height. Now we define the overlap (or overlap index) by the ratio of

$$\theta = \frac{\overline{A}_T}{\overline{A}_C} \qquad (6.12)$$

which is a measure of fuzziness between the two crisp points. Obviously when $\theta = 0$, we have two crisp points on the universe separated by empty space. Conversely when $\theta = 1$, the two crisp points are connected by a fuzzy region in the sense that all points between them have the possibility of 1. The most practical cases of overlap correspond to the cases in between b and d in Fig. 6.24 with overlap indexes $0 < \theta_d < \theta_b < 1$. Using piece-wise-linear membership functions and assuming symmetry, it can be shown that $\theta_d = 0.5$ and $\theta_b = 0.75$. It can also be shown that given the crisp points x_j and x_k of Fig. 6.24 and the overlap index θ, the coordinate of the overlap is given by

$$x_j^o = \frac{4x_j(1-\theta) + (x_k - x_j)}{4(1-\theta)}, \quad 0.5 \le \theta < 1 \qquad (6.13)$$

where x_j^o is the extension coordinate of the membership function μ_j to the right in Fig. 6.24.

Antecedent Variables

The conditional part of a rule uses antecedent fuzzy variables comprising membership functions defined over the corresponding universe of discourse. As described in the previous section, the shape of each individual membership function determines the degree of fulfillment on the conditional premises given external inputs. Regardless of the reason or knowledge behind the membership function shape design, we will examine the

Chap. 6 Membership Function Shape Analysis

effect of overlapping antecedent membership functions by considering a simple mapping process:

IF X IS Small THEN Y IS Small
IF X IS Large THEN Y IS Large

Suppose that this simple mapping process is designed as shown in Fig. 6.26, where the overlapping of antecedent membership functions are changed while the consequent membership functions are fixed. Three cases marked as b, c, and d are expected to produce different results.

For each case, we will numerically construct the mapping by entering input values changing from x_j to x_k. Also assume that this inference mechanism uses the Mamdani implication operator and the center-of-gravity aggregation method. When the implication operator and the aggregation method are changed, the output behavior will change. However, the relative effect of overlap from one case to another has consistent characteristics such that a principle can be formed. The transition between two points at the output space for each case of overlapping is shown in Fig. 6.27.

It is easy to interpret what is happening in Fig. 6.27. As shown in Fig. 6.26, case d has no overlapping (or $\theta = 0.5$). Thus, scanning the input entries from left to right produces DOF from the membership function *Small* until point d. During this time, there is no DOF production from

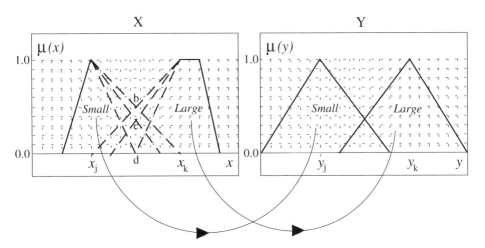

Figure 6.26 Overlap analysis of antecedent membership functions using two simple rules of mapping.

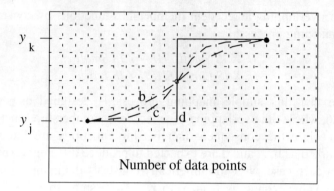

Figure 6.27 Transition between two points on the output space for different cases of overlapping of the antecedent membership functions.

the membership function *Large*. In other words, only the first rule is firing. After point d, the process reverses and only the second rule fires. Accordingly, the output behavior shown in Fig. 6.27 is a sudden switch from one output point to another. As overlapping increases (cases c and b), transition gets smoother due to partial cooperation between the two rules. Note that the two crisp points on the output space are reached in each case regardless of the differences in overlap. Also note that the sigmoidal behavior in case b ($\theta = 0.75$) is due to the center-of-gravity method.

Before stating a related principle, we need to clarify two things. First, there is no reason why the inputs will change in a scanning manner in the real world. This was employed here only to make a point. Second, the consequent membership functions are designed to be symmetric in this example, which produces a discrete switch in case d. Asymmetric design will change the point of switch for case d and for all others; however, the overall appearance of Fig. 6.27 will maintain its characteristics.

> Designing antecedent membership functions with overlap determines the speed of switch or the degree of cooperation between the corresponding rules. More overlap causes more cooperation and smoother transition between the two crisp points on the consequent space.

Most of the existing applications employ overlap between $\theta = 0.50$ and $\theta = 0.75$, which ranges from firing the rules one by one to firing them collectively with the minimum DOF value determined at the overlap coordinate. Thus, θ determines the minimum possibility value between x_j and x_k in Fig. 6.24. Overlap less than θ = 0.5 means undefined regions on the universe of discourse, which is a conceptual design error. Overlap more than θ = 0.75 means poor categorization or poor knowledge of the universe of discourse. Remember that all these arguments are only valid between the two crisply known points on the antecedent universe.

Consequent Variables

Unlike the previous case, in which the overlap analysis of an antecedent membership function on one side (of a crisp point) yielded conclusive results, the overlap analysis of consequent membership functions requires considering both sides. By both sides, we mean symmetry versus asymmetry. The two possible overlap designs are shown in Fig. 6.28 (a) and (b).

We will not consider the asymmetric overlap in consequent membership function design because the original location of the decomposed result (either using the center-of-gravity or any other defuzzification technique listed in this book) will shift to a new location. Thus, asymmetric overlap becomes the subject of "location design," which is related to de-

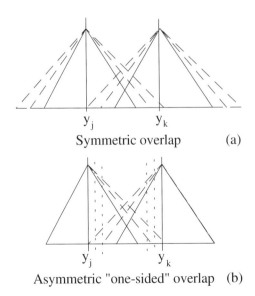

Figure 6.28 Symmetric (a) and asymmetric or one-sided (b) overlap design for consequent membership functions.

cision boundary analysis as described in Chapter 5. For the symmetric case, we will consider the same simple mapping rules:

IF X IS Small THEN Y IS Small
IF X IS Large THEN Y IS Large

Suppose that this mapping process is designed as shown in Fig. 6.29 in which the overlapping of both antecedent and consequent membership functions are changed. Nine cases emerging from three-by-three combinations marked as b, c, d (for antecedents) and e, f, g (for consequents) are intuitively expected to produce different results. Note that one fixed antecedent overlap would not be enough to characterize the output behavior; therefore, we have to look at all nine cases. Again, our objective in such an analysis is to extract useful design principles.

For each case, we will numerically construct the mapping by entering input values changing from x_j to x_k. Also assume that this inference mechanism uses the Mamdani implication operator and the center-of-gravity defuzzification method. When the implication operator and defuzzification method are changed, the output behavior will change. However, the relative effect of overlap from one case to another has consistent characteristics such that a principle can be formed. The transition between two points at the output space for each case of overlapping is shown in Fig. 6.30.

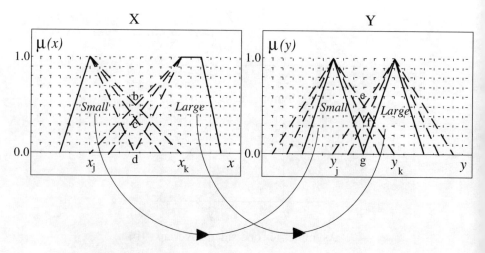

Figure 6.29 Overlap analysis of consequent membership functions using two simple rules of mapping.

Chap. 6 Membership Function Shape Analysis

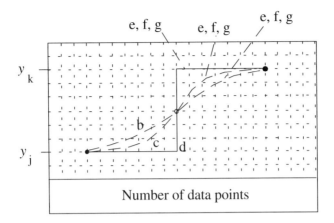

Figure 6.30 Transition between two points on the output space for different cases of overlapping of the antecedent (cases b, c, d) and consequent (cases e, f, g) membership functions.

These results are very similar to those illustrated in Fig. 6.27 (i.e., differences are hard to visualize for this particular example). In other words, symmetric overlapping of all of the membership functions in a consequent fuzzy variable does not produce significant effect. Particularly, the effect diminishes when the antecedent membership functions are not overlapped $\theta = 0.50$ (case d). This leads to the following design principle:

> When designing a consequent fuzzy variable, symmetric changes of overlap in all of its membership functions do not produce significant changes in the output behavior of the fuzzy inference engine; thus symmetric changes of overlap are equivalent for all practical purposes.

To visualize this property, we have drawn equivalent designs in Fig. 6.31 that would produce a similar transition between two crisp points on the output space.

The principle stated above, when remembered during design, will eliminate unnecessary considerations of symmetric global overlap. Now, let's consider individual overlap using the same pair of rules. This time, only one of the consequent membership functions (*Small*) is overlapped, while the neighboring membership function *Large* is fixed to a constant

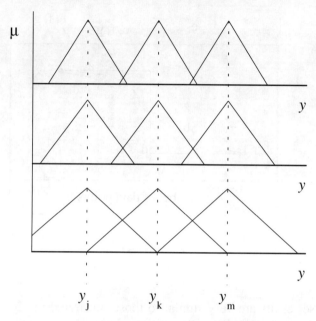

Figure 6.31 Consequent membership functions with different symmetric overlap ($0.5 \leq \theta \leq 0.75$) are equivalent for all practical purposes.

shape as shown in Fig. 6.32. We call this case the proportional-symmetric-overlap indicating a general case in which each membership function is symmetrically widened at different rates.

For each case, we will numerically construct the mapping by entering input values changing from x_j to x_k. Out of nine possible combinations between the antecedent and consequent overlap cases, the transitions from one crisp point to another on the output space are shown in Fig. 6.33. Starting from case d ($\theta = 0.5$) at the top, switching suddenly from one rule to another (imposed by the antecedent overlap) yields a transition profile in which the consequent overlap becomes ineffective. This leads to the following principle.

Overlapping consequent membership functions becomes ineffective when the overlapping antecedent membership functions in the corresponding rule do not allow any cooperation ($\theta = 0.5$). In such a design, the output behavior is a sudden switch from one crisp point to another.

Chap. 6 Membership Function Shape Analysis

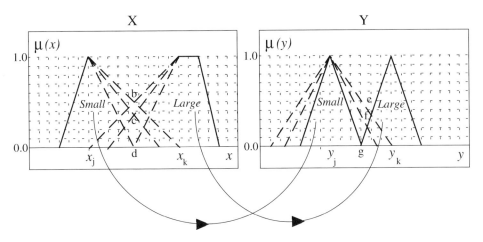

Figure 6.32 Proportional-symmetric overlap of the consequent membership function *Small*.

This principle merely suggests that there is no point in designing antecedent overlap with θ = 0.5 if the consequent universe includes fuzzy regions defined by noncrisp membership functions.

Between case c and case b, it is apparent that overlapping antecedent membership functions tend to flatten the transition toward a somewhat more linear profile. However, we have already analyzed this effect in Fig. 6.30. Examining the consequent overlap cases e, f, and g in each figure, we can see that widening the overlap causes the transition to shift towards the crisp value of the widened membership function (compare the lines f and e with the center point in Fig. 6.33). This suggests that the widened membership function, which has a larger area, is dominant in the fuzzy region between two crisp points.

> Symmetric widening of a consequent membership function while keeping its neighborhood membership functions unchanged results in a dominating effect in transition from its crisp value to that of its neighbor.

This principle is directly useful in design. However, the design challenge is to apply this principle by accurately estimating (or by computing) the desired dominance. This principle also brings out perhaps the

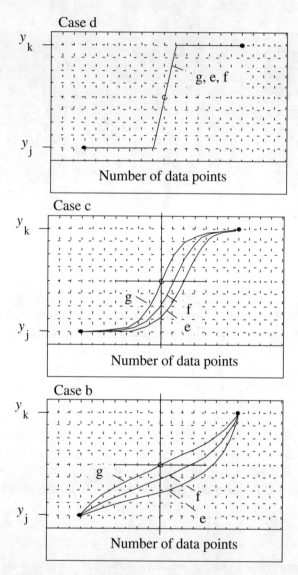

Figure 6.33 Transition between two crisp points on the output space affected by both antecedent (cases d, c, b) and consequent (cases e, f, g) overlaps.

most important characteristic difference between the antecedent and consequent membership function design. When an antecedent membership function is widened, widening of its base-width means increased uncertainty. The same fundamental concept applies to the consequents. However, increased uncertainty in a consequent membership function dominates the results as suggested by the last principle. This sounds like a paradox, but it is not. By enlarging the spread of a consequent membership function, we are increasing the number of possible values that represent the category. This causes a dominating effect in comparison to another membership function, which is represented by a smaller number of possible points on the universe of discourse.

Antecedent Versus Consequent Membership Functions

Adjustments on the antecedent domain can yield more profound changes in the output behavior than adjustments on the consequent domain. This is illustrated in Fig. 6.34 for an arbitrary modification on one side of the membership function. The top figure (a) illustrates the effect (gray area) produced by different slopes applied to the antecedent membership func-

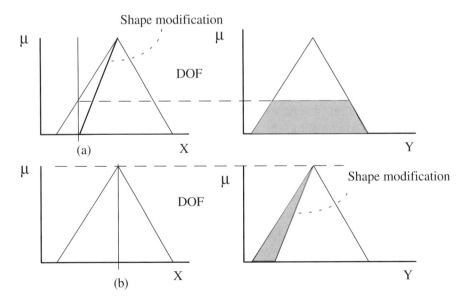

Figure 6.34 The effect of shape modification in the antecedent membership function (a) versus the consequent membership function (b).

tion. The bottom figure (b) illustrates the effect produced by applying the same slope difference to the consequent membership function. The results using the Mamdani implication operator shown by the gray areas are obtained for instantaneous inputs a and b, respectively. These inputs represent the worst case deviations. Note the difference in sizes between the two gray areas that represent the contributions to the aggregated output caused by shape modifications. For different forms of shape modification and different implication operators, this discussion holds in different proportions. However, antecedents generally produce more profound effects when their shapes are modified.

This discussion leads to the following design principle.

> In general, shape modifications in antecedent membership functions produce more significant effects on the output behavior than those produced by the same shape modification in consequent membership functions.

6.5 SUMMARY

The shape of a membership function describes the transition between two crisp points (also called *prototypes*) on a product space. Accordingly, this chapter describes the events taking place on fuzzy surfaces in the vicinity of crisp points. The location and the granularity design previously examined in Chapter 5 determine the overall performance of the basic fuzzy inference algorithm, whereas the shape design determines local performances. In other words, the shape design can be considered as fine tuning in comparison to the location and granularity designs. If the locations and granularity are poorly determined, the shape design cannot compensate for the poor performance. The shape design (or the fine tuning) may sometimes be unnecessary, as several applications in the industry indicate. Especially for problems like control, in which the regions beyond the crisp points are rarely known, the simple partitioning technique described in Section 5.2.1 is often adequate. In a broader spectrum however, the shape design can affect the performance of a fuzzy inference engine. Therefore, shapes should not be formed arbitrarily.

The principles formed in this chapter involve three fundamental issues: height, line style, and overlap. The principles related to the height issues suggest that the designer must pay special attention when the heights of the membership functions are designed to be different from 1.0. The membership function line style is a way to characterize uncertainty based on linguistic criteria such as conservatism, optimism, or high standards. When there is no criterion to apply as such, the line style design becomes difficult and sometimes impossible. If there is no criterion to apply, the line style should be determined by techniques like equal-distance partitioning. The overlap design of membership functions for antecedent and consequent fuzzy variables produces different effects. In the antecedent domain, the amount of overlap determines the degree of cooperation among the rules contributing to the same consequent. In the consequent domain, overlap is not significantly effective on the output behavior of the basic fuzzy inference algorithm when all membership functions are symmetrically expanded or shrunk. Changing the overlap of a single consequent membership function symetrically, on the other hand, changes the strength of its contribution to the aggregated result, thus its effects are easily traceable and significant.

PROBLEMS

6.1 Suppose a survey revealed the following statistics in finding dangerous substances in the drinking water over the acceptable ppm levels: 35 samples out of 100 taken from a reservoir, 85 samples out of 100 taken from a river, and 5 samples out of 100 taken from a well. What would be the most suitable design for the membership functions (called *reservoir*, *well*, and *river*) of the antecedent variable *water_contamination* in a fuzzy inference engine that estimates a *contamination index* over an area involving several other variables?

6.2 Using the Mamdani implication operator and centroid defuzzification, what are the two possible mechanisms to create a dead-band in control design? (Hint: The output from a dead-band controller does not change for certain inputs.)

6.3 How would you characterize the membership functions *Temperature_high* and *Temperature_medium* if 100 people agreed *Temperature_high* to be between 20° and 45° C and 25 people agreed *Temperature_medium* to be between 15° and 25° C in the context of outdoor weather? How would you incorporate the number of agreement votes into the membership function design?

6.4 Between the two equal base-width membership functions described below, which one expresses conservatism, pessimism, precision, or high standards? What is the maximum discrepancy (i.e., maximum level of possibility deviation given the same input) point on the universe of discourse?

$$A(x) = 0/0 \cup 0.2/2 \cup 0.5/4 \cup 1.0/6 \cup 0.5/8 \cup 0.2/10 \cup 0/12$$
$$B(x) = 0/0 \cup 0.4/2 \cup 0.8/4 \cup 1.0/6 \cup 0.8/8 \cup 0.4/10 \cup 0/12$$

6.5 Considering a rule such as IF X is A THEN Y is A, where A is

$$A(x) = 0/0 \cup 0.5/1 \cup 1.0/2 \cup 1.0/3 \cup 0.5/4 \cup 0.0/5$$

what would be the ratio of the effects produced by modifying this membership function to

$$A'(x) = 0/0 \cup 0.33/1 \cup 0.66/2 \cup 1.0/3 \cup 0.5/4 \cup 0.0/5$$

in antecedent domain versus in consequent domain? (Hint: Apply inputs that cause maximum deviation; use the Mamdani implication operator.)

6.6 When a fulfillment table is provided to the designer by experts, the overlap between the membership functions is already fixed. Given the overlap $\theta = 0.7$, construct a fulfillment table and design three adjacent membership functions on the universe of discourse [0–1]. Is it possible to repeat this process for $\theta = 0.75$ while not changing the locations of the crisp points on the same universe? (Hint: Consider triangular and trapezoidal shapes.)

6.7 What is the minimum possibility (DOF) value for the overlap index $\theta = 0.75$ between any two membership functions (piece-wise-linear with heights equal to 1.0)?

Chapter 7

Composing Fuzzy Rules

> Starting from the basic logic operators, this chapter describes rule composition techniques and provides several examples. Rule formation per inference type, rule composition strategies, and paradoxical cases are among the discussions included in this chapter. The final topic is the effects of membership function design in relation to rule composition.

7.1 INTRODUCTION

Composing fuzzy rules is a process natural to every human brain, because the underlying IF-THEN structure is part of our daily language and natural logic. Fuzzy system design in this context assumes that the rules are already determined by experts or are easily extractable from a data set by examination. The title of this chapter, "Composing Fuzzy Rules," mostly refers to the translation of fuzzy rules into the calculus of the basic fuzzy inference algorithm and to some organizational matters that define design strategies.

We begin this chapter with the design of logic operators that are used in composite rules. The two most commonly employed operators, AND and OR, are actually not designable because their roles are well established. We have also included other—rather experimental—logic operators that can be designed or selected as a design choice. Next are the basic forms of rule composition suitable for the basic fuzzy inference algorithm. The section titled "Rule Composition Strategies" describes some of the most commonly encountered methods of rule organization for a given or implied objective. Included are the competitive, cooperative, weighted, prioritized, hierarchical, and adaptive rule formation strategies. Following this section, a few practical issues are discussed, such as how to handle a large amount of evidence in systems sensitive to the amount of evidence and outliers. The last part of this chapter includes discussions related to membership function design and the interaction between rule formation and membership functions. These discussions are centered around the possible unexpected consequences emerging from the formation of composite rules.

7.2 BASIC LOGIC OPERATORS

The basic logic operators are also referred to as aggregation, algebraic, compensatory, or composition operators in the literature. The underlying mechanism is to compute the outcome of a logical proposition that includes more than one truth (membership) value. There are two basic logic operators: AND and OR. In Boolean logic, they correspond to intersection and union operations. In fuzzy set theory, they also correspond to intersection and union operations; however, their algebraic formulation is not unique due to a theoretically infinite number of possibilities when operating among fuzzy sets. Mathematicians have formulated the triangular norm (briefly *t-norm*) and triangular conorm (also known as *s-norm*) concepts through which different algebraic formulations of AND and OR can be derived [1]. Basic logic operators AND and OR take the NAND and NOR forms using the negation operator NOT. There are several operators in the fuzzy logic literature; however, we will only elaborate on some of the most commonly deployed logic operators along with some experimental ones.

7.2.1 AND Operator

The AND logic operator corresponds to the intersection operation in fuzzy set theory and is formulated as the minimum operation between the truth (membership) values. The result of AND logic via the min operation between two truth values is shown in Fig. 7.1.

Chap. 7 Composing Fuzzy Rules

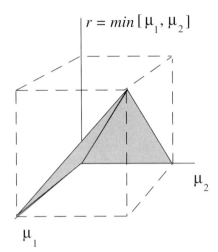

Figure 7.1 AND logic operator implemented as intersection via the min operation.

7.2.2 Product AND Operator

Formulation of AND as the product of truth values is used in various applications, and it produces an effect different from that of the standard AND operator. The result of product AND logic between two truth values is shown in Fig. 7.2.

Comparing Figs. 7.1 and 7.2, we see that the product operation yields a conservative AND effect in the sense that the fulfillment of both conditions for low truth values is suppressed. For example, consider the following fuzzy rule:

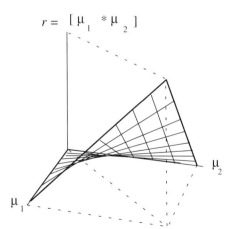

Figure 7.2 Product AND operation.

IF Weather IS Cold AND Wind IS High THEN We WILL Stay_home

Assume that the truth values of *Weather IS Cold* and *Wind IS High* are 0.5 and 0.3 at a given moment, respectively. Using the standard AND operator, the degree of fulfillment of the left-hand-side conditions is 0.3. Using the product AND operator, the degree of fulfillment is 0.15, half of the previous case. A lower level of fulfillment results in a weaker justification for us staying home. Therefore, the product AND operation is conservative in that it requires a more strict fulfillment of the conditions when firing the corresponding rules.

7.2.3 OR Operator

The OR logic operator corresponds to union operation in fuzzy set theory and is formulated as the maximum operation between the truth (membership) values. The result of *OR* logic via max operation between two truth values is shown in Fig. 7.3.

7.2.4 Yager Operators

Intersection and union operations are formulated by Yager [2] in a parameter-dependent manner in which the noncrisp evaluations of these operators can be changed. The intersection formulation is shown in

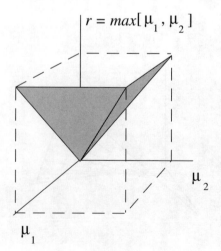

Figure 7.3 OR logic operator implemented as union via the max operation.

Fig. 7.4. The Yager's AND (intersection) operator pushes the diagonal intersection towards higher truth values and inhibits operations between low truth values. This is similar to product AND operation; however, the conservatism can be adjusted by k.

$$r = 1 - min\{1, ((1-\mu_1)^k + (1-\mu_2)^k)^{1/k}\}$$

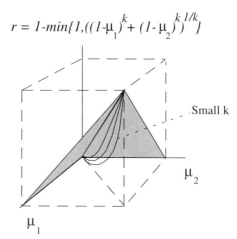

Figure 7.4 Yager's formulation of the intersection operator.

Similarly, union operation is expressed in terms of a class transformation parameter k as shown in Fig. 7.5. Just like the intersection, the union expression, when used as an OR logic operator, cannot be easily associated with a linguistic equivalent, and the adjustment of k is rather empirical.

$$r = min\{1, (\mu_1^k + \mu_2^k)^{1/k}\}$$

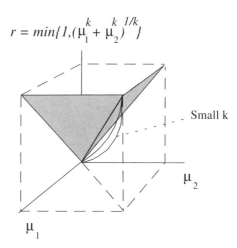

Figure 7.5 Yager's formulation of the union operator.

7.2.5 Consistency Operator

This operator functions in a unique manner, as shown in Fig. 7.6. There are different versions of this type of operation in the literature. One suggested by Dubois and Prade uses product AND instead of standard AND in addition to a class parameter k in the denominator [3]. The basic idea behind dividing minimum by maximum is that the outcome is always 1.0 when both truth values are equal regardless of how small or large they are. In other words, this operator searches for equality between two (or more) truth values. Therefore, its linguistic equivalence is consistency among fuzzy statements. For example, let's consider the following fuzzy statements:

> *Woman's_legs ARE Long*
> *Woman's_arms ARE Long*

Assume that woman's measurements yield truth (membership) values of 0.8 and 0.8 for the membership functions *Long_legs* and *Long_arms*, respectively. Now let's apply the consistency operator between these two fuzzy statements in a hypothetical fuzzy rule:

IF Woman's_legs ARE Long CONSISTENT_WITH Woman's_arms ARE Long THEN ...

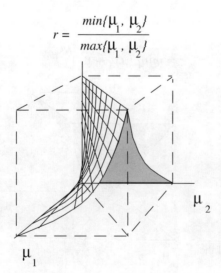

$$r = \frac{min\{\mu_1, \mu_2\}}{max\{\mu_1, \mu_2\}}$$

Figure 7.6 Consistency operator.

Min{0.8, 0.8}/Max{0.8, 0.8} yields 1.0 meaning that the two fuzzy statements are consistent. Note that we would have the same result if the truth values were 0.3, 0.3 or 0.0 , 0.0 for which, from the consistency point of view, the result is correct. As one can easily infer, the consistency operator checks the membership function design in a collective manner; that is, it checks the consistency between two or more membership functions that are logically bound to be consistent with each other (see Section 7.2.2 for more information on this operator and its use). Note that logical consistency is not a general requirement, but it may be a specific one depending on the problem.

7.2.6 Mean Operator

Mean, interpolation, or average between two or more truth values yields an outcome that defines a geometrical middle surface between the intersection and union surfaces. This is shown in Fig. 7.7.

It is not easy to find a linguistic equivalence for this operator because it represents the midpoint between AND and OR logic. However, it may be considered as a tolerant AND operator or a conservative OR operator. In some particular contexts, the mean operator can be viewed as the AND/OR logic. More discussion about this operator is included in the next section.

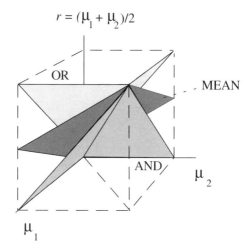

Figure 7.7 Mean operator.

7.3 LOGIC OPERATOR DESIGN ISSUES

Logic operators, when considering the general category of aggregation operators found in the literature, can be formed in a number of ways, some of which are described in the previous section. As suggested in the literature, a set of design criteria can be applied in general terms [4]. These include the following:

- *Axiomatic strength:* The less limiting the axioms it satisfies, the better an operator is.
- *Emprical fit:* An operator must be appropriate for emprical testing.
- *Adaptability:* Referring to parametrization, an operator must have the ability to change its behavior by an adjustable parameter.
- *Compensation:* An operator is preferred if its outcome can be reformed by different combinations of its arguments. For example, a min operator is not compensatory since its outcome cannot be reformed by more than one combination.
- *Susceptibility:* An operator must be susceptible to the number of its arguments if each argument is a piece of evidence, and if the amount of evidence is a variable of interest.

We will see in this section that the general criteria established for aggregation operators are applicable to the design of logic operators in a practical way. In the context of the basic fuzzy inference algorithm, design means selecting appropriate logic operators from those listed previously or from those not included in this book but reported in the literature. Let's first clarify how designing logic operators affects the performance of a fuzzy inference engine. Consider the generalized rule formation shown below.

$$X_i \bullet P_{i,x} \ \Theta \ X_j \bullet P_{j,x} \ \Theta \ldots \Theta \ X_k \bullet P_{k,x} \rightarrow \ldots \quad (7.1)$$

Fuzzy variables are denoted by X and predicates by P with subscripts indicating infinite degrees of freedom in forming a rule. The logic operators denoted by Θ are the same kind and there is no nested structure for simplicity. Because logic operators are used in the composition of conditions, we are only concerned with the left-hand side of each rule in the rule base. Logic operators appearing on the right-hand side with respect to a THEN operator are treated differently, as explained in Chapter 8. Let's recollect some of the concepts in the form of a design principle.

> Logic operator design determines the degree of fulfillment (DOF) of the left-hand-side conditions. DOF is a measure of how much the corresponding rule will fire, how strongly its implication result will be, or how strongly its right-hand-side actions will contribute to the aggregated result.

The application of the linguistic design criteria discussed in Chapter 4 to logic operator design does not require any special treatment. That is, if conservatism is to be exercised, logic operators will be selected accordingly. However, incorporating conservatism, for example, by means of logic operator design creates a more profound effect than that created by means of membership function design. This is obvious in Eq. (7.1) such that single design selection for Θ will affect N fuzzy statements at once. Accordingly, applying a design criterion by designing logic operators is a more practical (short cut) avenue than modifying each membership function. However, we must remember that applying conservatism to all the elements of a rule yields a somewhat uniform effect, which is often dysfunctional. This was the opening argument in Section 6.1. The fuzzy system designer may choose this avenue to distribute different nuances among the rules at the expense of designing different logic operators effective in each rule.

> Some of the linguistic design criteria can be applied to the design of logic operators instead of membership functions for simplicity. This practice is meaningful only when a linguistic design criterion is applied to an entire rule to assign a different importance relative to the other rules.

This is not an easy principle to implement because membership function modification is intuitive (i.e., involves interpretation of possibilities) whereas logic operator design is more like a trial-and-error process. Besides, distribution of nuances among the rules can be accomplished by importance weights in a relatively easier way.

7.3.1 Designing Product AND and Yager AND Operators

Let's elaborate on designing linguistically nonstandard logic operators starting from the product AND operator. The implementation of this operator can be tricky. For example, if there is a rule with many left-hand-side conditions, this logic operator will eventually diminish the DOF value by multiplying several real numbers smaller than 1.0. This is explained by Example 7.1.

EXAMPLE 7.1 PRODUCT AND VERSUS STANDARD AND

Let's consider a simple case using three fuzzy statements. Assume that the three fuzzy statements shown below are connected by the AND operator, and the truth values for a given input set are 0.5, 0.7, and 0.3.

OPERATORS	LHS OF A COMPOSITE RULE					DOF
	IF TEMPERATURE IS LOW AND FLOW IS MEDIUM AND HEAT IS HIGH THEN . . .					
AND	0.5	Min	0.7	Min	0.3	0.3
Product AND	0.5	*	0.7	*	0.3	0.105

Between the two different formulations of the intersection (AND) operation, we see that DOF results can be significantly different. In fact, if the number of fuzzy statements becomes higher than three above, we will get DOF values even less than 0.105 for any truth value less than 1.0.

This leads to the following design principle, which is valid only for noncrisp truth values (i.e., the region in which fuzzy inference engines operate most of the time).

> The product AND operator is sensitive to the number of fuzzy conditions on the premises such that a high number of fuzzy conditions inhibits the firing capability of the corresponding rule. Thus, this operator can be used when the outcome of fuzzy inference is desired to be sensitive to the degree of fuzziness characterized only by the *number* of fuzzy conditions.

Let's iterate this principle more from the rule importance point of view. Consider a case where the strength of a decision depends on the number of conditions to be met. Assume that you are buying a house and the conditions you impose are many (i.e., nice scenery, high ceilings, large garage, etc.). You will have a more difficult time finding such a house compared with someone else who is only looking for a shelter from rain and cold. The standard AND operator will perform its expected function in both cases, however it will settle on the minimum condition met. The product AND operator will be much more conservative and it will reflect the fact that you are more picky than the other person because you impose more conditions. It will settle on the combination of partially satisfied conditions always more suppressed than the minimum condition met.

The Yager AND operator functions in a similar manner to that of the product AND but with an additional capability of adjustment by the class transformation parameter k. Compared to the product AND operator, it is even more difficult to associate its mathematical function to a semantic criterion. Smaller k values produce a more strict or conservative DOF value. The Yager AND operator is also sensitive to the number of fuzzy conditions as the number of truth values produce a cumulative effect.

7.3.2 Designing the Consistency Operator

Consistency-based operations where minimum truth is divided by maximum truth can be useful during the design stage to improve the membership function allocation and shapes. This mode of utilization requires a set of data representing the expected domain of operations. The functionality of this type of logic operator is described in Example 7.2.

EXAMPLE 7.2 CONSISTENCY OPERATOR

Let's return to the previous case study, which includes the following simple fuzzy rule. Suppose that this fuzzy rule will be used in an inference engine with the AND logic operator. We first try the consistency operator (denoted as CONSISTENT_WITH) to test the fuzzy rule.

IF Woman's_legs ARE Long CONSISTENT_WITH Woman's_arms ARE Long THEN ....

Assume that we have membership functions for long legs and long arms as shown below. The question is whether these designs are adequate. Our criterion to make such a decision is to test the rule by a set of input data and make sure that this rule fires significantly at least once.

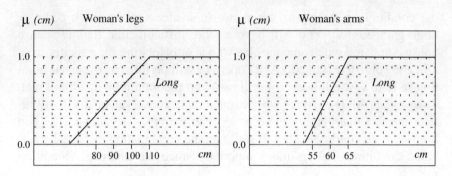

Now suppose that we have a set of field data that represents the domain of operations.

DATA SET	LEGS	ARMS
1	90 cm	58 cm
2	108 cm	61 cm
3	78 cm	58 cm
4	82 cm	60 cm
5	103 cm	65 cm

Evaluating the rule by each data set (1,2,...,5) produces the following truth values.

DATA SET	μ LONG-LEGS	μ LONG-ARMS
1	0.55	0.45
2	0.98	0.63
3	0.3	0.45
4	0.34	0.6
5	0.81	1.0

Now applying the consistency operator between them produces the following DOF values.

DATA SET	CONSISTENCY	DOF
1	0.45/0.55	0.81
2	0.63/0.98	0.64
3	0.30/0.45	0.66
4	0.34/0.60	0.56
5	0.81/1.00	0.81

The DOF values range from 0.56 to 0.81. Now let's change the membership function design for long arms and apply the same procedure.

Chap. 7 Composing Fuzzy Rules

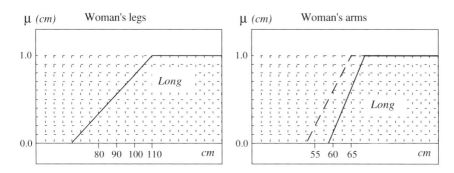

The new truth values for the same input data set are shown below.

DATA SET	μ LONG-LEGS	μ LONG-ARMS
1	0.55	0.01
2	0.98	0.22
3	0.3	0.01
4	0.34	0.20
5	0.81	0.60

The new DOF values are shown below.

DATA SET	CONSISTENCY	DOF
1	0.01/0.55	0.018
2	0.22/0.98	0.22
3	0.01/0.18	0.055
4	0.20/0.22	0.90
5	0.60/0.81	0.74

Compared to the previous range of DOF values, we realize that the new membership function is not only less effective, but also not responsive to 40 percent of the data. In other words, if this rule is used in a fuzzy inference engine with an AND operator, it would not significantly fire 40 percent of the time (for the representative data set).

In this example, we assumed that the membership functions were developed linguistically (without knowledge extraction from data) and the data set is used for testing only. As one can infer from this example, the consistency operator can be used systematically to allocate membership functions in their most suitable locations consistent with the semantic reasoning behind the fuzzy rule.

The most drastic case to be avoided by this procedure is that the expert mistakenly articulates a rule using the AND operator (and designs antecedent membership functions) that will never fire significantly in actual implementation because the two (or more) truth values will never have significantly large values simultaneously. The practicality of this argument can be understood more easily when we consider more complex fuzzy conditions (i.e., conditions that are not easily understood using common sense) than that of a woman's arms and legs. For example, in a rule with fuzzy statements such as *Temperature IS High*, and *Flow IS Medium*, it becomes quite important not to compose rules using the AND operator that will never fire (or never fire significantly). If a rule never fires during actual implementation, we call it inconsistent and the design is poor.

One possible extrapolation of the ongoing argument is to use the consistency operator in an adaptive fashion. In such a design, every rule composed with the AND operator will have its equivalent with the consistency operator.

IF X IS A AND Y IS B THEN Z IS C
IF X IS A CON Y IS B THEN A IS Good AND B IS Good

The consequent variables in the second rule qualify how good the membership functions are. By keeping a record for every specified interval of operations, we create a training data set. Then, a best fit algorithm can be devised (either in an exhaustive or directional manner) to allocate membership functions based on the training data set.

7.3.3 Designing the Mean Operator

The mean operator is applicable to cases where fuzzy rules include AND/OR type reasoning. For example, consider the following rule:

IF Leaves_on_trees ARE Brownish AND/OR Leaves_on_the_ground ARE Plenty THEN It IS Autumn.

It is not absolutely true that autumn has arrived when only one of the conditions are satisfied (OR operation). On the other hand, it is not absolutely necessary to have both conditions satisfied (AND operation) to declare autumn. This example illustrates that fuzzy logic does not always refer to logic of fuzzy measurements, but it may also refer to fuzziness in the way measurements are aggregated. The mean operator can be used to express this relaxed state of logic among the fuzzy conditions on the premises.

Chap. 7 Composing Fuzzy Rules

EXAMPLE 7.3 COMPARISON OF AND, OR, AND MEAN OPERATORS

Let's consider the previous example by completing the rule using the mean operator.

IF Woman's_legs ARE Long AND/OR Woman's_arms ARE Long THEN Her_size IS
(possibly) Large

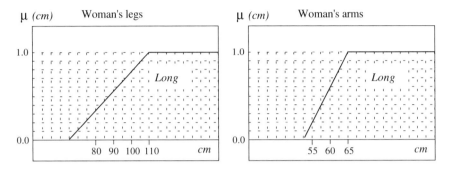

Now suppose that we have a set of field data that represents the domain of operations.

DATA SET	LEGS	ARMS
1	90 cm	58 cm
2	108 cm	61 cm
3	78 cm	58 cm
4	82 cm	60 cm
5	103 cm	65 cm

Evaluating the rule by each data set (1,2,...,5) produces the following truth values.

DATA SET	μ LONG-LEGS	μ LONG-ARMS
1	0.55	0.45
2	0.98	0.63
3	0.30	0.45
4	0.34	0.60
5	0.81	1.00

Now applying the AND, OR, and MEAN operators between them produces the following DOF values.

DATA SET	AND	OR	MEAN
1	0.45	0.55	0.50
2	0.63	0.98	0.80
3	0.30	0.45	0.37
4	0.34	0.60	0.47
5	0.81	1.00	0.90

As illustrated in this example, there are three sets of DOF values that will produce three sets of outputs yielding different levels of commensurateness to size *Large*. Because the underlying logic is fuzzy, the fuzzy system designer may employ the mean operator as an equal distance choice to each extreme logic. In each case, the fifth data set is computed to be the best description of size *Large* and the third data set is the worst one. Therefore, the mean operator retains the upper and lower bound selections of AND and OR operators, but at different DOF levels.

7.4   RULE FORMATION PER INFERENCE TYPE

We will present rule formation in this section for different inferencing schemes described earlier in Chapter 2. The three types of inference schemes listed below cover most of the practical decision-making problems in engineering, business, medicine, law, and so forth. The formation of rules is not determined based on the problem type (e.g., control, estimation, prediction, forecasting, classification, patterns recognition, modeling, etc.). Rule formation depends on the nature of the problem at hand as explained in the following sections. Also note that all mathematical equivalencies represented by fuzzy rules are—in a way—like models of reasoning useful for the solution of practical problems, but are not models that explain approximation skills of the human brain.

7.4.1   Relational Inference

We have defined relational inference in Chapter 2 to be the most straightforward form of fuzzy inference that consists of associations between the elements of two or more fuzzy sets. This basic structure, which defines a fuzzy relation, allows the development of higher-level inference mechanisms such as the compositional inference. It also constitutes the basic building block of inference mechanisms developed analytically, without linguistic articulation of expertise in the form of conditional IF-THEN rules. Analytically derived inference mechanisms are many in numbers and are somewhat problem dependent. One example is presented at the end of this section.

Relational inference is often expressed by unconditional rules or fuzzy statements such as *Fast cars ARE Expensive* or *X IS In_the_vicin-*

ity_of_Y. Normally, relational inference is linguistically monotonic. In other words, there is often a set of phrases repeated by changing the objects of associations or the nature of associations such as:

X IS In_the_vicinity_of_Y
W IS Very_close_to_Z
X IS Quite_similar_to_Z

There are no linguistically composed statements using logic operators. In addition, there are no direction assignments—often implied by conditional rules such as *IF A THEN B*—that allow forward and backward computations through the same representation. Relational fuzzy inference is often represented by the directed fuzzy graph method, as shown in Fig. 7.8. In this figure, the elements of fuzzy set X are connected to the elements of fuzzy set Y by different grades of elementhood often characterized by an expert. The connections are equivalent to an *N*-dimensional membership function. The formation of unconditional rules, fuzzy phrases, or directed graphs is based on collecting all necessary relationships describing the problem at hand. Almost every aspect of the knowledge acquisition process applies to relational fuzzy inference.

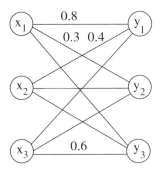

Figure 7.8 Directed fuzzy graph representation.

EXAMPLE 7.4 RELATIONAL INFERENCE FOR DIAGNOSTICS

The best example for the use of relational inference is a diagnostics problem where a set of faults, symptoms, reasons, and solutions are related to each other with different degrees of association. Let's consider a typical car-diagnostics problem as shown below. Boxes represent objects (fuzzy or crisp) that are connected to related categories.

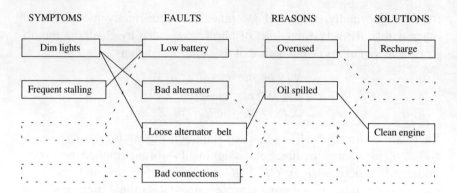

To examine relational inference, we must zoom in on one of the connections. Consider the *Dim_lights* and *Low_battery* pair. Because the degrees to which "lights are dim" and "battery is low" are fuzzy concepts, the relationship between two fuzzy sets is described by a two-dimensional membership function on the Cartesian product space as shown below. Obviously, zero light intensity corresponds to zero power battery.

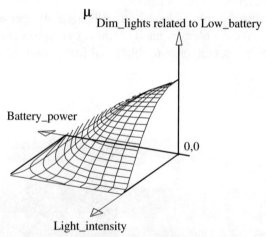

Given a *Light_intensity* value, this relationship yields a family of *Battery_power* values each corresponding to a different possibility value (i.e., a new fuzzy set) or vice versa. When both measurements are available, we have a single membership value describing the level of agreement or similarity. Considering the entire diagnostics problem domain, the space of operation is described by many surfaces (fuzzy relations) combined together from one end to another. Considering only the first two columns of the diagnostics problem, we define symptoms and faults as fuzzy sets:

Faults: F = { Low Battery, Bad Alternator, Loose Belt, Bad Connections, ... }
= { $Fi\,(i = 1,..,m)$ }
Symptoms: S = { Dim Lights, Frequent Stalling, ... }
= { $Sj\,(j = 1,..,n)$ }

Chap. 7 Composing Fuzzy Rules

A typical diagnostics algorithm using both the *modus ponens* and *modus tollens* is expressed by

$$(R_{i,j} \, oF_i) \to S_j$$
$$S_j \to \vee_i (R_{i,j} \, oF_i)$$

where R is the relational inference expressed by a unconditional rule such as *Fault F is casually related to symptom S*. When fuzzy relations are represented by fuzzy sets, the inference computation reduces to

$$(R_{i,j}^\alpha \wedge F_i^\alpha)_u \equiv (S_i^\alpha)_u + (\neg P_{i,j}^\alpha)_u \wedge 1 \quad \forall \alpha \in [0,1]$$
$$\vee_i (R_{i,j}^\alpha \wedge F_i^\alpha)_l \equiv (S_j^\alpha)_l - (\neg P_j^\alpha)_u \vee 0$$

where $\neg$ means negation, P_{ij} is the certainty level of *modus ponens*, P_j is the certainty level of *modus tollens*, and α is the alpha cut determining the upper (u) and lower (l) bounds for certainty levels. By finding a common solution between these two equations, the fault F is obtained. This requires the availability of observed symptoms S, fuzzy relation R_{ij}, and the knowledge of certainty levels subject to alpha cuts. The heart of the problem is the formation of certainty levels in the form of membership functions and the determination of a set of α values that guarantees the existence of at least one solution. Different algorithms exist (especially in the medical diagnostics field) each of which operates based on context-related assumptions.

7.4.2 Proportional Inference

Also called *monotonic reasoning*, the inference mechanism is basically the same as relational inference, in which the entire problem domain is defined by a surface topology. The difference between relational and proportional inference is that the linguistic articulation of the problem in terms of complex conditional rules is possible. In its simplest form, an output is obtained by applying the degree-of-fulfillment level of the conditions to the decomposition of the output fuzzy set without an implication process. Let's consider the following rule:

$$X \bullet P_x \to Y \bullet P_y \tag{7.2}$$

As shown in Fig. 7.9, an input value of x produces a scalar output y over the membership function contours. Obviously, the output domain membership functions are restricted to having monotonically increasing or decreasing shapes to avoid multiple answers, or there should be a selection criterion. Another restriction is that each consequent cannot be used in more than one rule, otherwise there has to be a mechanism to reduce mul-

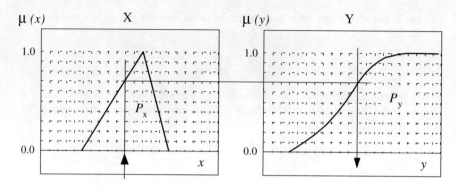

Figure 7.9 Proportional inference.

tiple scalar outputs to one answer. Note that a collection of already decomposed outputs does not constitute an aggregation problem as defined in the fuzzy logic literature because aggregation is mainly defined in the possibility domain between fuzzy sets. In the case of multiple scalar outputs, the fuzzy system designer may use one of the classical expert system solutions such as the Dempster-Shafer's rules of evidence method [5, 6].

There is no special requirement for the formation of conditional fuzzy rules except that multiple use of consequents will necessitate the treatment of multiple scalar outputs, as explained before. All aspects of the knowledge acquisition process discussed in Chapter 4 apply. For applications in which rules are formed based on knowledge extraction from data, this process is identical to inverse mapping for crisp points. Between the crisp points, the proportional inference is a possibilistic interpolation method—interpolation surfaces defined by possibility rather than employing straight line linear interpolation.

Although simple in terms of the implication process, which is decomposition of the output fuzzy set by the degrees of fulfillment, we cannot judge how suitable this approach is when applied to complex problems or to different problem types. This is simply because the single decomposed output will map as intended; thus, it cannot be a measure of performance. Its suitability is therefore measured by whether the problem at hand can be represented easily by the overall approach or not. The fuzzy system designer will chose this approach based on the limitations described here.

> As a method of direct mapping, proportional fuzzy inference may be used provided that the problem at hand suggests (1) contributions to each consequent come from one rule only, and (2) all consequent membership functions are monotonically increasing or decreasing.

Although not very common, there are a few applications of this type of inferencing especially in the business world. A typical example is the fuzzy financial risk assessment system [6].

7.4.3 Compositional Inference

Compositional inference is the most generalized form of fuzzy inferencing that allows linguistic expression for the articulation of the problem at hand. Mathematically speaking, it was shown in Chapter 2 that composability requires the identification of relationships among fuzzy sets as well as the relationships between the elements of each fuzzy set. In more practical terms, composition must connect individual relational inferences through a linguistic framework. The inference mechanism includes composition operators such as max-min or max-star, an implication process such as Mamdani or Zadeh, aggregation, and a defuzzification method such as the centroid method. The inference structure based on the generalized *modus ponens*, which constitutes the essence of the basic fuzzy inference algorithm, is summarized in Fig. 7.10.

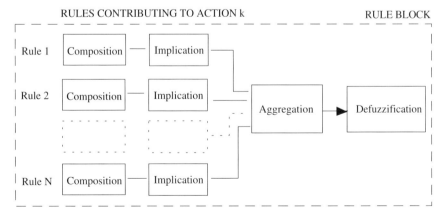

Figure 7.10 Compositional inference.

Remember the difference between the composition operators (max-min, max-*) and logic operators (AND, OR). Also note that composing rules means a creation process just like composing music, whereas compositional rule of inference refers to the structure shown in Fig. 7.10. This figure only shows a segment of a typical rule base; that is, rules contributing to one consequent fuzzy variable (action), which is called a rule block. A collection of rule blocks define a rule base and the fuzzy inference algorithm. The symbolism corresponding to a rule block is given below.

$$(X_i \bullet P_{i,j}) \Theta (X_k \bullet P_{k,l}) \Theta \ldots \Theta (X_m \bullet P_{m,n}) \rightarrow Y_1 \bullet P_{1,a}$$
$$\ldots \qquad\qquad \ldots \qquad\qquad \rightarrow Y_1 \bullet P_{1,a}$$
$$\ldots$$
$$(X_I \bullet P_{I,J}) \Theta (X_K \bullet P_{K,L}) \Theta \ldots \Theta (X_M \bullet P_{M,N}) \rightarrow Y_1 \bullet P_{1,D}$$
(7.3)

The subscripts on the left-hand side imply an infinite degree of freedom in selecting the antecedent fuzzy variables of the problem. However, the consequent variable on the right-hand side is fixed. The consequent predicates $a, b, \ldots, D$ do not have to be equally distributed and they can repeat. A rule block is linguistically expressed as shown below.

IF Current IS High AND Flow IS Low AND .... AND Pressure IS Medium
THEN Valve IS Open
.... *THEN Valve IS Half*
.... *......*
IF Current IS Medium AND Flow IS Low AND .. AND Pressure IS High
THEN Valve IS Closed

Many variations exist in forming the left-hand side of each rule. This is explained next.

Formation of the Individual Rules

There are two distinct methods for conditional rule formation. The first one, which is entirely based on context-specific expert opinion or common sense, takes variety of forms depending on the strategy and logic. Strategic choices are explained in Section 7.5. We will examine a few possible formations here. A rule is called homogeneous if all the logic operators are of the same kind. This is illustrated in Fig. 7.11 using a fuzzy AND gate (min operation).

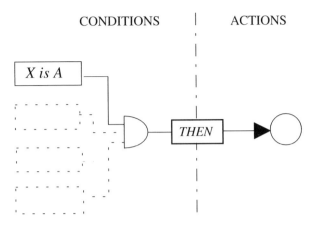

Figure 7.11 A homogeneous rule with fuzzy AND operator.

The linguistic example given for Eq. (7.3) consists of homogeneous rules. A nonhomogeneous rule is the one with mixed logic operators arranged in a nested structure. An example is shown in Fig. 7.12 using both fuzzy AND and OR operators. Nested structures (or groups) are denoted by extra brackets in the equations for simplicity.

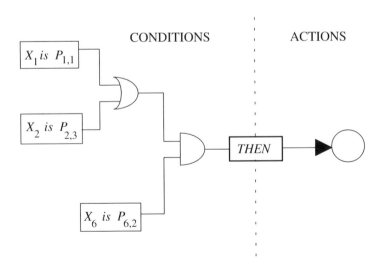

Figure 7.12 A nonhomogeneous rule with fuzzy AND and OR logic operators.

For example, the left-hand side of the rule in Fig. 7.12 is represented by

$$((X_1 \bullet P_{1,1}) \Theta_{max} (X_2 \bullet P_{2,3})) \Theta_{mim} (X_6 \bullet P_{6,2}) \rightarrow \ldots \quad (7.4)$$
IF (Current IS High OR Flow IS Low) AND Pressure IS Low THEN ....

where the grouping of max and min operations are distinguished by an extra pair of brackets. Such brackets have no meaning in fuzzy logic, but they are meaningful for fuzzy algorithms. In the example above, the extra bracket ensures that the min operation will not be performed between X_2 and X_6. When grouping different operations there is one principle to remember that has roots in classical logic and has also been seen in conventional programming languages; that is, putting different operators in the same group yields an ambiguous proposition. Removing the extra bracket in Eq. (7.4) would result in such a case.

The right-hand-side rule structure may include additional THEN statements as shown below.

$$X_4 \bullet P_{4,3} \rightarrow ((X_1 \bullet P_{1,1}) \Theta_{max} (X_2 \bullet P_{2,3})) \Theta_{min} (X_6 \bullet P_{6,2}) \rightarrow \ldots \quad (7.5)$$
IF System IS Running_slow THEN
   IF (Current IS High OR Flow IS Low) AND Pressure IS Low THEN ....

Normally, the right-most conditional proposition is considered to be the basis of the rule. This is shown above as the second line. The first statement functions like a navigator in the sense that if the first condition is not satisfied, the rest of the rule may not be evaluated, as shown in Fig. 7.13.

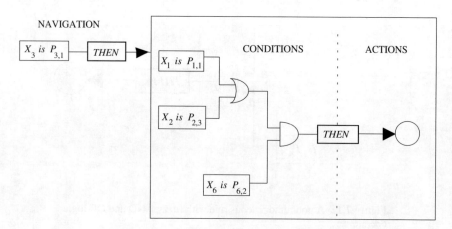

Figure 7.13 Navigation among the rules by additional THEN statements.

This is easily done in crisp logic in which the navigating condition (also called Boolean flag) is either true or false. The first THEN operator can also be replaced by the AND operator.

In fuzzy logic, however, navigation requires additional considerations due to partial truth. There are two options the fuzzy system designer may consider. The first option is to use a threshold on the navigational condition such that if the condition is not satisfied above a certain value then the rest of the rule is not evaluated. The second option is to treat the degree of fulfillment of the navigational condition as an importance weight to be incorporated during the implication process. Both approaches affect the results in a similar (but not identical) way.

EXAMPLE 7.5 HOMOGENOUS, NONHOMOGENOUS RULES, AND NAVIGATION

As an example to rule formation based on the articulation of expert knowledge, we will examine a decision-making problem from the social sciences. In a group home for behaviorally disturbed developmentally disabled adolescents, the staff must decide a course of action in response to a wide variety of problematic behaviors. An incorrect response may reinforce the incorrect behavior, unnecessarily infringe upon the rights of the individual, or result in injury. A fuzzy logic system is considered to simulate possible disturbance scenarios to adjust policies and regulations in addition to providing a training tool for the new members of the response team. By the articulation of the expert, the overall system has the following structure:

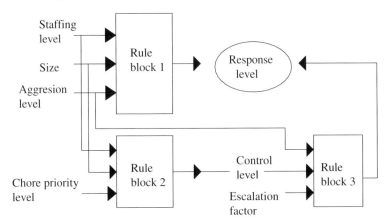

Most of the variables of the problem are fuzzy and can only be measured by human interpretation. Let's examine them briefly. Staffing level and size are the only two fuzzy variables that can be numerically measured. Staffing level means the number of personnel at the site of the disturbance; size refers to physical stature. Aggression level, chore priority level, and escalation factors are entirely linguistic variables and are quantified by a normalized universe. Control level indicates how much the situation is under control by the staff, whereas the response level determines how much to interfere. After the knowledge acquisition process, the variables are designed in the following manner.

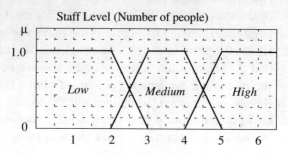

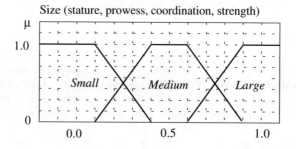

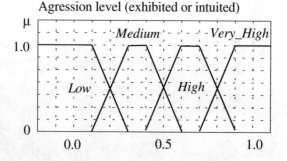

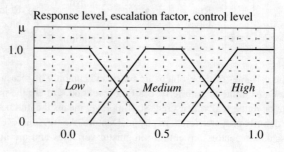

The key factor is the translation of the predicates *Low, Small, Medium, High*, and so on into behavior measures and vice versa. Such measures have been studied in social science. A book written by George Edmunds [7] was used to quantify linguistic variables and to represent them numerically. For example, the determination of aggression level uses the following dictionary:

BEHAVIOR	LEVEL	AGGRESSION CATEGORY
Calm, silent	0.1	Low
Calm, articulate	0.2	Low–medium
Nervous, articulate	0.3	Medium
Nervous, no speech	0.4	Medium
Nervous, bad speech	0.5	Medium–high
Physical restlessness	0.6	High
Physical threats	0.7	High
Holds back aggression	0.8	High–very high
Mild aggression	0.9	Very high
Strong aggression	1.0	Very high

Symptoms in the list above are combined with an OR operator, so the maximum aggression level is obtained from multiple symptoms. Other linguistic variables are treated in the same manner. The number of fuzzy rules in this particular example is too high to examine them all. Three segments from the rule base are shown below to illustrate the variations employed in this system.

IF Staff_level IS High AND Aggression_level IS Low THEN Response_level MUST BE Low

IF Staff_level IS High AND Aggression_level IS Medium THEN Response_level MUST BE Medium

IF Staff_level IS High AND Aggression_level IS High THEN Response_level MUST BE High

....

IF Escalation_Level IS Low THEN

IF Staff_level IS High AND Aggression_level IS High THEN Response_level CAN BE Medium

....

IF Staff_level IS Low AND {Size IS Small OR Size IS Medium} AND aggression_level IS Medium THEN Response_level IS Low

....

The first segment at the top is an example of homogeneous rules using the AND operator only. The second segment illustrates fuzzy navigation. The response level is reduced by the condition that the situation is not escalating. The third segment shows the use of the mixed logic operators AND and OR.

In this example, we have seen that compositional inference is applicable to situations where measurements are linguistic rather than numerical. Although a conversion table is utilized in this example, fuzzy inference is suitable for linguistic input entries as shown in Chapter 4. Second, we realize that design (rule formation) by knowledge acquisition is entirely driven by the expert in the field who designs fuzzy variables, then composes conditional rules. The granularity of the solution, which is determined by the number of membership functions per variable, determines the complexity of the rules. Third, the concept of rule blocks is clearly seen; this is discussed in the next section.

The second method of rule formation is governed by the process of knowledge extraction from data and is always composed by homogeneous rules using the AND logic operator. In contrast to the previous case, rule formation is pretty much fixed by data in the manner explained by the following example.

EXAMPLE 7.6 FORMATION OF THE CONTROL RULES FROM DATA

Suppose we are designing a tank liquid level controller and we have a data set of the variables of the problem including exemplified control actions. We will extract control knowledge from data and automate the control action by means of a fuzzy controller. The process, as shown below, is in thermodynamic equilibrium between two phases and any heat input will decrease the subcooled height (level) due to increased boiling rate (just like boiling water in a tea kettle). The control objective is to complete a heat increase maneuver while maintaining the level constant. Note that there is a constant liquid feed and the valve position for the outlet flow maintains mass balance at steady-state operation.

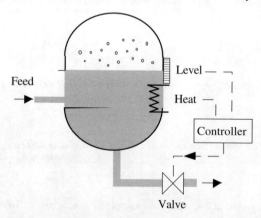

The data describing the intended operation is shown below. On the left, we have a transient profile with the horizontal axis representing the sampling time. Assume that this is the highest sampling rate from this process, and the points in between are not explicitly observed. On the right, we have the point membership functions for each data point developed by the fuzzy system designer. The horizontal axis is the universe of discourse, whereas the vertical axis is possibility. Membership functions are labeled by letters for simplicity, and later defined using proper linguistic terms during rule formation.

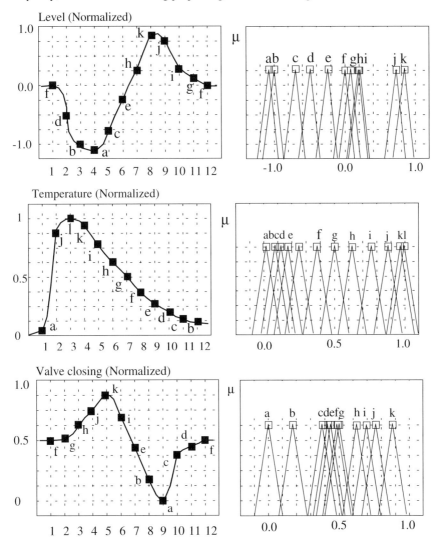

Note that the number of point membership functions in the first and last variables is 11 instead of 12 because some data points exactly coincide. Given the 12 data points, this is the most precise (highest granularity) clustering of the mapping spaces. Exhaustive rule formation is performed by considering each data point on the plots (left). Because *Valve* is the control variable, it ends up on the right-hand side after the THEN operator.

1- IF Level IS f AND Temperature IS a THEN Valve IS f
2- IF Level IS d AND Temperature IS j THEN Valve IS g
3- IF Level IS b AND Temperature IS l THEN Valve IS h
4- IF Level IS a AND Temperature IS k THEN Valve IS j
5- IF Level IS c AND Temperature IS i THEN Valve IS k
6- IF Level IS e AND Temperature IS h THEN Valve IS i
7- IF Level IS h AND Temperature IS g THEN Valve IS e
8- IF Level IS k AND Temperature IS f THEN Valve IS b
9- IF Level IS j AND Temperature IS e THEN Valve IS a
10- IF Level IS i AND Temperature IS d THEN Valve IS c
11- IF Level IS g AND Temperature IS c THEN Valve IS d
12- IF Level IS f AND Temperature IS b THEN Valve IS f

Note that letters of each fuzzy variable refer to different membership functions. For example *a* of *Temperature* is not the same as *a* of *Level*. This system will exactly map the crisp *Valve* points regardless of the choice of implication operators (see Chapter 8); however, the transition between crisp points will be determined by the design choices.

Now let's assume that the fuzzy system designer wants to simplify the rules by combining some of the point membership functions located very close to each other. Starting from *Level*, combining a and b yields new membership function m, and combining f, g, h, and i yields a new membership function n. Also notice that the overlaps between the membership functions m and c, and n and j are increased to avoid dead zones (see Chapter 5). Similar approximations are applied to *Temperature* and *Valve* as shown below.

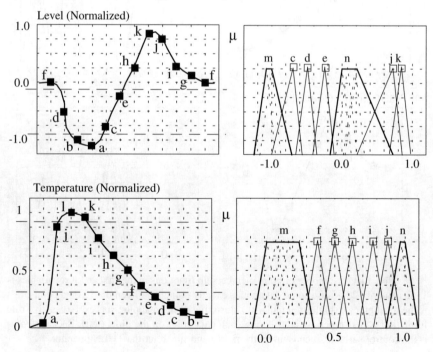

Chap. 7 Composing Fuzzy Rules

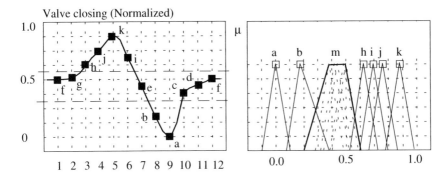

Now, new rules are formed in the same manner as before. However, a more simple method is to rewrite the rule block and replace the new membership function labels.

1- IF Level IS n AND Temperature IS m THEN Valve IS m
2- IF Level IS d AND Temperature IS j THEN Valve IS m
3- IF Level IS m AND Temperature IS n THEN Valve IS h
4- IF Level IS m AND Temperature IS n THEN Valve IS j
5- IF Level IS c AND Temperature IS i THEN Valve IS k
6- IF Level IS e AND Temperature IS h THEN Valve IS i
7- IF Level IS n AND Temperature IS g THEN Valve IS m
8- IF Level IS k AND Temperature IS f THEN Valve IS b
9- IF Level IS j AND Temperature IS m THEN Valve IS a
10- IF Level IS n AND Temperature IS m THEN Valve IS m
11- IF Level IS n AND Temperature IS m THEN Valve IS m
12- IF Level IS n AND Temperature IS m THEN Valve IS m

As seen above there are four identical rules: 1, 10, 11, and 12. So they can be reduced to one.

1- IF Level IS n AND Temperature IS m THEN Valve IS m
2- IF Level IS d AND Temperature IS j THEN Valve IS m
3- IF Level IS m AND Temperature IS n THEN Valve IS h
4- IF Level IS m AND Temperature IS n THEN Valve IS j
5- IF Level IS c AND Temperature IS i THEN Valve IS k
6- IF Level IS e AND Temperature IS h THEN Valve IS i
7- IF Level IS n AND Temperature IS g THEN Valve IS m
8- IF Level IS k AND Temperature IS f THEN Valve IS b
9- IF Level IS j AND Temperature IS m THEN Valve IS a

Rules 3 and 4 appear to conflict in Boolean logic; however they do not conflict in fuzzy logic (see Chapter 5). Rule formation is not finalized yet. The fuzzy system designer must simulate the system performance and evaluate the other design options employed. Once the performance is satisfactory, the rules can be finalized by assigning linguistic terms for the labels so that the underlying semantic reasoning becomes transparent:

LEVEL:	TEMPERATURE:	VALVE:
m - Extremely_negative	m - Very_low	a - Completely_open
c - High_negative	f - Low	b - Little_open
d - Medium_negative	g - Somewhat_low	m - Medium_open
e - Small_negative	h - Medium	h - About 60%_closed
n - Zero	i - Medium_high	i - About 70%_closed
j - Medium_positive	j - High	j - About 80%_closed
k - High_positive	n - Extremely_high	k - Almost_closed

The final form of the rules is:

1- IF Level IS Zero AND Temperature IS Very_low THEN Valve IS Medium_open
2- IF Level IS Medium_negative AND Temperature IS High THEN Valve IS Medium_open
3- IF Level IS Extremely_negative AND Temperature IS Extremely_high
THEN Valve IS About_60%_closed
4- IF Level IS Extremely_negative AND Temperature IS Extremely_high
THEN Valve IS About_80%_closed
5- IF Level IS High_negative AND Temperature IS Medium_high THEN Valve IS
Almost_closed
6- IF Level IS Small_negative AND Temperature IS Medium THEN Valve IS
About_70%_closed
7- IF Level IS Zero AND Temperature IS Somewhat_low THEN Valve IS Medium_open
8- IF Level IS High_positive AND Temperature IS Low THEN Valve IS Little_open
9- IF Level IS Medium_positive AND Temperature IS Very_low THEN Valve IS
Completely_open

We have seen a few things in this example. First, the rule formation depends on the granularity of the mapping process. Lower granularity results in fewer rules. This is true in general for all types of development, including design by knowledge acquisition from an expert. However, it is much more vivid when we design by rule extraction from data. Second, the whole process, if it can be automated, is a method for extracting rules from a data set—a valuable adaptive fuzzy system structure. The challenge is to employ an appropriate clustering method to reduce the number of point membership functions and to form rules that do not conflict. Third, it is obvious that assigning linguistic terms, when the number of membership functions is high, can be a difficult task. In many cases, rules extracted from a data set are not as informative as the rules articulated by an expert.

Formation of Rule Blocks

There are three types of rule block architecture. The first type is the parallel architecture in which none of the outputs affect each other. As shown in Fig. 7.14, each block receives an input set, which may include common elements, to produce distinct outputs.

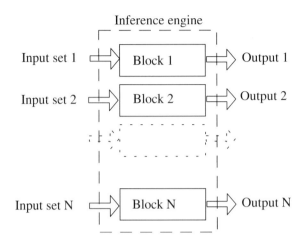

Figure 7.14 Parallel rule block architecture.

The second type is the cascade architecture shown in Fig. 7.15, in which the output from one rule block becomes the input to another block. Some of the outputs from the first line of blocks may also be observed at the output. This is depicted as Output k in Fig. 7.15.

The architecture in Example 7.5 includes cascade elements. The two types discussed above are easily sortable. The basic fuzzy inference algorithm described in Chapter 3 is a block inference engine that can eas-

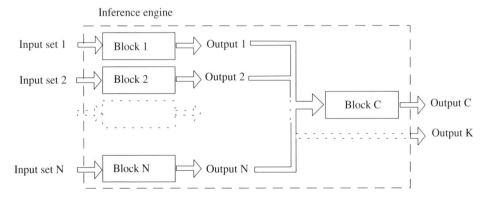

Figure 7.15 Cascade rule block architecture.

ily be expanded to accommodate arrangements as shown in Figs. 7.14 and 7.15. The last case is the combination of parallel and cascade structures. Consider the structure shown in Fig. 7.16.

Notice the difference between Fig. 7.15 and Fig. 7.16. In Fig. 7.16, the rule block B receives external inputs like a parallel block as well as internal inputs like a cascade block. In Fig. 7.15, only block C receives internal inputs as a cascade block. Normally, this type of architecture indicates syllogism in the original solution. Consider the following pair of rules:

$$\begin{aligned} X_1 \bullet P_{1,x} \ominus X_2 \bullet P_{2,x} &\rightarrow X_3 \bullet P_{3,k} \\ X_3 \bullet P_{3,m} \ominus X_2 \bullet P_{2,x} &\rightarrow X_4 \bullet P_{4,x} \end{aligned} \quad (7.6)$$

Subscript x means that it can be anything, whereas subscripts k and m are specific ones. Syllogism due to X_3 causes a problem in the basic fuzzy inference algorithm. The problem is how to distinguish between input X_3 in the second rule from the output X_3 in the first rule. In calculus, such a pair of equations can be reduced to one by substitution. Here, we can employ substitution only if $k = m$; that is, if the predicates (membership functions) are the same. This is illustrated below.

IF Temperature IS High AND Flow IS Low THEN Valve IS Half_open
IF Valve IS Half_open AND Temperature IS Low THEN Pressure IS Low
⇓
IF Temperature IS High AND Flow IS Low THEN Valve IS Half_open
IF Temperature IS High AND Flow IS Low AND Temperature IS Low
THEN Pressure IS Low

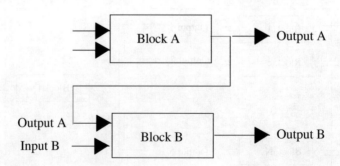

Figure 7.16 Combination of parallel and cascade arrangements of rule blocks.

The first rule is inserted into the second rule by substituting for *Valve IS Half_open*. The first rule can also be kept just for observation. This treatment would not be possible for $k \neq m$ as shown below.

IF Temperature IS High AND Flow IS Low THEN Valve IS Completely_open
IF Valve IS Half_open AND Temperature IS Low THEN Pressure IS Low

Valve IS Half_open cannot be substituted for *Valve IS Completely_open*. In such a case, the only solution is to separate these two rules and place them in different rule blocks in a cascade arrangement. This may not be an easy task, as discussed in Section 5.6.5.

7.5 RULE COMPOSITION STRATEGIES

Basically, rule composition is governed by the knowledge acquisition process. When the knowledge is acquired in the form of articulated expertise, the rule formation is fixed by the expert. When knowledge is extracted from a data set, normally homogeneous-AND rules are composed. In both cases, the fuzzy system designer makes sure that paradoxical problems are not encountered and unintended outcomes are avoided. We realize that rule formation is not a function of problem type since all problems can use the same rule structure. In terms of design criteria, precision/cost and robustness/sensitivity considerations apply to rule formation as shown earlier. Now, we will examine rule composition strategies. A strategy is defined as the style of approach to problem solving that is related to the objective behind a given problem.

7.5.1 Competitive Rules

Competitive rule formation strategy is best understood by modeling a decision-making process similar to one in real life. Consider a classroom full of students, each asked to recognize a partially photographed object either as a frog or as a lizard. The photograph under examination is labeled as evidence in Fig. 7.17. Assume that each student expresses an opinion and confidence level in the scale of 0–1. In this analogy, every student is a fuzzy rule contributing to the same consequent fuzzy variable.

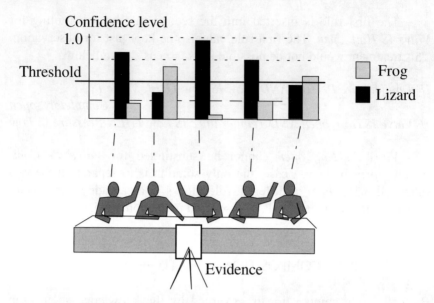

Figure 7.17 Competitive decision-making scenario in which students are examining evidence and interpreting it according to their own skills.

To make a final decision out of the responses shown in Fig. 7.17, competitive strategy employs a threshold to discriminate between the purely guessed low-quality decisions and the educated or experienced guesses of high quality. Elimination of the guesses with low confidence level often improves the possibility distribution of the overall result. In some cases, there may be only one qualified answer, or none. As seen in Fig. 7.17, the decision is more easily visible above the threshold than without a threshold. This operation may also be viewed as removing noise (low-quality information) from the decision-making process. Note that determination of the threshold level is the key challenge, and that it is often achieved through trial and error for each specific problem at hand.

Competitive rule formation in fuzzy system design assumes that all rules in the same rule block are intended to fire at full capacity or close to full capacity. Weak associations are not tolerated. Competitive rule formation is normally designed by applying one threshold level to the entire rule block as depicted in Fig. 7.18. Rules producing degrees of fulfillment less than the threshold are discarded.

Because the same threshold is applied to the entire block, the threshold can be viewed as the property of the consequent variable. As

Chap. 7 Composing Fuzzy Rules

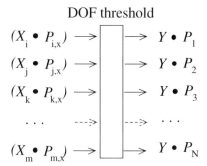

Figure 7.18 Competitive rule formation using block threshold.

expected, a higher threshold produces jumpier outcome, which may be considered a selective behavior.

Competitive rule formation can also be employed across rule blocks by applying one threshold to the entire fuzzy inference engine. This is depicted in Fig. 7.19 for only two rule blocks.

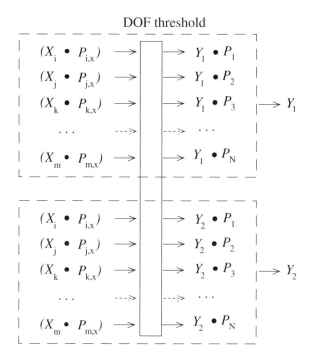

Figure 7.19 Competitive rule formation by applying a DOF threshold across all rule blocks.

As one can infer from above, threshold across all rule blocks results in selecting different numbers of rules from each block at different times. However, selected rules only contribute to their own rule block aggregation in the standard fuzzy inference algorithm. An important point in competitive strategy is that the consequents must be related to each other in a competitive or comparative manner. Spatial relationship between output classes in classification problems may be considered as such. Similarly, decision-making problems with many alternative decisions to choose from fall into this category.

> Competitive inference strategy can be applied to problems in which the consequents have the same units and are related to each other in a competitive or comparative manner.

The DOF threshold function can be designed in a number of ways:

A: Apply a single DOF threshold. The result is zero if none of them exceeds the threshold.

B: Accept the maximum DOF value if all of them are below the threshold.

C: Take the first N maximum DOFs without applying a single threshold.

D: Take the first N maximum DOFs among those which exceed the threshold.

E: Take the winner DOF.

These options are not always clear to an expert in the field, and they are often incorporated through interaction between the fuzzy system designer and the expert during the knowledge acquisition process.

EXAMPLE 7.7 FINGERPRINT CLASSIFICATION

Computerized recognition of fingerprints is a time-consuming process due to the abundance of records and the exhaustive comparison nature of the existing technology. A fuzzy logic-based classification system is designed to reduce the computation time of pattern recognition problems. This is possible due to the fact that all fingerprints can be classified by a number of easily identifiable characteristics. Once accomplished, the classification reduces the exhaustive search domain by a factor of N, N being the number of classes. In other words, if N is 6, the recognition time is roughly expected to be reduced by a factor of 6 (i.e., from one week to one day). The part of the problem that will be examined here includes the following class definitions: *Arch, Left_loop, Right_loop, Tented_arch,* and *Whorl*. The three antecedents of the problem are obtained from a Poincare delta/core finder that produces confidence measures for the number of cores, number of deltas, and the difference between the horizontal coordinates of cores and deltas. Here, core and delta are the shapes encountered in a fingerprint. This system employs proportional inference; thus, the degrees of fulfillment of the conditions determines the classification. A simplified schematic of the problem is shown below.

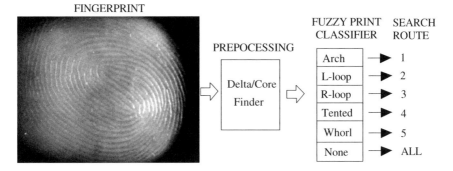

The three antecedents described above are normalized by a universe of discourse between 0 and 1. Now, let's examine the rules articulated by the expert.

IF Number_of_core IS Large AND Number_of Delta IS Large THEN It IS Whorl
IF Number_of_core IS Medium AND Number_of Delta IS Medium
              AND Diff_coordinates IS Negative THEN It IS Right_loop
IF Number_of_core IS Medium AND Number_of Delta IS Medium
              AND Diff_coordinates IS Positive THEN It IS Left_loop
IF Number_of_core IS Small AND Number_of Delta IS Small THEN It IS Arch
IF Number_of_core IS Small AND Number_of Delta IS Small THEN It IS Tented

The last two rules above indicate that the classes *Arch* and *Tented* arch are detected in the same manner. The number of cores and deltas each have three membership functions, whereas the difference in coordinates has two (either negative or positive). Now let's iterate the inference process using one set of test input data for each rule formation strategy. Assume that we have the following truth values:

$\mu\ Large\_number\_of\_cores = 0.65$
$\mu\ Medium\_number\_of\_cores = 0.59$
$\mu\ Small\_number\_of\_cores = 0.11$
$\mu\ Large\_number\_of\_deltas = 0.82$
$\mu\ Medium\_number\_of\_deltas = 0.62$
$\mu\ Small\_number\_of\_deltas = 0.19$
$\mu\ Positive\_diff\_coordinates = 0.35$
$\mu\ Negative\_diff\_coordinates = 0.75$

These values are selected such that the solution is right around the class boundaries. DOF results are shown below for each strategy.

Rule	Left-hand-side	DOF	A T = 0.50	B	C N = 3	D N = 3	E
1	0.65 ∧ 0.82	0.65	0.65	-	0.65	0.65	0.65
2	0.59 ∧ 0.62 ∧ 0.75	0.59	0.59	-	0.59	0.59	-
3	0.59 ∧ 0.62 ∧ 0.35	0.35	-	-	0.35	-	-
4	0.11 ∧ 0.19	0.11	-	-	-	-	-
5	0.11 ∧ 0.19	0.11	-	-	-	-	-

Among the five different competitive strategies, we have three different sets of solutions. The winner strategy (E) selects the first rule with the answer of *Whorl* supported by 0.65. The standard threshold strategy (A) yields *Whorl* and *Right_loop* with DOF values 0.65 and 0.59. The first N ($N = 3$) maximum DOF method (C) adds a third solution *Left_loop* with DOF value 0.35.

Because proportional inference is employed, DOF results indicate which route is to be searched directly. The third column lists all DOF results, which indicates the order of search. The appropriateness of a strategy is determined by the cost of implementing it. The cost is the computation (search) time determined by the data length (in Terra Bytes TB).

Route 1: 2000 TB   Route 2: 155,000 TB   Route 3: 144,000 TB
Route 4: 778 TB   Route 5: 1033 TB

There are several considerations to implement the best search. First, if the classification is not successful, we want the system to switch to *Reject* mode as soon as possible so that other methods can be applied before wasting too much time. Strategy (A) serves this purpose. Second, when the reject mode is selected we still want to know which route was the best candidate so that other methods may use this information. Strategy (B) serves this purpose. We want the system to be sure before selecting Route 2 or Route 3 because they are the largest data segments. So if the DOF values for other routes are close to that of 2 or 3, we may first select small data segments (gambling) before trying the large ones. The combination of other DOF selection strategies (C, D, E) makes it possible to implement a route selection inference process. Considering the fact that data segments in each route are updated regularly, their size becomes another fuzzy variable. This extension of the system is shown below.

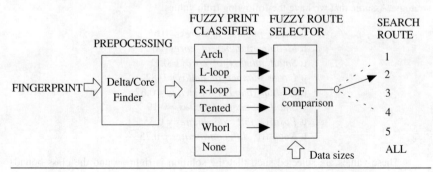

7.5.2 Cooperative Rules

Referring to the previous classroom analogy, let's assume that students are asked to recognize a partially displayed object either as a frog or as a lizard. In this case, each student examines a different photograph displaying different parts of the entire figure. This is depicted by labels a, b, c, d, and e in Fig. 7.20. Again, each student expresses an opinion and confidence level in the scale of 0–1. In this analogy, every student is a fuzzy rule contributing to the same consequent fuzzy variable.

To make a final decision, cooperative strategy makes use of every bit of information provided by each student. The strategy is mainly determined by the objective function, which requires the collection of different evidences to form a final conclusion. There may be individual thresholds for each student if their guessing practices significantly vary.

Cooperative rule formation in fuzzy system design is employed in the same manner as described above. Although some α-cut thresholds may be applied, the cooperative strategy does not employ a block DOF threshold. All evidence is evaluated fully if the rules are composed based on collecting evidence. Otherwise, all partial truth values are utilized in the implication process. This strategy may be viewed as the opposite of the competitive strategy; thus, individual elements are treated as important parts of the overall decision-making process. Most of the existing fuzzy control applications fall into this category. In the following exam-

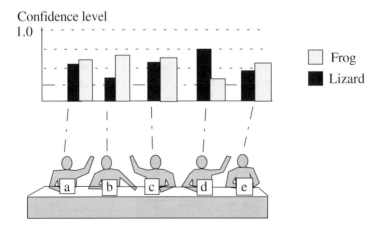

Figure 7.20 Cooperative decision-making scenario in which each student is examining a different property to contribute to the final decision.

ple, we will examine a fuzzy control application for a waste treatment facility, in which the control performance is improved by incorporating as much information as possible.

EXAMPLE 7.8 CONTROL OF A WASTEWATER TREATMENT PROCESS

The activated sludge wastewater treatment system involves the production of an activated mass of microorganisms (biomass) that is capable of biodegrading a waste stream into simple end-products such as carbon dioxide and water. An aerobic environment must exists by means of diffused or mechanical aeration. The system schematic is shown below.

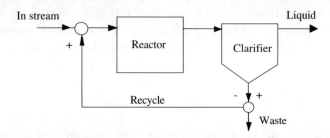

The control problem is the determination of the rate of recycled biomass back to the system, or equivalently, the amount of biomass to be wasted. If all biomass is recycled, microorganism population increases out of control. If too little is recycled then the effluent standards established in environmental regulations may not be met. The antecedent variables of the control problem are *bode* (biological oxygen demand), *sse* (suspended solids effluent), *mlss* (mixed liquor suspended solids), *do* (dissolved oxygen), *qr* (recycled biomass flow), and *q* (biomass flow). All the units are parts per million (ppm). The consequents, which are the control actions, are do' and qr'—same as *do* and *qr*—but they belong to the next time step. Thus, the overall control approach is a feedback method.

The control law implemented in the table below is based on the settling flux theory—an analytical method for establishing optimal state values—that is in agreement with the observed dynamics of the plant under operator control. Every row in this table is a fuzzy rule composed by the AND operator.

	INPUTS						OUTPUTS	
CASE	BODE	SSE	MLSS	DO	QR	Q	DO$'$	QR$'$
1	Small	Small	Medium	Medium	Medium	Medium		None
2	Small	Small	Medium	Large			Lrgneg	
3	Small	Small	Medium	Small			Smlpos	
4	Small	Small	Medium		Large			Smlneg
5	Small	Small	Medium		Small			Smlpos

	INPUTS						OUTPUTS	
CASE	BODE	SSE	MLSS	DO	QR	Q	DO'	QR'
6	Small	Small	Medium			Large		Lrgpos
7	Small	Small	Medium			Small		Lrgneg
8		Medium						Smlneg
9		Large						Lrgneg
10			Large					Smlpos
11			Small					Smlneg
12			Large		Large			Smlpos
13			Small		Small			Smlneg
14	Medium	Small						Smlpos
15	Large	Small						Lrgpos
16						Large		Smlpos
17						Small		Smlneg

This control system is reported to have potential for further improvement incorporating new variables such as temperature, odor, foam, color, algal growth, effluent clarity, bubbles, floating material, and turbulence. Notice that the new variables, like odor, are not measurable; instead, they are observable by the operator, who can make a qualitative judgment and define them as *high, low, very blurry*, and so on. Design incorporates such variables via fuzzy rules articulated by the operator(s). Linguistic variables of this nature are often unitless, and they require the development of a dictionary of predicates and linguistic hedges. Once such a capability exists (i.e., resides in the memory), then the fuzzy inference engine can accept external linguistic inputs comprised of one or more of the words in the memory.

To illustrate the cooperative nature of the development, let's consider the following fuzzy rules articulated by an experienced plant operator.

IF CASE-1 AND Temperature IS Medium AND Odor IS Mild THEN Recycled_biomass IS None
IF CASE-8 AND Temperature IS High THEN Dissolved_oxygen IS Smlpos
IF CASE-8 AND Temperature IS Low THEN Recycled_biomass IS Smlpos
IF CASE-8 AND Odor IS Strong THEN Dissolved_oxygen IS Smlpos
IF Odor IS Strong THEN Dissolved_oxygen IS Lrgpos

Here, the operator has articulated operational experience based on new antecedents, *Temperature* and *Odor*. This set of rules introduces more knowledge into the control law; thus, they cooperate rather than compete. Similarly, more rules are proposed by plant operators, including the new antecedents *color, bubbles,* and *foam*.

IF Color IS Medium_dark OR Quite_dark THEN Dissolved_oxygen IS Small
IF Color IS Brown OR Little_dark THEN Dissolved_oxygen IS None
IF Bubbles IS Plenty THEN Recycled_biomass IS Lrgpos
IF Bubbles IS None THEN Recycled_biomass IS None
IF Foam IS Light_billowy THEN Recycled_biomass IS Lrgpos

7.5.3 Weighted Rules

In most of the decision-making situations in real life, contributing factors often have different importances. Particularly, problems in social sciences, economics, and business require some sort of importance distribution among conditional rules. Formation of rules in a weighted manner is often encountered in inference engines of evidence collection or voting type. Referring back to the previous classroom analogy, we now assign different weights to different students based on their abilities and performance levels in making individual decisions. This is shown in Fig. 7.21. Again, each student corresponds to one fuzzy rule in a fuzzy inference system.

In fuzzy system design, importance weights can be assigned to fuzzy rules if (1) the antecedents involved have different importances in terms of their contribution to the final decision, and/or (2) the designers of the rules are different experts with varying levels of expertise. In the basic fuzzy inference algorithm, weights are numerically incorporated

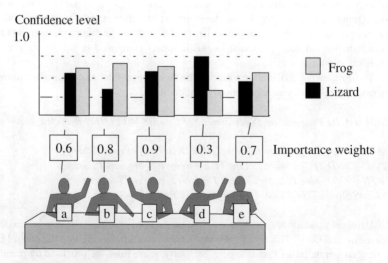

Figure 7.21 Weighted decision-making scenario in which each student's ability is taken into account.

during the implication process, as explained in Chapter 8. The following example describes a typical case in which a set of rules are generally accepted or known by commonsense. However, determining their importance distribution is what makes each solution an expert solution.

EXAMPLE 7.9 STOCK PICKING

We will examine a fuzzy inference system that picks stocks for the profitable growth of stock portfolios. A combination of expert knowledge is used for rule formation and fuzzy variable design. Experts have employed the so-called momentum investment method. The rule formation strategy is an evidence collection type with small thresholds when necessary. The antecedents of the problem are explained as follows:

> *Screening variables:*
> Alpha = Positive price movement above index movement.
> Reward_risk_ratio = Alpha divided by monthly standard deviation.
> Quarterly_earnings = Reported earnings of the firm every 3 months.
> Profit_margin = Latest quarterly profit margin.
> Delta_profit_margin = Latest quarterly profit margin minus annual profit margin.
> Beta = Systematic volatility in relation to general stock market.
> Average_daily_volume = (As stated)
> 52_week_high = Maximum price during the last one year.
> Industry_relative_strength = Index from reported sources.

The only consequent variable is the purchase decision with three decision categories (membership functions) labeled as *Must_buy, Consider_to_buy, Don't_buy*. The universe of discourse is unitless and ranges between 0.0 and 1.0. A defuzzified output is interpreted as 80 percent *Must_buy* and 20 percent *Consider_to_buy*. This inference engine, and many others designed in this manner, is useful when there are many stocks under consideration such that the differences between them can be assessed. In other words, the inference engine determines which stock looks the most profitable among several under consideration. Without such a comparison, an implication result such as 80 percent *Must_buy* is not as informative.

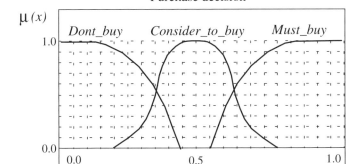

The membership functions of each antecedent are designed by experts. For example, the membership functions of the variable *Reward_risk_ratio* are shown below.

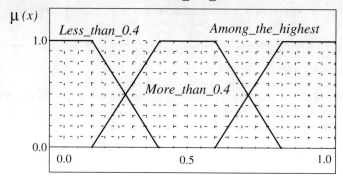

The entire rule base is too broad to be fully covered in this book. However, we will examine a segment that constitutes an example for the weighted strategy. The segment we will examine defines *Must_buy* cases. There are other rules for the other consequent categories not shown below. Accordingly, the right-hand side of the rules shown below repeats *THEN Stock IS A Must_buy*.

1- IF Industry_relative_strength IS Top_quartile THEN .....
2- IF Quarterly_earnings IS More_than_50%_from_previous_year THEN ...
3- IF 52_week_high IS Observed OR 52_week_high IS Nearly_observed THEN ...
4- IF Average_daily_volume IS At_least_1000_shares_per_day THEN ....
5- IF Alpha IS More_than_5% THEN .....
6- IF Reward_risk_ratio IS Among_the_highest OR Reward_risk_ratio IS
   More_than_0.4 THEN..
7- IF Profit_margin IS Among_Top_ three THEN ...
8- IF Profit_margin IS Near_inflection_point AND Delta_profit_margin IS High
   THEN ....
9- IF Beta IS Low AND Alpha IS High THEN ...
10- If Stock IS Technology_stock THEN If Beta IS About_1.7 AND Alpha IS High
   THEN ...

The 10 rules stated above are assigned different weights by different experts as shown below. Notice the differences between two experts. For instance, the first expert thinks the fifth and sixth rules are equally important, whereas the second expert thinks the ninth and tenth rules are absolutely irrelevant.

Rules	1	2	3	4	5	6	7	8	9	10
Expert 1	0.6	0.8	0.4	0.9	1.0	1.0	0.7	0.7	0.5	0.5
Expert 2	0.5	0.9	0.6	0.9	0.8	1.0	0.7	0.7	0.0	0.0

The implication process employs multiple aggregation in this application (see Chapter 8), therefore contributions to the *Must_buy* category by these 10 rules are different based on their importance weights. Note that these 10 rules are not uniquely composed by individual experts because their design only requires commonsense reasoning and a moderate level of market knowledge. The uniqueness or expertise is not in the rules but is embedded in the weight distribution, which reflects experience, style, and belief.

7.5.4 Prioritized Rules

The priority distribution among fuzzy rules determines which rule will fire first, second, third, and so on. In other words, this strategy is employed when the firing sequence of the rules is important for some specific reason. Firing sequence is only important if, at some point, the next rule in the line is not fired because the previous firings have satisfied some criterion. Considering the classroom analogy previously hypothesized, this strategy corresponds to asking students a question and continuing to ask until a satisfactory answer is given by at least one student, then moving to the next question.

The basic fuzzy inference algorithm does not give importance to which rule is fired first, second, third, and so forth. Although rules are fired one at a time due to the sequential processing nature of existing CPUs, all rules are treated equally regardless of how they are ordered. Therefore, to implement a prioritized strategy, the mechanics of the basic fuzzy inference algorithm must be modified. Such a modification consists of a termination condition based on a specified DOF threshold. Consider N rules contributing to the same consequent variable Y as shown in Fig. 7.22.

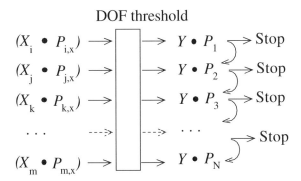

Figure 7.22 Prioritized rule structure with the capability of terminating the process.

Every rule in this rule block is evaluated for its left-hand-side fulfillment, and the evaluation of the rule block is terminated as soon as a fulfillment threshold is exceeded. Then, the next rule block is evaluated. Thus, the order of the rules from top to bottom is important in this strategy. Prioritized rule formation is useful basically for two purposes: (1) the order of rule firing is in agreement with the nature of the problem, and/or (2) the rule base is too large and prioritized firing saves computation time. An example is given below that directly serves the first purpose. An example for the second purpose will have the same characteristics as the one given below.

EXAMPLE 7.10 AUTOMATIC ALARM SYSTEM IN A CHEMICAL FACILITY

Consider an alarm system in a chemical facility that prevents explosion due to wrong mixing of chemicals. As shown in the schematic below, trucks deliver chemicals Y and Z from the loading ports which are mixed with the chemical X supplied from a permanent tank. Four valves are used to allow proper mixing through the blending manifold and two valves are used to deliver the product to the outgoing trucks. A total of six valves determine the overall operation.

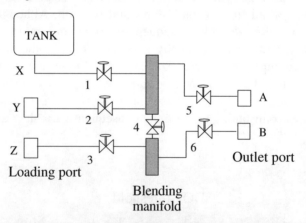

Devising an alarm system to prevent all modes of faulty operations requires the identification of a fault tree.[1] Two top events that we will analyze here are the excessive mixtures of XZ and YZ caused by faulty valve operations and/or wrong truck loading. Because the entire fault tree is too large to consider here, only the relevant part is illustrated below.

---

[1] Event trees and fault trees are standard tools in reliability analysis and they consist of a schematic of logical occurrences leading to the top event. A fault tree is also useful in alarm system design.

Chap. 7 Composing Fuzzy Rules

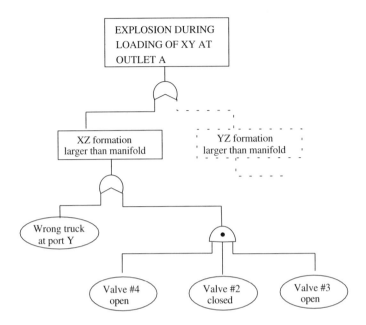

The system is equipped with digital flow meters, port video cameras, and electronic truck ID readers. Each instrument is connected to a centralized control unit and each has a different sampling/downloading time characteristic. Now, consider the following fuzzy rules that activate the alarm system based on the truck misplacement fault.

IF ID_at_port_Y IS_NOT Y_truck THEN Alarm_level IS Full
IF Image_at_port_Y IS_NOT Similar_to_Y_truck THEN Alarm_level IS Alert

The antecedent of the first rule *ID_at_port_Y* is a crisp variable that takes a linguistic value such as *Y_truck*. The consequent *Alarm_level* is also a crisp variable with linguistic values *Full* and *Alert*. An electronic ID recognizer at the port detects the ID signal mounted on each incoming truck and sends a response signal to the control center where the fuzzy algorithm resides. Using this signal, the algorithm is activated for the first rule. If DOF level is satisfactory, which is a matter of Yes/No evaluation for crisp rules, then alarm is turned on. The second rule uses a response signal from an image processing system. The image processing system analyzes a video image of the truck's license plate and tries to recognize the color-coded tag. The antecedent *Image_at_port_Y* is a fuzzy variable.

Because the first rule is given a higher priority, the fuzzy inference algorithm does not wait for the inputs of the second rule. Even if all the inputs are received at the same time, the first rule is activated first and an alarm is turned on without activating the second rule if the conditions of the first rule are met. The second rule in this application is designed as redundancy in case of measurement anomalies in the ID recognizer. The *Alert* category prompts the field maintenance team to check the operation. The *Full* category of alarm activates a number of interlocks to mitigate the likelihood of continued operation.

Similar strategy is applied to the following fuzzy rules, which cover the right leg of the fault tree. The valve openings are detected by flow measurements.

IF Valve4 IS Open AND Valve3 IS Open AND Valve2 IS Closed
THEN Alarm_level IS Full
IF Valve4 IS Medium_open AND Valve3 IS Medium_open AND Valve2 IS Medium_closed
THEN Alarm_level IS Alert

Again, the priority is given to the first rule. If its DOF value is above a specified threshold level then the second rule is not evaluated.

7.5.5 Hierarchical Rules

A chain of investigative questions fall into this category of decision making in which a navigation is conducted through the rule base according to each answer received. Most of the large-scale decision-making tasks require some form of hierarchical structure with rule blocks connected to each other in a cascade or parallel form. A simple example, depicted in Fig. 7.23, involves a decision about whether the economy is in recession. The deciding committee relies on the decisions made by subcommittees, each focusing on one particular indicator. In this example, subcommittees represent separate rule blocks or inference engines.

The rule formation strategy for this purpose requires the determination of the input-output structure between different layers of decision

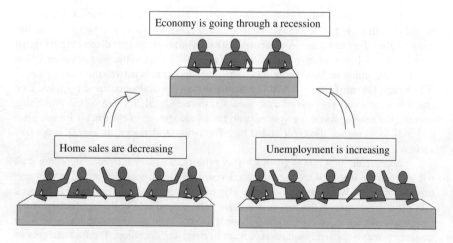

Figure 7.23 Hierarchical decision-making scenario.

making, then the building of each layer by composing rules using the identified inputs and outputs. In practical terms, hierarchical rule formation is nothing but building multiple inference engines with appropriate connections. Although hierarchical structures are often seen in large-scale decision-making systems, we will examine a simplified example next.

EXAMPLE 7.11 ON-LINE PERFORMANCE MONITORING IN A LARGE-SCALE PLANT

A successful maintenance program in a power plant depends on the on-line monitoring capability of each subsystem. The key idea is to detect anomalies before the failure of a subsystem so that preventive actions can be taken in advance.[2] Performance monitoring of a nuclear reactor heat removal system consists of monitoring hundreds of components via measurements and inspections throughout the coolant loop. This primarily includes valves, pipes, and pumps. The pump system we examine in this example is a part of the heat removal system and its operational performance is one of the determining factors in the performance of the entire heat removal system.

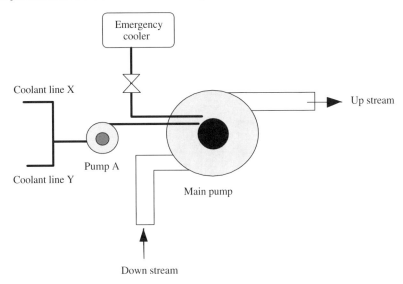

The operation of the main pump relies on cooling of the pump bearings via an auxiliary pump called PUMP A. In the case of its failure, an emergency cooling system is activated by opening the valve. The emergency cooling valve is tested regularly to ensure its operation when needed. The main pump operation is evaluated by a set of fuzzy rules as listed below.

---

[2] Preventive maintenance and predictive maintenance are two interrelated disciplines that make use of on-line monitoring of plant components.

IF Main_pump_RPM IS Steady THEN Performance IS Good
IF Main_pump_RPM IS Slightly_oscillatory THEN Performance IS Mediocre
IF Main_pump_RPM IS Quite_oscillatory THEN Performance IS Poor
IF Emergency_cooling_valve HAS Never_been_stuck THEN Performance IS Good
IF Emergency_cooling_valve IS Sometimes_stuck THEN Performance IS Risky

IF Pump-A_RPM IS Steady THEN Performance IS Good
IF Pump-A_RPM IS Slightly_oscillatory THEN Performance IS Mediocre
IF Pump-A_RPM IS Quite_oscillatory THEN Performance IS Poor
IF Pump-A_RPM IS Sometimes_zero THEN Performance IS Risky

The antecedents involving RPM measurements use average readings over a period of time to compare deviations from time to time. Valve status is determined by test results over a period of time. The consequent *Performance* in this rule block can be *Good, Mediocre, Poor,* and *Risky*.

Considering the three main pumps of the plant, the overall pump performance is determined by a high-level inference engine that uses the performance outputs of the three lower-level inference engines as depicted below.

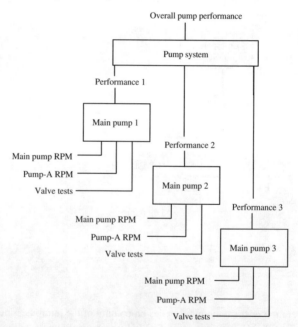

The high-level inference engine consists of the following fuzzy rules:

IF Performance-1 IS Good AND Performance-2 IS Good AND Performance-3 IS Good
  THEN Pump_overall_performance IS Good
IF Performance-1 IS Mediocre OR Performance-2 IS Mediocre OR Performance-3
  IS Mediocre
  THEN Pump_overall_performance IS Mediocre

IF Performance-1 IS Poor OR Performance-2 IS Poor OR Performance-3 IS Poor
　　　　　　　　　　　　　THEN Pump_overall_performance IS Poor
IF Performance-1 IS Risky OR Performance-2 IS Risky OR Performance-3 IS Risky
　　　　　　　　　　　　　THEN Pump_overall_performance IS Risky

On the highest level of hierarchy, the overall performance of the plant operation is determined by a number of inference engines (or rule blocks) connected in a hierarchical manner. This is shown next, including other elements not examined in this example.

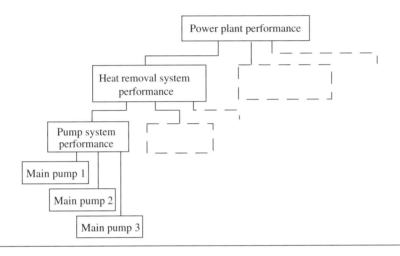

As clearly indicated in Example 7.11, hierarchical design is only a high-level structuring, and individual rule formation is not affected. Referring to the basic fuzzy inference algorithm presented in Chapter 3, such a structure can be built by developing as many inference algorithms as the number of layers.

7.5.6 Adaptive Rules

Adaptive rule formation strategy is more complex than the strategies discussed previously. This is due to the abundance of adaptable parameters the designer can choose. The main idea behind an adaptive fuzzy inference engine is to have the initial design capable of coping with the environmental[3] changes evolving in time. There are two distinct levels of

---

[3] Environment in this context refers to external system disturbances as described in Chapter 4.

adaptive design: (1) adapting a set of selected parameters of an inference engine, and (2) automatically generating new rules to cope with the changes. The second level of adaptive systems, which also includes adaptive learning functions using neural networks, is at the experimental stage; its theory has not been fully established yet.

The most straightforward adaptation technique is to change the shape of the membership functions by some criterion. This requires additional expertise or knowledge acquisition to determine how such a change will be designed. Consider a fuzzy rule of the form

$$X_j \bullet P_{j,k} \to Y_m \bullet P_{m,l} \tag{7.7}$$

where the membership function $P_{j,k}$ and/or $P_{m,l}$ are to be modified when the circumstances change considerably and the underlying logic of the fuzzy rule prevails. There are two ways of developing an adaptive algorithm to accommodate such a function.

- Design one or more adaptive rules with consequents directly controlling the coordinates of the membership functions of the original rule.
- Design one or more adaptive rules with consequents producing linguistic hedge effects to modify the shape of the membership functions of the original rule.

Also assume that the adaptive rule or rules are as simple as the original rule. The antecedent of the adaptive rule W_j is a variable that represents the environmental changes in time. The consequent Z_m is an output variable that controls the shape of the membership function(s) of the original rule as shown in Fig. 7.24.

Let's consider the case in which Z_m is changing the coordinates. There are several ways to incorporate adaptive functions depending on the shape of the membership function to be modified. Assume that the membership function to be modified is defined in piece-wise-linear form, such as triangles or trapezoids.

The first method is to shift the entire membership function without changing its profile. If there are four break points describing a trapezoid as shown in Fig. 7.25, the output variable Z_m can be used to determine the amount of shift.

The shift determined by Z_m changes the horizontal coordinates of the four break points defining the original membership function.

Chap. 7 Composing Fuzzy Rules

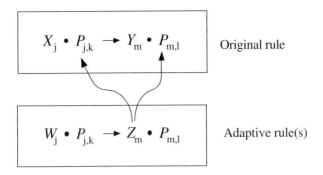

Figure 7.24 A simple adaptive fuzzy rule structure.

$$\begin{aligned} Z_m &= \Delta x \\ x_1^{new} &= x_1 + \Delta x \\ x_2^{new} &= x_2 + \Delta x \\ x_3^{new} &= x_3 + \Delta x \\ x_4^{new} &= x_4 + \Delta x \end{aligned} \qquad (7.8)$$

This adaptive rule, therefore, must be designed to determine the rate of shift based on selected observations. The consequent Z_m may be designed as shown in Fig. 7.26, such that deviations are categorized as posi-

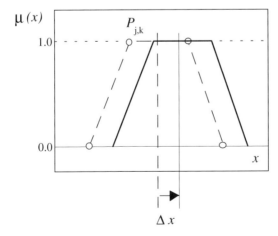

Figure 7.25 Shifting the location of a membership function via adaptive rules.

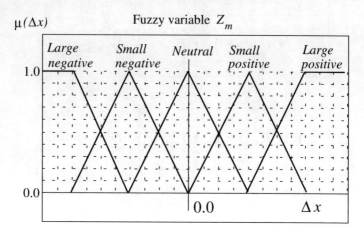

Figure 7.26 Deviation categories in an adaptive rule.

tive, negative, and neutral. Shifting a membership function location is a direct attempt to modify the original decision boundaries of the input-output mapping process, thus it significantly affects the performance of the fuzzy inference engine.

Another method is to reduce or increase the width of the membership function without changing its original location, as shown in Fig. 7.27. As a rule of thumb, reducing the width results in a crisper (or more certain) membership function shape. Width modification, when applied to consequent membership functions, does not change the performance drasti-

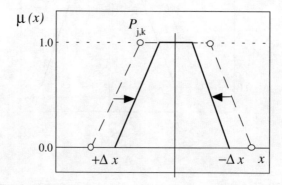

Figure 7.27 Reducing or increasing the width of a membership function.

cally; however, it produces a variable weight effect in the aggregation-defuzzification process. When applied to the antecedent membership functions, width modification changes the overlap among membership functions, thus producing effects as described in Chapter 6.

$$\begin{aligned} Z_m &= \Delta x \\ x_1^{new} &= x_1 + \Delta x \\ x_2^{new} &= x_2 + \Delta x \\ x_3^{new} &= x_3 - \Delta x \\ x_4^{new} &= x_4 - \Delta x \end{aligned} \qquad (7.9)$$

The horizontal coordinates in Eq. (7.9) are ordered from left to right. Obviously, the adaptive fuzzy algorithm must ensure that reduction does not exceed beyond certain limits where the shape becomes ambiguous.

There are numerous other possibilities for modifying a membership function shape, ranging from height adjustments to modifying curvatures like the linguistic hedge effects described in Chapter 5. All such modifications primarily serve one purpose; that is, reducing the number of static membership functions by designing a few dynamic ones. Otherwise, all possible shapes can be incorporated statically, which may potentially become too cumbersome and tedious to design.

An adaptive fuzzy algorithm in which the membership functions of one set of fuzzy rules are modified by another set of fuzzy rules potentially eliminates the cumbersome task of designing many static membership functions by designing a few dynamic ones.

Besides modifying the membership function shapes, there are a few other alternatives an adaptive algorithm can be based on. All adaptable parameters[4] are listed in Table 7.1 along with the effects they produce. These effects are discussed in Chapters 5 and 6 in detail.

The most important question at this point is how to design adaptive rules. Just like designing a set of original rules, the knowledge required to

---

[4] Other adaptable parameters can be devised, depending on the level of sophistication and available knowledge of the problem.

Table 7.1 Elements of an Adaptive Fuzzy Inference Algorithm

ADAPTABLE ELEMENTS	EFFECTS ON
Membership function location	Decision boundaries
Membership function width	Overlap, robustness, sensitivity
Membership function height	Rule importance, paralysis
Membership function line style	Transition between crisp points
Importance weights	Implication results
Thresholds	Decision boundaries

design adaptive fuzzy rules comes from the knowledge acquisition process and from related techniques discussed in Chapter 4. Contrary to traditional adaptive control theory, there is no single piecemeal solution applicable to all fuzzy inference engines. In other words, the adaptive knowledge is problem specific.

> The design knowledge for fuzzy adaptive rules is always problem specific, and is obtained via one of the knowledge acquisitions techniques.

The following example is a simple one, but it clearly illustrates the fundamental approach of adaptive fuzzy inference engines.

EXAMPLE 7.12 ADAPTIVE STRATEGY FOR CLIMATE CONTROL

Consider a typical climate control system in which the primary objective is to maintain room temperature at a constant level. The set point is called *ideal* temperature. The fuzzy climate control system operates by the following rules.

> IF Temperature IS Very_hot THEN AC_level IS High
> IF Temperature IS Little_hot THEN AC_level IS Medium
> IF Temperature IS Warm THEN AC_level IS Low
> IF Temperature IS Ideal THEN AC_level IS Zero
> IF Temperature IS Ideal THEN Heater_level IS Zero
> IF Temperature IS Chilly THEN Heater_level IS Low
> IF Temperature IS Cold THEN Heater_level IS Medium
> IF Temperature IS Freezing THEN Heater_level IS High

The antecedent *Temperature* is designed as shown below. In application to a large office building located in a harsh climate zone, the occupants have repeatedly complained about the inefficiency of this climate control system, especially when the weather abruptly changed from one extreme to another.

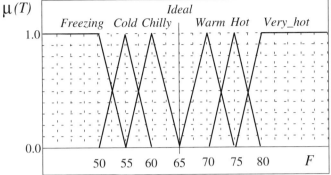

The problem was that the fuzzy climate controller did not consider the temperature outside the building, which drops and rises very rapidly, putting extreme demand on the air-conditioner/heater system. Apparently, it took a while to cool or heat the entire building during such extreme changes in weather. To cope with this problem, the designers had two choices: (1) increase the number of rules by including the outside temperature as a new variable, or (2) design an adaptive fuzzy algorithm to modify the existing rules. The second choice, designing a simple adaptive fuzzy algorithm, consisted of the following fuzzy rules.

IF Gradient IS positive THEN Shift IS negative
IF Gradient IS Zero THEN Shift IS Zero
IF Gradient IS negative THEN Shift IS positive

The antecedent *Gradient* is the difference between outside and inside temperatures whereas the consequent *Shift* is the amount of location shift of the membership functions of the original variable *Temperature*. *Shift* and *Gradient* are both designed in the following manner.

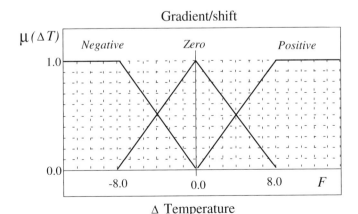

After a few trials and errors, the designers established an interval of 16° F, which both the *Gradient* and *Shift* operated on. The idea is simple. When the outside temperature drops suddenly, then *Gradient* becomes negative, producing a positive shift for all membership functions as illustrated above.

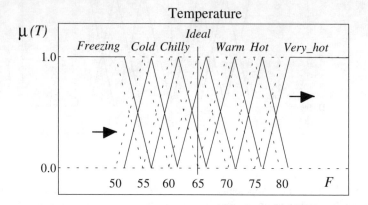

Accordingly, the ideal temperature definition along with all others becomes higher than the original design, thus causing the heater to overreact. In other words, a room temperature value, which is previously considered chilly, becomes cold after the shift, therefore requiring higher levels of heating action. This strategy is an anticipatory approach such that the heating action is taken in advance before it is too late to recover. Obviously, the same discussion holds for the cooling action. This design greatly reduced the number of complaints within the following months of operation.

As experienced readers will recognize, this example uses derivative control—an element of standard PID control law—as an adaptive mechanism. More importantly, the adaptive rules were designed based on the properties of the problem at hand rather than using a piecemeal strategy applicable to other problems. However, once the logic behind adaptive strategy is established, the adaptive fuzzy inference is applicable to any problem, not only to control.

7.6 PARADOXICAL CASES

The calculus of fuzzy IF-THEN rules found in the literature cannot accurately represent every variation in semantic reasoning. We will analyze a few important cases in this section that may drastically change the composition of fuzzy rules.

7.6.1 Amount of Evidence

The basic logic operators in fuzzy logic (AND and OR) may be viewed as filters through which the maximum outcome cannot exceed the maximum truth value. Any combination of AND and OR will yield an outcome smaller than the maximum truth value due to minimum operation. The question is whether they are adequate to formulate every possible logical statement. Consider the following fuzzy statements:

> *Animal HAS Long_fur*
> *Animal HAS Sharp_teeth*
> *Animal IS Quite_smaller_than_man*
> *Animal HAS Whiskers*

If we continue the description, it will get easier to recognize the cat behind the curtain. Although the recognition process is in the consequent domain, one wonders if the collection of partially satisfied conditions should result in a degree of fulfillment higher than the minimum single truth value; that is, if we combine all these fuzzy statements using the AND operator and form the conditional part of a fuzzy rule for recognition. Let's do it in a two-stage manner by first including only the two initial statements, then by including the four of them.

RULE 1:
IF Animal HAS Long_fur AND Animal HAS Sharp_teeth THEN Cat IS
Likely_answer

RULE 2:
IF Animal HAS Long_fur AND Animal HAS Sharp_teeth
AND Animal IS Quite_smaller_than_man AND Animal HAS Whiskers
THEN Cat IS Most_likely_answer

Now assume that for a given evidence, we have the following truth evaluations:

μ *Long_fur* $= 0.8$
μ *Sharp_teeth* $= 0.8$
μ *Quite_smaller_than_man* $= 0.8$
μ *Whiskers* $= 0.8$

We have selected equal truth values just to make a point. Using the truth values above, we see that the degrees of fulfillment of the first and second rules are both 0.8 due to minimum. In other words, both of these rules will equally support their distinct action. This seems contradictory to common sense because we know from daily experience of life that an increased amount of evidence (even with partial quality) improves our belief. Thus, we encounter a paradoxical situation when we use standard AND or OR logic operators such that the increased amount of evidence does not increase the level of support for the corresponding action.

The solution is trivial. To account for the amount of evidence, the fuzzy rule labeled RULE 2 should be broken into individual statements without AND or OR operators:

IF Animal HAS Long_fur THEN Cat IS Most_likely_answer
IF Animal HAS Sharp_teeth THEN Cat IS Most_likely_answer
IF Animal IS Quite_smaller_than_man THEN Cat IS Most_likely_answer
IF Animal HAS Whiskers THEN Cat IS Most_likely_answer

This manipulation solves the problem in a practical manner by shifting the problem from the left- to right-hand side. Note that the right-hand side of this fuzzy inference engine must be sensitive to multiple (additive) aggregation. This is explained in Chapter 8. The "amount of evidence" issue is often encountered in diagnostics systems. However, there are other circumstances where the collective occurrence of conditions (imposed by the AND operator) is represented suitably by intersection operation such as in pattern recognition problems. The fuzzy system designer must consider the following design principle:

> When the problem at hand is sensitive to the amount of evidence, all conditions composed by the AND operator in a rule can be broken into individual rules provided that the fuzzy inference algorithm employs multiple (additive) aggregation.

7.6.2 Outliers

Borrowed from statistics, the term *outlier* refers to a single item of evidence that kills all other evidence unfairly. This discussion is closely related to the previous one where the effects of collective evidence were examined. Let's reconsider the previous example.

Chap. 7 Composing Fuzzy Rules

IF Animal HAS Long_fur AND Animal HAS Sharp_teeth
AND Animal IS Quite_smaller_than_man AND Animal HAS Whiskers
THEN Cat IS Most_likely_answer

Now assume that the membership function evaluations have yielded the following truth values.

μ *Long_fur* = 0.8
μ *Sharp_teeth* = 0.8
μ *Quite_smaller_than_man* = 0.8
μ *Whiskers* = 0.1

In other words, we assume that one of the items of evidence is almost unrecognized due to some unfortunate reason, such as a measurement error. The AND composition yields 0.1 as the degree of fulfillment that drastically affects the outcome. The solution is again trivial and practical. This rule can be repeated by using the mean operator to replace AND. Now we have a pair of rules instead of a single rule.

RULE 1:
IF Animal HAS Long_fur AND Animal HAS Sharp_teeth
AND Animal IS Quite_smaller_than_man AND Animal HAS Whiskers
THEN Cat IS Most_likely_answer

RULE 2:
IF Animal HAS Long_fur AND/OR Animal HAS Sharp_teeth
AND/OR Animal IS Quite_smaller_than_man AND/OR Animal HAS
Whiskers THEN Cat IS Most_likely_answer

We have denoted the mean operator by AND/OR in the second rule. The degree of fulfillment of the second rule is 2.5 / 4 = 0.61. Comparing the degrees of fulfillment of both rules (0.1 versus 0.61) indicates a large difference (0.51). Thus, when there is a large difference between a pair of rules designed in this manner, we can detect a possible outlier. As a result, the design trick is to suppress the contribution of the outlier by a weight equal to 1 minus the difference in degrees of fulfillment between the rule *j* and its identical pair designed with the mean operator. This is called the DOF comparison method, which requires the conversion from a composite rule to its equivalent canonical rules.

$$B_j = 1 - |DOF_j^{AND} - DOF_j^{MEAN}| \qquad (7.10)$$

If the difference in DOFs is large in Eq. (7.10), the weight will decrease, as will the contribution of the outlier when the original rule is broken into many canonical rules as illustrated in Section 7.6.1. The reader will realize that an outlier problem is directly related to the robustness/sensitivity criterion described in Chapter 4. A single bad measurement drastically affecting the outcome is considered poor robustness. This leads to the following design principle:

> Composing a homogeneous rule using the AND operator among many conditional statements will decrease robustness against single measurement anomaly. Robustness can be increased by detecting the anomaly using the DOF comparison method.

This principle can be reversed for the sensitivity criterion. If there is a desire for the system to be sensitive to single measurement anomaly, the AND operator will just accomplish that and there is no need for the DOF comparison method. Note that the problem at hand does not have to fit the "evidence collection" category for this principle to apply.

7.6.3 Conflicts in Disguise

It is easy to fabricate a conflict in bivalent logic where every statement is true or false. The simplest form of conflict emerges from a condition producing both true and false consequences:

> IF Price IS Expensive THEN Buy A_car
> IF Price IS Expensive THEN Don't buy A_car.

Similarly, another conflicting form is when opposite conditions produce the same consequence:

> IF Price IS Expensive THEN Buy A_car.
> IF Price IS NOT Expensive THEN Buy A_car.

In expert systems based on crisp logic, there is no solution to this problem, and crisp algorithms may halt by detecting the conflict. In fuzzy logic, however, conflicts do not emerge from rule composition directly. Although semantically incorrect, conflicting rules will not halt the fuzzy inference process (due to excluded-middle laws). In the first set of rules shown above, the result will be indecision (located between buy and don't buy) for any price value. This shows that the basic fuzzy algorithm will respond to a conflicting rule correctly. In the second set of rules, the result will be "buy the car no matter what" which is again the correct interpretation of the conflict. This property of fuzzy algorithms is as valuable as it is misleading. In other words, a conflicting rule can be overlooked.

One easy method for detecting a conflict is the paralysis phenomenon. The outcomes of a pair of conflicting rules will tend to be the same regardless of the inputs. Thus, the fuzzy system designer must observe the outputs of the fuzzy inference engine during test runs to ensure that there is no stationary response. More complex forms of conflict arise from poor design of several elements. Conflicts are usually observed in the implication process, which is described in Section 8.6. The outcome may not be paralyzed in every case, especially when conflicting statements are mixed with nonconflicting statements in a composite rule. Thus, there is no absolute remedy to this paradox.

7.6.4 Conversion From Multiple Rules to OR Combination

This is a special case often encountered in rule blocks including homogeneous-AND rules. Although not a paradox per se, conversion from multiple rules to a single rule using the OR operator may produce different outcomes if the implication mechanism of the inference engine is not designed appropriately. This is shown in Fig. 7.28. Consider the following two rules:

IF Weather IS Cold AND It IS Raining THEN Wear A_raincoat
IF Sailing_in_the_Open_sea AND Wind IS Strong THEN Wear A_raincoat

Now consider the conversion:

IF (Weather IS Cold AND It IS Raining) OR (Sailing_in_the_Open_sea AND Wind IS Strong) THEN Wear A_raincoat

which is correct in classical logic and will produce identical results. However, outcomes can be different in fuzzy logic depending on the type of aggregation imposed on the consequents. For example, the first group of rules will produce a collective evidence effect via implication with multiple aggregation (see Chapter 8) whereas the combined version will produce a single implication result by the maximum degrees of fulfillment. To see this, let's assume the following partial truth values.

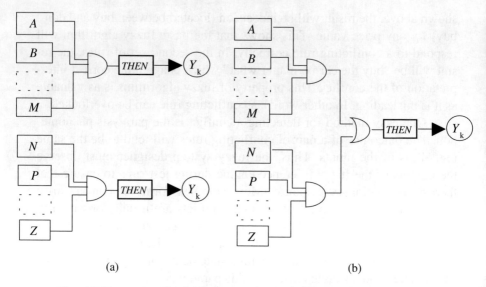

Figure 7.28 Two homogeneous rules with the same consequent (a) are combined using the OR operator (b).

$$\mu\_Cold\_weather = 0.7$$
$$\mu\_Raining = 0.2$$
$$\mu\_Sailing = 1.0$$
$$\mu\_Strong\_wind = 0.6$$

The first two rules produce the two DOF values 0.2 and 0.6. Accordingly, they will support the consequence twice by these values. On the other hand, the combined version will support the consequence once by DOF = 0.6. The difference is the 0.2 contribution in the first case. If the implication process employs multiple aggregation then this difference will affect the outcome. Otherwise, using the max aggregation operation instead of fuzzy additive method, both representations are equivalent. Multiple aggregation is described in Section 8.4.1.

7.6.5 Mutual Feedback in Block Formation

Consider the structure shown in Fig. 7.29. Normally, this type of architecture indicates a logical inconsistency in the original solution. In calculus, such a structure requires a simultaneous solution; however, in fuzzy inference this is not possible. Consider the following pair of rules:

IF Temperature IS High AND Flow IS Low THEN Valve IS Half_open
IF Valve IS Half_open AND Temperature IS Low THEN Flow IS High

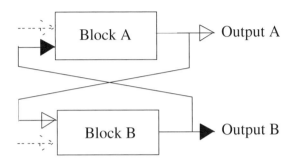

Figure 7.29 Mutual feedback arrangement of rule blocks.

Consequents *Valve* and *Flow* are mutually feeding back on each other. This type of rule formation is unresolvable for simultaneous execution, and it implies prioritized execution with a specified sequence. It is important for the designer to recognize such hierarchical structures among several rules composed by experts and to make sure that the rules are fired in a proper sequence.

7.6.6 Empty Input Space

During the composing of fuzzy rules, it is not always possible to formulate a solution using all the combinations among the antecedent predicates. For example, assume that there are two antecedent fuzzy variables each divided up into three categories (membership functions): *Level (Negative, Zero, Positive)*, and *Flow (Low, Medium, High)*. Assume that a consequent fuzzy variable *Valve* also has three categories: *Valve (Open, Half, Closed)*.

Suppose, a fuzzy inference engine is designed in the following manner:

IF Level IS Negative AND Flow IS Low THEN Valve IS Open
IF Level IS Zero AND Flow IS High THEN Valve IS Half
IF Level IS Positive AND Flow IS Medium THEN Valve IS Closed

This fuzzy inference engine can also be represented in a tabular form as shown in Table 7.2. Input categories are listed along the first column and first row, and actions are written inside the corresponding boxes. Note that all the definitions are used in this rule base at least once; however, there are combinations shown by shaded boxes that are undefined.

So the question is: What happens if *Flow* is *Medium* and *Level* is *Zero*, or *Flow* is *High* and *Level* is *Negative*? Since there is no rule to cover these cases, the inference results will be undefined. The challenging task the designer can face is to fill in such cases in the absence of expertise or acquired knowledge. However, it can be a tedious and impossible task depending on the type of problem at hand and available knowledge.

If there are k antecedents each with n categories, the total number of combinations is n^k using the AND formation. If the number of combinations that are specified by rules is N, then the ratio N/n^k is a measure of how much the input space is covered. For example, the coverage in Table 7.2 is $3/9 = 0.33$. In an ideal design this ratio is 1.0, which is also referred to as *full rank design*.

If $N < 1$, there are two cases to be considered during design. First, the rules composed using available knowledge must ensure that the uncovered combinations of inputs are not feasible in practice. Referring to the example in Table 7.2, this means that the combination *Flow* is *Medium* and *Level* is *Zero* will never occur in reality due to the physics of the problem. Second, if such cases may occur, the design must include rules for the uncovered combinations with a consequent variable called *Unknown_case*. Such rules often use negation such as *IF Flow IS NOT Low AND Level IS NOT Negative THEN*. Exercising this method ensures that the fuzzy inference engine is capable of detecting empty locations on the input space due to lack of knowledge in its design. The following principle summarizes the discussions above.

Table 7.2 A Tabular Representation of Fuzzy Inference Engine

LEVEL	FLOW		
	LOW	MEDIUM	HIGH
Negative	Open		
Zero			Half
Positive		Closed	

Chap. 7 Composing Fuzzy Rules 391

> When there are empty regions in the input space due to lack of knowledge to compose the corresponding rules, a fuzzy inference engine can be designed to detect its limits by (1) composing the unknown rules using a consequent called *Unknown_case*, and (2) programming techniques to detect the cases when all DOF values are zero.

7.7 MEMBERSHIP FUNCTION SHAPE EFFECTS

When composing fuzzy rules, it may not be clear to the designer what portion of the input space is becoming fuzzified and what effects are being produced. We will demonstrate in this section visual images of the input space to facilitate the reader's understanding of those effects caused by membership function shapes. Consider the LHS of a rule of the form:

$$(X \bullet P_{x1}) \Theta (Y \bullet P_{y1}) \xrightarrow{DOF} \ldots \qquad (7.11)$$

The degrees of fulfillment (DOF) produced by using triangular and trapezoidal membership functions are shown in Figs. 7.30 and 7.31, respectively. In each case, there are two plots, one for AND composition and one for OR. In Fig. 7.32, a combination of trapezoid and triangle is dis-

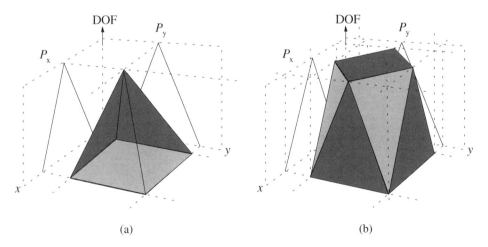

Figure 7.30 Triangle membership functions combined with (a) AND operator and (b) OR operator.

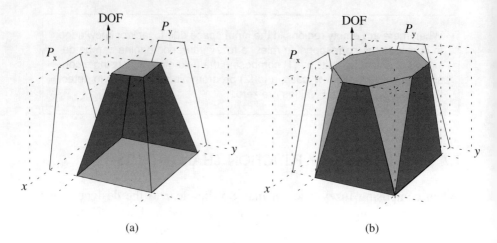

Figure 7.31 Trapezoidal membership functions composed with (a) AND operator and (b) OR operator.

played. As clearly illustrated by these figures, the input space for which degree of fulfillment is 1.0 varies depending on the membership function design, although using the same rule and the same logic. Also remember that the degree of fulfillment determines the strength of contribution this rule will bring to the overall decision. When DOF = 1.0, the contribution is maximum. What we are visualizing in these figures is the ability to

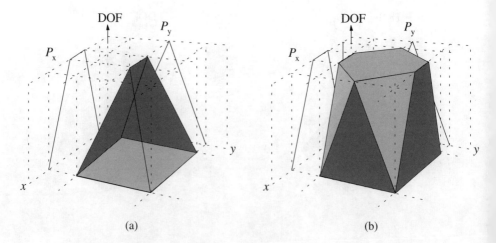

Figure 7.32 Trapezoidal P_x and triangular P_y membership functions composed with (a) AND operator and (b) OR operator.

contribute. Apparently, OR composition is always more contributive (i.e., flat top surfaces) as expected. There is an interesting outcome from AND composition, explained next.

7.7.1 Statistical Favoritism

Although there is no statistical or probabilistic mechanism embedded in Eq. (7.11), a fuzzy rule can become statistically biased due to its membership function shape. To examine this phenomenon, let's consider the input product space $X \times Y$ again. Assume that the likelihood of having any pair of X and Y values from the external world is equal. This is depicted in Fig. 7.33(a) for a large number of measurements. Now consider the bird's-eye view of the pyramid shown in Fig. 7.30 obtained from AND composition. This is illustrated in Fig. 7.33(b). The center indicates the tip of the pyramid, and squares indicate equal height points on the fuzzy surfaces. Figure 7.33(c) shows the superimposition of the two figures on the left.

As one can easily deduct from Fig. 7.33(c), the likelihood of receiving an input pair increases as we depart from the center and go towards the edges. That is because the area (between two squares) gets larger moving outwards. Conversely, the likelihood of receiving an input pair right on the tip of the pyramid is very low. Thus, the likelihood of a rule composed with the AND operator and triangular membership functions to have a DOF = 1 and to contribute to the overall decision at full strength is extremely low. Comparison with a trapezoidal combination as shown

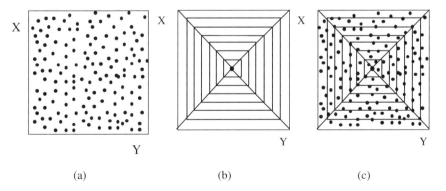

Figure 7.33 (a) Equal likelihood of measurements on the product space; (b) bird's-eye view of the pyramid in Fig. 7.30; and (c) superimposing measurements on the pyramid surfaces.

in Fig. 7.34 indicates that rules with triangular membership functions operate more on the fuzzy surfaces than those with trapezoidal membership functions.

The effect produced by the shape of membership functions along with randomly distributed data assumption is examined next for two separate cases.

7.7.2 Dominance Among Rules

The behavior in case of randomly distributed input data is examined here using the same fuzzy rule composed in two different forms. Between the two forms of design—one using triangles the other trapezoids—the degree of their contribution to the final aggregated result will be different on the average.

using triangular membership functions

$$(X \bullet P_{x1}) \Theta (Y \bullet P_{y1}) \xrightarrow{DOF_1} Z \bullet P_{z1}$$

using trapezoidal membership functions

$$(X \bullet P_{x2}) \Theta (Y \bullet P_{y2}) \xrightarrow{DOF_2} Z \bullet P_{z2}$$

(7.12)

The rule with trapezoidal membership functions will dominate[5] the result compared to the rule with triangular membership functions. On the average $DOF_2 > DOF_1$ will hold for Θ = AND. Note that if such two rules are in the same rule block (i.e., they have the same consequent vari-

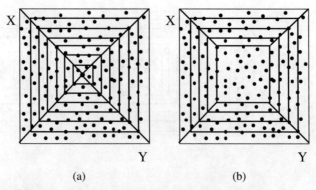

Figure 7.34 Triangular (a) versus trapezoidal (b) combination of membership functions superimposed with randomly distributed data.

---

[5] In this context, dominance means the capability to effect the outcome.

able Z), then this design has inadvertently incorporated a favoritism for (or put more importance on) the second rule, which will dominate most of the time.

> Between the two rules in a rule block, the one composed with trapezoidal antecedent membership functions will dominate the other composed with triangular antecedent membership functions when the input data are randomly distributed on the universe and when the logic operators are AND.

EXAMPLE 7.13 DOMINANCE DUE TO MEMBERSHIP FUNCTION DESIGN

Consider a pair of simple decision-making rules that define the likelihood of recession based on two measures of the economy.

> IF Unemployment IS High AND House_sales ARE Low THEN Recession IS Possible
> IF Unemployment IS Low AND House_sales ARE High THEN Recession IS Unlikely

Considering the left-hand-side statements, assume that the membership functions in the first rule are all triangles, whereas the membership functions in the second rule are all trapezoids.

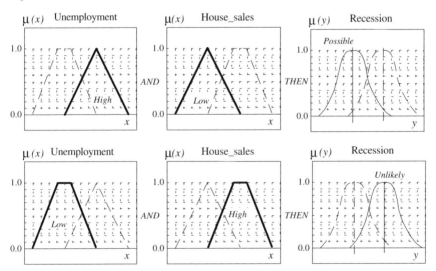

The rightmost column above shows consequent membership functions that are not part of the ongoing discussion. Using the standard center-of-gravity and union aggregation methods, the performance of this simple inference mechanism is shown below in terms of the normalized frequency of recession diagnostics. Note that only the interval between the center of the trapezoids and tip of the triangles is used to construct the following histogram, because the rest of the universe will not reflect the difference between the two geometrical shapes due to the implication process.

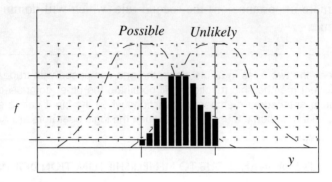

The vertical axis is normalized frequency. The inputs were hypothetically selected with random distribution (i.e., equal likelihood of occurrence). The frequency distribution between the two diagnostic states indicates that the second rule using trapezoids determines the answer more frequently than the first rule with triangles.

The average behavior is reversed in a dynamic system with feedback effects in which the inputs are affected by the outputs generated in the previous step. The behavior is reversed such that a rule with triangular membership functions is effective more frequently than that of a rule with trapezoidal membership functions. Because the rule with trapezoidal membership functions reaches full firing capacity faster, this bounces the system (via feedback) back to the domain of the rule with triangular membership functions. It takes longer for the rule with triangular membership functions to reach full firing capacity. Thus, the system spends more time in that domain. Note that the mechanism is the same, yet its effects are reversed due to feedback.

> Between the two rules in a rule block, the one composed with triangular antecedent membership functions will dominate the other composed with trapezoidal antecedent membership functions when (1) the logic operators are AND; and (2) inputs are affected by outputs via a feedback mechanism.

EXAMPLE 7.14 DOMINANCE IN DYNAMIC SYSTEMS DUE TO MEMBERSHIP FUNCTION DESIGN

Consider a car navigation problem. The objective is to steer the car through an opening without colliding against the obstacles. Depending on the initial position of the car, the

Chap. 7 Composing Fuzzy Rules   397

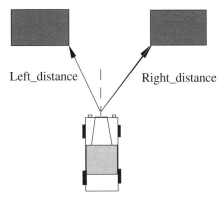

two variables *Left_distance* and *Right_distance* can have any values; thus, there is no statistically predictable initial value for them. There are two simple rules for steering control:

IF Left_distance IS Large AND Right_distance IS Small THEN Turn Left
IF Left_distance IS Small AND Right_distance IS Large THEN Turn Right

Considering the left-hand-side statements, assume that the antecedent membership functions in the first rule are all triangles whereas the antecedent membership functions in the second rule are all trapezoids.

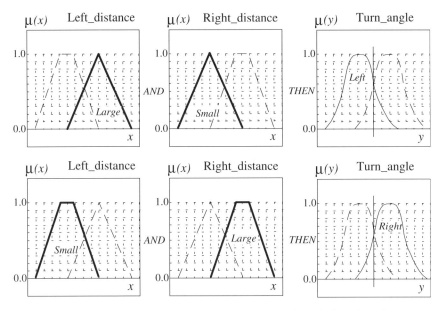

The rightmost column above shows consequent membership functions that are not part of the ongoing discussion. Using the standard center of gravity and union aggregation methods, the performance of this simple inference mechanism is shown in terms of tracks generated by the car motion in the following illustration. Out of many trials with randomly selected initial positions, six are shown that represent the general trend.

As indicated by left and right arrows, there is a difference in turning behavior. The car turns right in a much sharper manner than it turns left. This is because the action of turning right is fully supported by the second rule more easily than by the first rule. *More easily* means that the plateaus of the trapezoids are used more often than the tips of the triangles. The first rule operates more on the fuzzy surfaces, thus generating a somewhat smoother and slower turn.

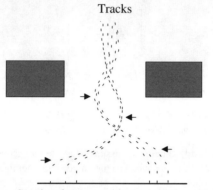

Starting from an arbitrary initial position

In the statistical sense, the behavior is determined by turning to the left more than turning to the right because right turn is sharper and quicker. The system spends more time on a left turn.

The example above is not presented to compare control performances between triangles and trapezoids, because such a comparison would not be correct. Instead, it shows that, when both of them are used in such a manner, the rule with triangles will be more dominant than the rule with trapezoids for systems with feedback. This conclusion also agrees with the discussions in Chapter 5. The last two design principles are warning messages rather than design steps to be followed.

7.7.3 Dominance Among the Components of a Rule

A similar "dominance" phenomenon is encountered within the same fuzzy rule when both triangular and trapezoidal membership functions are used. Assume that P_x is a triangle membership function whereas P_y is a trapezoid.

Chap. 7 Composing Fuzzy Rules 399

$$(X \bullet P_x) \ominus (Y \bullet P_y) \xrightarrow{DOF_1} .... \qquad (7.13)$$

The corresponding "likelihood" discussions are one-dimensional in this case for obvious reasons. Because a triangular membership function operates more on the fuzzy regions compared with a trapezoidal function, the DOF values are determined by the triangular membership function on the average. In this case, the fuzzy statement with a triangular membership function dominates because of the minimum operation (AND operator). To understand this behavior, let's consider the following example.

EXAMPLE 7.15 DOMINANCE AMONG THE COMPONENTS OF A RULE DUE TO MEMBERSHIP FUNCTION DESIGN

Consider a hypothetical fuzzy rule:

IF Wind IS Strong AND Temperature IS Icy_cold THEN Flight IS Canceled

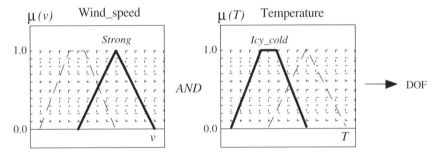

The membership functions of the antecedents are shown above. The difference between the components of this rule can be visualized by superimposing both membership functions as shown below. The shaded area between the trapezoid and triangle is what makes the trapezoid dominate in truth values, thus making the triangle dominate the DOF values via minimum operation.

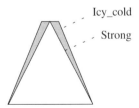

Therefore, wind speed becomes a more strict (or determining) requirement as compared to temperature in the rule stated above just by the shape of the membership functions. The normalized frequency diagram shown below indicates the likelihood of obtaining DOF values determined by trapezoids and triangles. The input space was randomly sampled for the construction of the histogram.

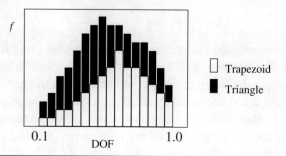

7.8 SUMMARY

This chapter focused on the rule formation issues within the framework of the basic fuzzy inference algorithm. The first topic, logic operator design, describes the role of the two common operators AND and OR along with more advanced and somewhat experimental operators such as product AND, Mean, Yager AND, Yager-OR, and CONSISTENCY. The logic operators AND and OR are adequate to form almost any composite rule. The designer might find it useful to consider other operators when the overall behavior of the fuzzy inference engine or part of it needs to be modified (either off-line or adaptively) based on some linguistic design criteria. Especially, product AND and Yager operators are useful for this purpose.

The relational inference, which is defined by unconditional fuzzy statements in this book, is useful for navigating among the different rules or rule blocks. The basic fuzzy inference algorithm is normally designed by conditional fuzzy statements either in the simple form of proportional inference or in the more complex form of compositional inference as shown in Chapter 3. Compositional inference is the most commonly applied approach in the industry and includes composite rules, implication operation, aggregation, and defuzzification. A combination of cascaded and parallel rule blocks is the most complex structure the designer can implement.

Most of the problems in practice can be modeled in one or more of the six strategic forms presented in Section 7.6. The strategies include competitive, cooperative, weighted, prioritized, hierarchical, and adaptive rule formation. The examples show that each strategy is distinct. However they can be mixed. These strategies are determined by an objective (either given or implied) in the original solution (either articulated by experts or extracted from data). The design challenge is to identify a strategy and modify the basic fuzzy inference algorithm accordingly rather than just putting the rules together.

In some situations, which are called paradoxical cases in this book to emphasize the discrepancy between the articulated logic and its translation into fuzzy IF-THEN rules, the designer must pay special attention not to undermine the implied objective. Problems sensitive to the amount of evidence and outliers must be treated as described in this section of the book. Other problematic issues include conflicts, conversion from multiple rules to OR logic, mutual feedback, and the holes in the input product space.

The last part of this chapter is focused on the interaction between composite rule formation and membership function design. When the membership functions are designed per variable basis, the formation of composite rules using these membership functions may cause unexpected favoritism. This is true for systems that receive inputs from the external world in a random manner (i.e., flat probability distribution). The designer must make sure that the combination of membership functions of different shapes does not create inadvertant favoritism or dominance. It is recommended that the same considerations are also applied to systems receiving biased input to examine the most likely region of operations and to estimate the consequences.

REFERENCES

[1] Kruse, R., J. Gebhardt, and F. Klawonn. *Foundations of Fuzzy Systems,* Chichester: John Wiley & Sons, 1994.

[2] Yager, R. R., and L. A. Zadeh. *An Introduction to Fuzzy Logic Applications in Intelligent Systems,* Boston, MA: Kluwer, 1992.

[3] Dubois, D., and H. Prade. "A Class of Fuzzy Measures Based on Triangular Norms." Int. Journ. of General Systems 8 (1982): 43–61.

[4] Zimmerman, H. J. *Fuzzy Set Theory.* Boston, MA: Kluwer Academic Publishers, 1990.

[5] Shafer, G. A. *A Mathematical Theory of Evidence.* Ph.D. diss., Princeton University, 1976.

[6] Cox, E. *The Fuzzy Systems Handbook.* New York: AP Professional, 1994.

[7] Edmunds, G. *The Measurement of Human Aggression.* Chichester: Ellis Horwood Ltd., 1980.

SELECTED BIBLIOGRAPHY

Katsuno, H. and A. O. Mendelzon. "Proportional Knowledge Base Revisions and Minimal Change." *Artificial Intelligence* 52, (1991): 263–294.

Saaty, T. L. "Exploring the Interface Between Hierarchies, Multi-objectives, and Fuzzy Sets." *Fuzzy Sets and Systems* 1 (1978): 57–68.

Tong, R. M. "Some Properties of Fuzzy Feedback Systems." *IEEE Trans. on Systems, Man, and Cybernetics* 10 (1980): 327–331.

Zadeh, L. A. "Test-Score Semantics as a Basis for a Computational Approach to the Representation of Meaning." *Literary and Linguistic Computing* 1 (1986): 24–35.

Zadeh, L. A. "Fuzzy Algorithms." *Information Control* 19 (1969): 94–102.

Chapter 8

Implication Process

> The implication process yields the output of fuzzy inference. Although the results obtained using different implication methods are invariant in principle, there are differences mainly in approaching the desired output behavior on fuzzy surfaces. This chapter is committed to the discussion of implication operators along with all other computations related to the right-hand side of composite fuzzy IF-THEN rules. Included are the output aggregation and defuzzification methods and their design.

*The best weather instrument yet devised is a pair
of human eyes.
Harold M. Gibson
Chief meteorologist, NYC Weather Bureau*

8.1 INTRODUCTION

For composite rules, the right-hand-side (RHS) computations of the basic fuzzy inference algorithm includes implication relation, output aggregation, and defuzzification steps. In Chapter 7, the compositional inference was presented to include the implication process as shown in Fig. 8.1. The

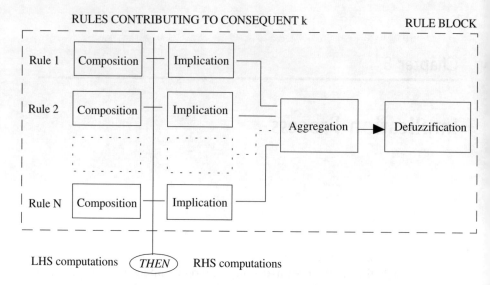

Figure 8.1 RHS and LHS computations in a rule block.

border between LHS and RHS computations are marked by the THEN operator, which appears only once in a regularly composed fuzzy rule. Multiple use of the THEN operator is discussed in Chapter 7. The implication process starts by obtaining the DOF values from the LHS computations. As illustrated in Chapter 3, this structure is simplified for canonical rules.

In Chapter 7, we have shown that a rule block is identified by a set of fuzzy rules contributing to a single consequent variable. Although an implication relation exists for every rule, an aggregation process is only possible for a set of rules that belong to the same rule block. The aggregation result is the result produced by an inference engine in the form of possibility distribution. Defuzzification (also known as decomposition) is the final treatment to obtain a scalar that is often suitable for the operation of systems in practice.

The design issues related to the implication process are discussed in this chapter following the order of computations from left to right as shown in Fig. 8.1. The selected implication operators are analyzed by a typical case study and their surface topologies are illustrated. These illustrations, although they depend on the shape of membership functions, are quite enlightening because the reader can visualize their behavior on fuzzy surfaces and their relative differences. In each subsection, there is an accompanying illustration that shows the clipping effect of the DOF value on the consequent membership functions. The clipping mechanism is the practical (or algorithmic) implementation and is called the *geometric approach*. Thus, the reader is provided with both fuzzy set operations and the geometric approach to implication computation. Their differences were emphasized in Chapter 3. Following these arguments, we introduce behavior analysis using the same case study. The design principles stated in this section highlight some of the most significant behavioral properties that may be associated with the linguistic design criteria described in Section 4.6 or with the designer's own criteria.

8.2 SELECTING IMPLICATION OPERATORS

A list of available implication operators [1–4] were given in Table 2.2. The selection of an appropriate implication operator is unfortunately one of the most confusing tasks a designer can face. Also consider that the final output is not only determined by the implication operator but also by the accompanying aggregation operator and the defuzzification method. This triplet yields more than a hundred combinations to examine when considering the different methods found in the literature. Due to limited space and scope, we will restrict our focus to some commonly used combinations.

There are two important points to consider when evaluating an implication operator on an individual basis. The first is how well the operator represents a decision or how accurately it maps. The second point, which is often more important, is how it behaves from one initiation step to another. However, regardless of the differences between them, all of the implication operators yield the same result at crisp points (i.e., when DOF = 1.0, the left-hand-side conditions are completely satisfied). The differences between implication operators emerge in operations on fuzzy surfaces (i.e., when DOF < 1.0, the left-hand-side conditions are partially satisfied).

The only significant design (selection) criterion between the different implication operators is their mapping properties during partial fulfillment of the left-hand-side of the rules; that is, when the rules operate on fuzzy surfaces. Crisp points of mapping between the input and output spaces are not affected by the type of the implication operator.

This is a very important design principle to remember, because fuzzy rules are primarily designed for their ideal cases (crisp points of mapping). Thus, implication operator selection does not affect ideal cases. In other words, we don't compose fuzzy rules differently for different implication operators. A hypothetical transition is illustrated in Fig. 8.2 in a curve-fitting analogy that shows the effects of using different implication operators.

Let's first examine each operator individually by examining the decision surfaces they generate. Consider a rule of the form *IF A THEN B* where new evidence A' is to be evaluated through a standard GMP inference. Suppose the membership functions implied by A and B are as shown in Fig. 8.3 with a new evidence A' represented by a singleton.

The membership functions drawn above will be used to draw decision surfaces using different implication operators in the following sections. The trapezoidal thickness in B will help the reader to distinguish X × Y spaces in the figures to come. Note that these membership functions are quite standard and they represent a generalized case in which certain conclusions can be easily made. Although the decision surfaces shown in later sections strictly belong to A and B shown above, different convex shapes of A and B will not change the basic arguments and the principles set forth. Also note that we employed a discrete (fuzzy set) representation for A and B instead of a continuous one to give the reader the opportunity to trace the computations.

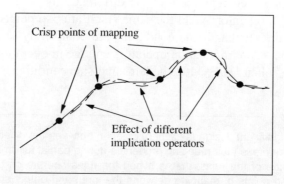

Figure 8.2 Effects of using different implication operators.

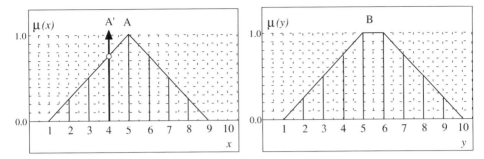

Figure 8.3 Two arbitrarily drawn membership functions implied by *IF A THEN B* and a new evidence A′ given as a singleton.

The fuzzy sets A and B are given as follows:

$$A = \bigcup_{i=1}^{10} \mu_i(x)/x_i$$
$$= 0/1 \cup 0.25/2 \cup 0.5/3 \cup 0.75/4 \cup 1.0/5 \cup 0.75/6 \cup 0.5/7 \cup 0.25/8 \cup 0/9 \cup 0/10$$

$$B = \bigcup_{j=1}^{10} \mu_j(y)/y_j$$
$$= 0/1 \cup 0.25/2 \cup 0.5/3 \cup 0.75/4 \cup 1.0/5 \cup 1.0/6 \cup 0.75/7 \cup 0.5/8 \cup 0.25/9 \cup 0/10$$

$A' = 1.0/4$

The implication relation is given by Eq. (8.1), which includes one of the implication operators listed in Table 2.2.

$$R(x,y) = \phi[\mu_A(x), \mu_B(y)] \quad (8.1)$$

The implication result B′ given the new evidence A′ is computed by

$$B'(y) = A'(x) \circ R(x,y) \quad (8.2)$$

where the symbol ∘ denotes the max-min composition operator. Note that the implication result B′ is a new fuzzy set. This operation is suitable for canonical rules, whereas the geometric representation (described in Chapter 3) is suitable for composite rules.

8.2.1 Mamdani Operator

The Mamdani implication operator yields an implication relation expressed by

$$R(x,y) = \phi[\mu_A(x),\mu_B(y)] = \mu_A(x) \wedge \mu_B(y) \qquad (8.3)$$

producing the following data (x rows y columns).

.00	.00	.00	.00	.00	.00	.00	.00	.00	.00
.00	.25	.25	.25	.25	.25	.25	.25	.25	.00
.00	.25	.50	.50	.50	.50	.50	.50	.25	.00
.00	.25	.50	.75	.75	.75	.75	.50	.25	.00
.00	.25	.50	.75	1.00	1.00	.75	.50	.25	.00
.00	.25	.50	.75	.75	.75	.75	.50	.25	.00
.00	.25	.50	.50	.50	.50	.50	.50	.25	.00
.00	.25	.25	.25	.25	.25	.25	.50	.25	.00
.00	.00	.00	.00	.00	.00	.00	.00	.00	.00
.00	.00	.00	.00	.00	.00	.00	.00	.00	.00

The implication relation surfaces are shown in Fig. 8.4. The final result B′ for given A′ is a slice shown by the arrow that corresponds to the 4th data row above.

As can be seen from Fig. 8.4, the Mamdani implication operator generates a nonbinary decision surface in which elements vary between 1 and 0. The result B′ is shown in Fig. 8.5.

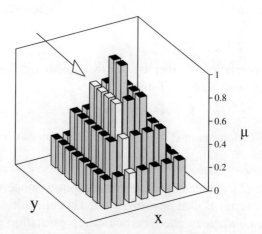

Figure 8.4 A Mamdani implication relation between a trianglular and a trapezoidal membership function.

Chap. 8 Implication Process

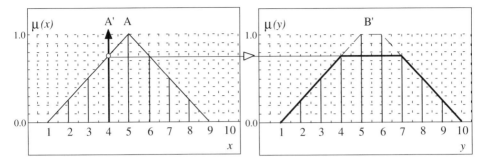

Figure 8.5 Inference result B' given evidence A' using the Mamdani implication operator.

As the evidence A' moves toward the ideal case (one step to the right), the implication result B' becomes equal to B. This can be viewed as a complete match when the ideal case is reached on the conditional part of the rule. In the opposite direction, when the evidence moves to the edges, B' is the area under the DOF line imposed on B. At the extreme edge points, B' becomes zero, a null set. The area of B' is directly proportional to the DOF.

Among the 10 implication operators selected in this book, the behavior described above constitutes one of the two distinct behaviors. When the total area of B' is directly proportional to the DOF, we call the implication operator a *first category operator*. Otherwise (e.g., if B' is inversely proportional to the DOF), the implication operator is a *second category operator*. This terminology will make recognition easy for the reader in the remainder of the book and has no further connotations.

The Mamdani implication operator is the most commonly used operator, especially in control applications. One reason is that its behavior is somewhat regular (see Fig. 8.5), which makes it easy to anticipate the outcome of the modeled reasoning. Like most of the first category implication operators, the Mamdani operator is based on a "match-seeking" paradigm.

8.2.2 Larsen Operator

The Larsen implication operator yields an implication relation expressed by

$$R(x,y) = \phi[\mu_A(x), \mu_B(y)] = \mu_A(x) \cdot \mu_B(y) \tag{8.4}$$

producing the following data (x rows y columns).

0	0	0	0	0	0	0	0	0	0
0	0.06	0.13	0.19	0.25	0.25	0.19	0.13	0.06	0
0	0.13	0.25	0.38	0.5	0.5	0.38	0.25	0.13	0
0	0.19	0.38	0.56	0.75	0.75	0.56	0.38	0.19	0
0	0.25	0.5	0.75	1	1	0.75	0.5	0.25	0
0	0.19	0.38	0.56	0.75	0.75	0.56	0.38	0.19	0
0	0.13	0.25	0.38	0.5	0.5	0.38	0.25	0.13	0
0	0.06	0.13	0.19	0.25	0.25	0.19	0.13	0.06	0
0	0	0	0	0	0	0	0	0	0
0	0	0	0	0	0	0	0	0	0

The implication relation surfaces are shown in Fig. 8.6. The final result B′ for given A′ is a slice shown by the arrow that corresponds to the fourth data row above.

As can be seen from Fig. 8.6, the Larsen implication operator generates a nonbinary decision surface in which elements vary between 1 and 0. The result B′ is shown in Fig. 8.7.

As the evidence A′ moves toward the ideal case (one step to the right), the implication result B′ becomes equal to B. This can be viewed as a complete match when the ideal case is reached on the conditional part of the rule. In the opposite direction, when the evidence moves to the edges, B′ keeps the shape of B with its height scaled by DOF. At the

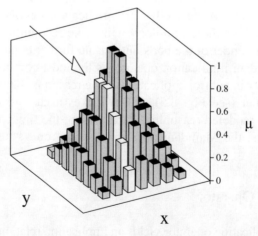

Figure 8.6 A Larsen implication relation between a triangular and a trapezoidal membership function.

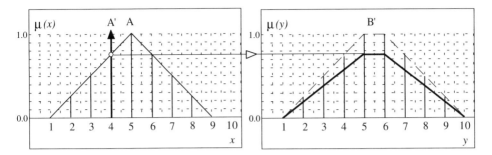

Figure 8.7 Inference result B' given evidence A' using the Larsen implication operator.

extreme edge points, B' becomes zero. The area of B' is directly proportional to the DOF; thus we classify this operator as a first category operator.

The Larsen implication operator is also one of the most commonly used operators in control applications. Its behavior is also regular, which makes it easy to anticipate the outcome of the modeled reasoning. Like most of the first category implication operators, the Larsen operator may be viewed as a "match-seeking" paradigm.

8.2.3 Lukasiewicz Operator

The Lukasiewicz implication operator (also known as the *arithmetic product*) yields an implication relation expressed by

$$R(x,y) = \phi[\mu_A(x),\mu_B(y)] = 1 \wedge (1 - \mu_A(x) + \mu_B(y)) \quad (8.5)$$

producing the following data (x rows y columns).

1	1	1	1	1	1	1	1	1	1
0.75	1	1	1	1	1	1	1	1	0.75
0.5	0.75	1	1	1	1	1	1	0.75	0.5
0.25	0.5	0.75	1	1	1	1	0.75	0.5	0.25
0	0.25	0.5	0.75	1	1	0.75	0.5	0.25	0
0.25	0.5	0.75	1	1	1	1	0.75	0.5	0.25
0.5	0.75	1	1	1	1	1	1	0.75	0.5
0.75	1	1	1	1	1	1	1	1	0.75
1	1	1	1	1	1	1	1	1	1
1	1	1	1	1	1	1	1	1	1

The implication relation surfaces are shown in Fig. 8.8. The final result B' for given A' is a slice shown by the arrow that corresponds to the fourth data row above.

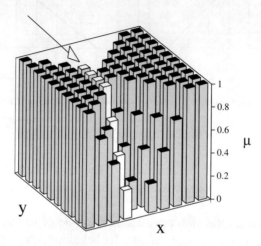

Figure 8.8 A Lukasiewicz implication relation between a triangular and a trapezoidal membership function.

As can be seen from Fig. 8.8, the Lukasiewicz implication operator generates a nonbinary decision surface in which elements vary between 1 and 0. The result B' is shown in Fig. 8.9.

As the evidence A' moves toward the ideal case (one step to the right), the implication result B' becomes equal to B. This can be viewed

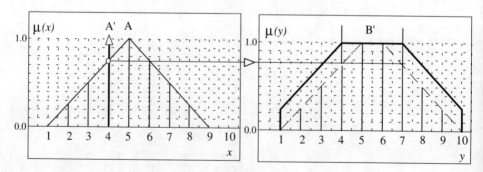

Figure 8.9 Inference result B' given evidence A' using the Lukasiewicz implication operator.

as a complete match when the ideal case is reached on the conditional part of the rule. In the opposite direction, when the evidence moves to the edges, B' keeps enlarging sideways. At the extreme edge points, B' becomes a complete rectangle. The area of B' is inversely proportional to the DOF; thus we classify this operator as a second category operator.

Although the ideal case B' is a complete match of B, this is not a match-seeking type implication operation, because departure from the ideal case increases the fuzziness or uncertainty of B', and partial matching (like that encountered using Mamdani or Larsen operators) does not apply. Thus, the Lukasiewicz implication operation may be viewed as a search for the designed uncertainty that only occurs at the ideal case.

8.2.4 Kleen-Dienes Operator

The Kleen-Dienes implication operator (also known as the Boolean operator) yields an implication relation expressed by

$$R(x,y) = \phi[\mu_A(x), \mu_B(y)] = (1 - \mu_A(x)) \vee \mu_B(y) \tag{8.6}$$

producing the following data (x rows y columns).

1	1	1	1	1	1	1	1	1	1
0.75	0.75	0.75	0.75	1	1	0.75	0.75	0.75	0.75
0.5	0.5	0.5	0.75	1	1	0.75	0.5	0.5	0.5
0.25	0.25	0.5	0.75	1	1	0.75	0.5	0.25	0.25
0	0.25	0.5	0.75	1	1	0.75	0.5	0.25	0
0.25	0.25	0.5	0.75	1	1	0.75	0.5	0.25	0.25
0.5	0.5	0.5	0.75	1	1	0.75	0.5	0.5	0.5
0.75	0.75	0.75	0.75	1	1	0.75	0.75	0.75	0.75
1	1	1	1	1	1	1	1	1	1
1	1	1	1	1	1	1	1	1	1

The implication relation surfaces are shown in Fig. 8.10. The final result B' for given A' is a slice shown by the arrow and corresponds to the fourth data row above.

As can be seen from Fig. 8.10, this implication operator generates a nonbinary decision surface in which elements vary between 1 and 0. The result B' is shown in Fig. 8.11.

As the evidence A' moves toward the ideal case (one step to the right), the implication result B' becomes equal to B. This can be viewed as a complete match when the ideal case is reached on the conditional

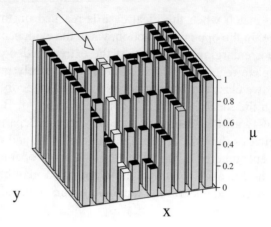

Figure 8.10 A Kleen-Dienes implication relation between a triangular and a trapezoidal membership function.

part of the rule. In the opposite direction, when the evidence moves to the edges, B' keeps the shape of B while adding (union) a rectangle of height (1–DOF). At the extreme edge points, B' becomes a complete rectangle of height 1.0. The area of B' is inversely proportional to the DOF; thus we classify this operator as a second category operator.

Just like the Lukasiewicz operator, the Kleen-Dienes implication operator searches for the designed uncertainty that only occurs at the ideal case. Departure from the ideal case increases the fuzziness of B'.

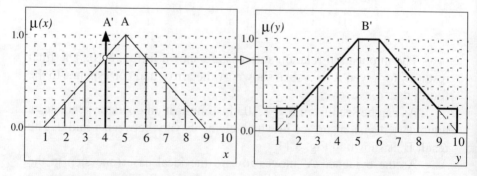

Figure 8.11 Inference result B' given evidence A' using the Kleen-Dienes implication operator.

8.2.5 Standard Sequence Operator

The standard sequence implication operator yields an implication relation expressed by

$$R(x,y) = \phi[\mu_A(x), \mu_B(y)] = \begin{bmatrix} 1 & if & \mu_A(x) \le \mu_B(y) \\ 0 & if & \mu_A(x) > \mu_B(y) \end{bmatrix} \quad (8.7)$$

producing the following data (x rows y columns).

1.00	1.00	1.00	1.00	1.00	1.00	1.00	1.00	1.00	1.00
.00	1.00	1.00	1.00	1.00	1.00	1.00	1.00	1.00	.00
.00	.00	1.00	1.00	1.00	1.00	1.00	1.00	.00	.00
.00	.00	.00	1.00	1.00	1.00	1.00	.00	.00	.00
.00	.00	.00	.00	1.00	1.00	.00	.00	.00	.00
.00	.00	.00	1.00	1.00	1.00	1.00	.00	.00	.00
.00	.00	1.00	1.00	1.00	1.00	1.00	1.00	.00	.00
.00	1.00	1.00	1.00	1.00	1.00	1.00	1.00	1.00	.00
1.00	1.00	1.00	1.00	1.00	1.00	1.00	1.00	1.00	1.00
1.00	1.00	1.00	1.00	1.00	1.00	1.00	1.00	1.00	1.00

The implication relation surfaces are shown in Fig. 8.12. The final result B′ given new evidence A′ is a slice shown by the arrow that corresponds to the fourth data row above.

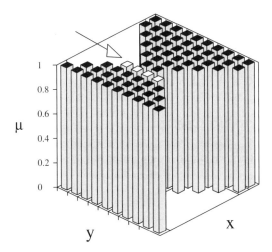

Figure 8.12 A standard product implication relation between a triangular and a trapezoidal membership function.

As can be seen from Fig. 8.12, the standard sequence implication operator generates a binary decision surface in which all elements are either 1 or 0. The result B' is shown in Fig. 8.13.

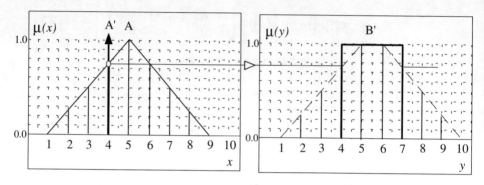

Figure 8.13 Inference result B' given evidence A' using the standard sequence implication operator.

As the evidence A' moves toward the ideal case (one step to the right), the implication result B' becomes a thin stripe located on the maximum possibility value of B. This can be viewed as removing the fuzziness from B when the ideal case is reached on the conditional part of the rule. In the opposite direction, when the evidence moves to the edges, the thickness of B' increases, reaching toward a complete rectangle covering the entire consequent universe. This means the fuzziness of B is at maximum. Note that the total area is inversely proportional to the degree of fulfillment (DOF) of the LHS conditions.

8.2.6 Drastic Product Operator

The drastic product implication operator yields an implication relation expressed by

$$R(x,y) = \phi[\mu_A(x), \mu_B(y)] = \begin{bmatrix} \mu_A(x) & \text{if} & \mu_B(y) = 1 \\ \mu_B(y) & \text{if} & \mu_A(x) = 1 \\ 0 & \text{if} & \mu_A(x) < 1, \mu_B(y) < 1 \end{bmatrix}$$

(8.8)

producing the following data (x rows y columns).

Chap. 8 Implication Process

.00	.00	.00	.00	.00	.00	.00	.00	.00	.00
.00	.00	.00	.00	.25	.25	.00	.00	.00	.00
.00	.00	.00	.00	.50	.50	.00	.00	.00	.00
.00	.00	.00	.00	.75	.75	.00	.00	.00	.00
.00	.25	.50	.75	1.00	1.00	.75	.50	.25	.00
.00	.00	.00	.00	.75	.75	.00	.00	.00	.00
.00	.00	.00	.00	.50	.50	.00	.00	.00	.00
.00	.00	.00	.00	.25	.25	.00	.00	.00	.00
.00	.00	.00	.00	.00	.00	.00	.00	.00	.00
.00	.00	.00	.00	.00	.00	.00	.00	.00	.00

The implication relation surfaces are shown in Fig. 8.14. The final result B' given the new evidence A' is a slice shown by the arrow that corresponds to the fourth data row above.

As can be seen from Fig. 8.14, the drastic product implication operator generates a nonbinary decision surface in which elements vary between 1 and 0. The result B' is shown in Fig. 8.15.

As the evidence A' moves toward the ideal case (one step to the right), the implication result B' becomes equal to the entire shape of B. This irregular behavior can also be viewed from Fig. 8.14. In the opposite direction, when the evidence moves to the edges, the thickness of B' remains the same (a thin strip equal to the thickness of the trapezoid at the top) while its height decreases. Note that the total area is directly proportional to the DOF of the LHS conditions. Because of direct proportionality around the ideal case, we classify this implication operator as a first category operator.

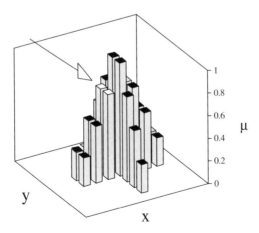

Figure 8.14 A drastic product implication relation between a triangular and a trapezoidal membership function.

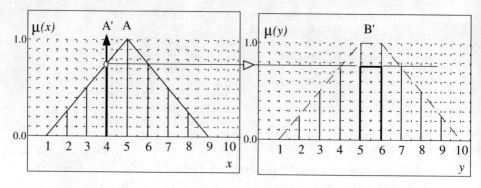

Figure 8.15 Inference result B' given evidence A' using the drastic product implication operator.

8.2.7 Zadeh Operator

The Zadeh implication operator yields an implication relation expressed by

$$R(x,y) = \phi[\mu_A(x),\mu_B(y)] = (\mu_A(x) \wedge \mu_B(y)) \vee (1 - \mu_A(x)) \qquad (8.9)$$

producing the following data (x rows y columns).

1	1	1	1	1	1	1	1	1	1
0.75	0.75	0.75	0.75	0.75	0.75	0.75	0.75	0.75	0.75
0.5	0.5	0.5	0.5	0.5	0.5	0.5	0.5	0.5	0.5
0.25	0.25	0.5	0.75	0.75	0.75	0.75	0.5	0.25	0.25
0	0.25	0.5	0.75	1	1	0.75	0.5	0.25	0
0.25	0.25	0.5	0.75	0.75	0.75	0.75	0.5	0.25	0.25
0.5	0.5	0.5	0.5	0.5	0.5	0.5	0.5	0.5	0.5
0.75	0.75	0.75	0.75	0.75	0.75	0.75	0.75	0.75	0.75
1	1	1	1	1	1	1	1	1	1
1	1	1	1	1	1	1	1	1	1

The implication relation surfaces are shown in Fig. 8.16. The final result B' for given A' is a slice shown by the arrow that corresponds to the fourth row above.

As can be seen from Fig. 8.16, the Zadeh implication operator generates a nonbinary decision surface in which elements vary between 1 and 0. The result B' is shown in Fig. 8.17.

As the evidence A' moves toward the ideal case (one step to the right), the implication result B' becomes identical to B. In the opposite di-

Chap. 8 Implication Process 419

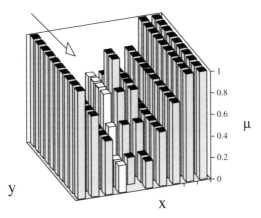

Figure 8.16 A Zadeh implication relation between a triangular and a trapezoidal membership function.

rection, when the evidence moves to the edges, the vertical thickness of B' increases while its horizontal thickness remains the same. This behavior continues until a critical DOF value (0.5), where the behavior changes afterwards. Below the critical DOF value, this operator acts like the inverse of the Mamdani operator. Notice the negation part in Eq. (8.5) (1-µ) which causes the area to be inversely proportional to the DOF. Because of inverse proportionality around the ideal case, we classify this implication operator as a second category operator.

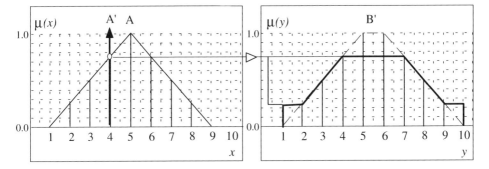

Figure 8.17 Inference result B' given evidence A' using the Zadeh implication operator.

8.2.8 Bounded Product Operator

The bounded product implication operator yields an implication relation expressed by

$$R(x,y) = \phi[\mu_A(x), \mu_B(y)] = 0 \vee (\mu_A(x) + \mu_B(y) - 1) \qquad (8.10)$$

producing the following data (x rows y columns).

0	0	0	0	0	0	0	0	0	0
0	0	0	0	0.25	0.25	0	0	0	0
0	0	0	0.25	0.5	0.5	0.25	0	0	0
0	0	0.25	0.5	0.75	0.75	0.5	0.25	0	0
0	0.25	0.5	0.75	1	1	0.75	0.5	0.25	0
0	0	0.25	0.5	0.75	0.75	0.5	0.25	0	0
0	0	0	0.25	0.5	0.5	0.25	0	0	0
0	0	0	0	0.25	0.25	0	0	0	0
0	0	0	0	0	0	0	0	0	0
0	0	0	0	0	0	0	0	0	0

The implication relation surfaces are shown in Fig. 8.18. The final result B' for given A' is a slice shown by the arrow and corresponds to the fourth data row above.

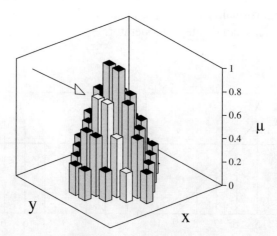

Figure 8.18 A bounded product implication relation between a triangular and a trapezoidal membership function.

Chap. 8 Implication Process

As can be seen from Fig. 8.18, the bounded product implication operator generates a nonbinary decision surface in which all elements vary between 1 and 0. The result B' is shown in Fig. 8.19.

As the evidence A' moves toward the ideal case (one step to the right), the implication result B' becomes equal to B. This can be viewed as a complete match when the ideal case is reached on the conditional part of the rule. In the opposite direction, when the evidence moves to the edges, B' keeps the shape of B scaled by DOF. At the extreme edge points, B' becomes zero. The area of B' is directly proportional to the DOF; thus we classify this operator as a first category operator.

The operator's behavior is regular, which makes it easy to anticipate the outcome of the modeled reasoning. Like most of the first category implication operators, a bounded product operator might be viewed as a match-seeking paradigm. The differences between the first category implication operators are explained later in this chapter.

8.2.9 Gougen Operator

The Gougen implication operator yields an implication relation expressed by

$$R(x,y) = \phi[\mu_A(x), \mu_B(y)] = \begin{pmatrix} 1 & if & \mu_A(x) \leq \mu_B(y) \\ \dfrac{\mu_B(y)}{\mu_A(x)} & if & \mu_A(x) > \mu_B(y) \end{pmatrix}$$

(8.11)

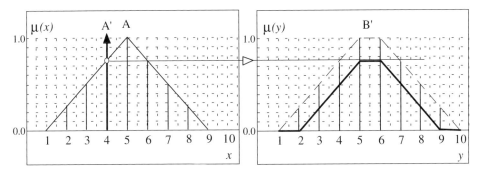

Figure 8.19 Inference result B' given evidence A' using the bounded product implication operator.

producing the following data (x rows y columns).

1	1	1	1	1	1	1	1	1	1
0	1	1	1	1	1	1	1	1	0
0	0.5	1	1	1	1	1	1	0.5	0
0	0.33	0.67	1	1	1	1	0.67	0.33	0
0	0.25	0.5	0.75	1	1	0.75	0.5	0.25	0
0	0.33	0.67	1	1	1	1	0.67	0.33	0
0	0.5	1	1	1	1	1	1	0.5	0
0	1	1	1	1	1	1	1	1	0
1	1	1	1	1	1	1	1	1	1
1	1	1	1	1	1	1	1	1	1

The implication relation surfaces are shown in Fig. 8.20. The final result B' for given A' is a slice shown by the arrow and corresponds to the fourth data row above.

As can be seen from Fig. 8.20, the Gougen implication operator generates a nonbinary decision surface in which elements vary between 1 and 0. The result B' is shown in Fig. 8.21.

As the evidence A' moves toward the ideal case (one step to the right), the implication result B' becomes equal to B. This can be viewed as a complete match when the ideal case is reached on the conditional part of the rule. In the opposite direction, when the evidence moves to the edges, the Gougen implication operator widens the original shape of B to

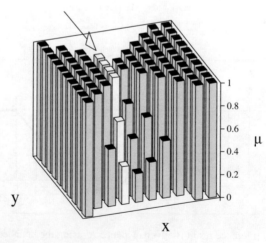

Figure 8.20 A Gougen implication relation between a triangular and a trapezoidal membership function.

Chap. 8 Implication Process

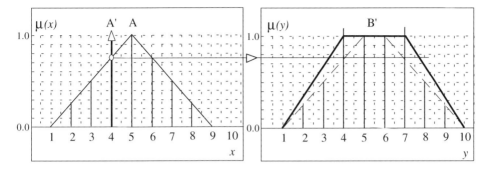

Figure 8.21 Inference result B' given evidence A' using the Gougen implication operator.

a trapezoid with the top two coordinates determined by the intersecting points between the DOF and B. At the extreme edge points, B' becomes a rectangle covering the entire universe. The area of B' is inversely proportional to the DOF; thus we classify this operator as a second category operator. The Gougen implication operator, like many other second category operators, searches for the designed uncertainty that only occurs at ideal case. Departure from the ideal case increases the fuzziness of B'.

8.2.10 Godelian Operator

The Godelian implication operator yields an implication relation expressed by

$$R(x,y) = \phi[\mu_A(x), \mu_B(y)] = \begin{pmatrix} 1 & if & \mu_A(x) \leq \mu_B(y) \\ \mu_B(y) & if & \mu_A(x) > \mu_B(y) \end{pmatrix}$$

(8.12)

producing the following data (x rows y columns).

```
1  1     1     1     1  1     1     1     1     1
0  1     1     1     1  1     1     1     1     0
0  0.25  1     1     1  1     1     1     0.25  0
0  0.25  0.5   1     1  1     1     0.5   0.25  0
0  0.25  0.5   0.75  1  1     0.75  0.5   0.25  0
0  0.25  0.5   1     1  1     1     0.5   0.25  0
0  0.25  1     1     1  1     1     1     0.25  0
0  1     1     1     1  1     1     1     1     0
1  1     1     1     1  1     1     1     1     1
1  1     1     1     1  1     1     1     1     1
```

The implication relation surfaces are shown in Fig. 8.22. The final result B' for given A' is a slice shown by the arrow and corresponds to the fourth data row above.

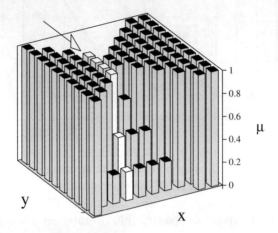

Figure 8.22 A Godelian implication relation between a triangular and a trapezoidal membership function.

As can be seen from Fig. 8.22, the Godelian implication operator generates a nonbinary decision surface in which all elements vary between 1 and 0. The result B' is shown in Fig. 8.23.

As the evidence A' moves toward the ideal case (one step to the right), the implication result B' becomes equal to B. This can be viewed as a complete match when the ideal case is reached on the conditional

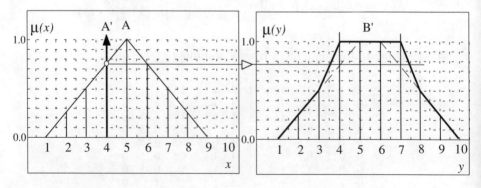

Figure 8.23 Inference result B' given evidence A' using the Godelian implication operator.

Chap. 8 Implication Process 425

part of the rule. In the opposite direction, when the evidence moves to the edges, B' widens outward to a concave trapezoid with the two top coordinates determined by the intersecting points between the DOF and B. At the extreme edge points, B' becomes a rectangle covering the entire universe. The area of B' is inversely proportional to the DOF; thus we classify this operator as a second category operator.

8.3 BEHAVIORAL PROPERTIES

We have pointed out earlier that one of the criteria for selecting an implication operator is the behavior in which a fuzzy system approaches toward the most ideal case from the least ideal case. In the *IF A THEN B* example, the ideal case occurs when the new evidence A' completely matches the antecedent predicate A, producing a possibility value of 1.0. This is shown in Fig. 8.24 by the fifth A'.

Note that the LHS of the rule examined here (*IF A*) could have been in a more complex form. However, the complexity of the LHS propositions does not affect the ongoing analysis because we are only interested in a range of degrees of fulfillment between 0 and 1 that will propagate to the implication operation. Such a range of values is assumed to exist.

Remember that the implication result B' is a new fuzzy set formed according to the selected implication operator [5]. The shape of B' and its

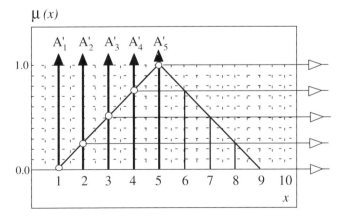

Figure 8.24 A series of new evidences (A') evaluated on the LHS of *IF A THEN B*.

location on the output universe represent a contribution to the final aggregated result. Contribution is characterized by the area of B', and its size determines the strength of contribution. The behavior refers to the rate of change of the size of B' with respect to new evidences presented. Note that locations are fixed via design and are not affected by the implication operation.

Illustrated in the last section was that the implication operators fall into two categories: one aggregated with the union operator, the other aggregated with the intersection operator. The size of B' is directly proportional to the amount of contribution when the first category of implication operators are used. The size of B' is inversely proportional to the amount of contribution when the second category of implication operators is used due to negation. Therefore, the behavior analysis is conducted separately.

8.3.1 Area Under the Curve Analysis

When the degree of fulfillment on the LHS of a rule approaches to the ideal case (possibility = 1.0), the implication result B' represents the biggest contribution to the aggregated result. Remembering this fact, let's configure the size behavior for the case shown in Fig. 8.24. The area under the B' curve for each A' is computed by[1]

$$Area(A') = \sum_{i=1}^{N} \mu_{B'}(y_i) \Delta y \qquad (8.13)$$

and plotted in Fig. 8.25 for the first category implication operators. This summation is the addition of all the elements (times the width Δy, which is 1) in each row in each data box presented in the previous sections.

Figure 8.25 indicates that the four implication operators (Mamdani, Larsen, bounded, and drastic) normally used with union aggregation produce the same area at the ideal case. Therefore, the contribution of B' to the aggregated result will be the same using any of these operators when the LHS condition is satisfied completely (crisp solution). However, on the fuzzy surfaces the contributions differ. As can be seen from Fig. 8.25, the order of Mamdani, Larsen, bounded, and drastic implication operators indicates a decrease in tolerance (i.e., Mamdani is the most

---

[1] The summation sign denotes algebraic addition, not the union operation.

Chap. 8   Implication Process

Area under the B' curve

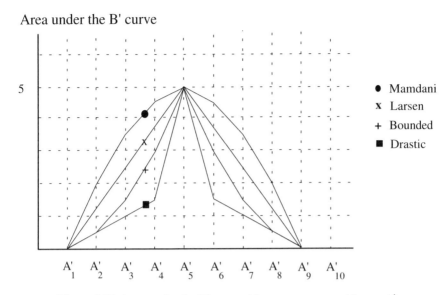

Figure 8.25 Areas under the B' curve with respect to new evidences A'.

tolerant and drastic is the least tolerant). The same order indicates an increase in conservatism. Tolerance means the degree of willingness to take an action (or to make a decision) given fuzzy information. The difference of tolerances is maximum at points A'_4 and A'_6 where different implication operators produce different areas for the same fuzzy evidence. These discussions lead to the following design principle.

> The choice among Mamdani, Larsen, bounded, and drastic implication operators, which are accompanied by the union aggregation, must be made by considering the level of tolerance they employ.

It is obvious from Fig. 8.25 that if the level of fuzziness a system may experience increases (or if a system operates on fuzzy surfaces more often) the effect of the choice of an implication operator becomes more important.

Although there is no closed-form formula to evaluate tolerances, it is intuitively known that higher tolerances improve the chance of finding at least one solution at the expense of imprecision. On the contrary, low tolerances improve precision at the expense of not finding a valid solution in some cases. The balance between tolerance versus conservatism is somewhat related to the decision boundaries in a mapping process, and therefore related to the design of the membership functions. If the membership functions are designed to yield poor mapping (i.e., if there is not adequate knowledge for design) then a tolerant implication operator tends to be more suitable.

> Tolerant implication operators such as Mamdani and Larsen yield a larger area of contribution to the aggregated result than do conservative implication operators such as bounded and drastic product for the same fuzzy evidence (input).

There is no direct relation between the tolerance characteristics of the implication operators and the type of problems (control, classification, etc.) to which they can be applied. In other words, there might be two different control problems, one requiring the most tolerant treatment and the other requiring the most conservative treatment. This situation can be encountered in other types of problems as well. Although this is a trivial conclusion from the fuzzy systems point of view, it is somewhat important for the designer to remember in the form of the following design principle.

> The choice of implication operators cannot be characterized by the type of problem at hand, but it can be related to the properties of the problem.

The second category of implication operators (based on intersection aggregation) exhibit a different behavior. The areas under the B' curves for all A's are plotted in Fig. 8.26.

The same discussions presented earlier can be applied to the implication operators shown in Fig. 8.26 in the reverse manner, such that a smaller area yields a larger contribution to the aggregated result due to intersection. This is depicted in Fig. 8.27 by assuming the center-of-gravity defuzzification method. Note that this property is not dependent on the defuzzification method of choice.

Chap. 8 Implication Process   429

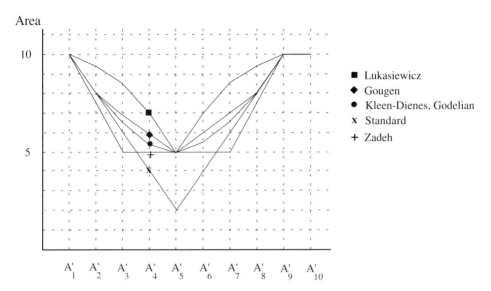

Figure 8.26 Areas under the B' curve with respect to new evidences A'

Among the implication operators shown in Fig. 8.26, the standard operator settles on a different ideal case compared to the other operators listed. Zadeh, Godelian, Kleen-Dienes, Gougen, and Lukasiewicz implication operators yield the same area at the ideal case. Note that the areas produced by the Kleen-Dienes and Godelian operators are equal; however, the shapes of those areas are different as shown in the previous sec-

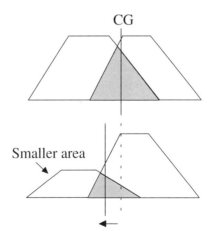

Figure 8.27 When aggregation is done via intersection, a smaller area contributes more to the aggregated result.

tion. Thus, their contributions are different. The order of Zadeh, Godelian/ Kleen-Dienes, Gougen, and Lukasiewicz implication operators indicate a decrease in tolerance and an increase in conservatism.

8.3.2 Rate of Change Analysis

The second criterion is the rate of change (i.e., the speed) of the implication result B' from one item of evidence A' to another. This criterion only applies to events or processes that are sequentially correlated. For example, the control actions of steering a bicycle are sequentially correlated because one move after another makes a difference. On the other hand, insurance risk assessment from one applicant to another is not a sequentially correlated event.

> Implication operator selection based on "rate of change of total area" criterion only applies to sequentially correlated events.

The rate of change of the areas under the B' curve is computed by taking the differences of areas per unit change in A'.

$$\frac{\Delta Area(A')}{\Delta A'} = \left(\sum_{i=1}^{N} \mu_{B'}(y_i)\Delta y\right)_{A'_k} - \left(\sum_{i=1}^{N} \mu_{B'}(y_i)\Delta y\right)_{A'_{k+1}} \quad (8.14)$$

Because approaching the ideal case from left or from right possesses the same properties, we will only look at the left approach for simplicity. All implication operators are marked in Fig. 8.28, indicating their speed in approaching to the ideal case (crisp solution). On the upper half of the plot, the first category implication operators are grouped where the rates are positive. Among them, the rate of change of area using the Mamdani operator slows down when approaching to the crisp solution. On the other extreme, the rate of change of area using the drastic operator speeds up. The second category implication operators shown in the bottom half have negative rates approaching from the left. Among them, the rate of change of area using the Zadeh, Gougen, Kleen-Dienes, and Godelian operators slows down, whereas that of the Lukasiewicz operator speeds up toward the ideal case.

Chap. 8 Implication Process 431

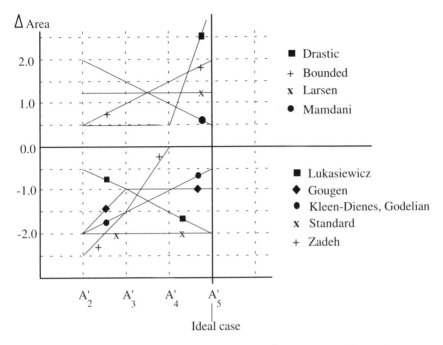

Figure 8.28 Rate of change of areas under the B' curve approaching to the ideal case (crisp solution) from the left.

It is apparent that the rate-of-change property of each implication operator is another expression of the tolerance/conservatism property discussed earlier. However, there is no established theory that directly relates the behavioral properties of the implication operators to the problem types. Considering the difficulty of such a task, a theory might never be developed or it might never be possible. Nevertheless, practical experience and intuitive reasoning give us some guidelines to follow.

Control of feedback systems using a rule-based fuzzy control law is mainly a mapping task between the state variables of the system and control inputs to the system. State variables, which form the antecedent variable space, are defined by crisp points on the antecedent space filled with fuzzy regions among them. In most of the applications, crisp points represent the best knowledge whereas fuzzy regions represent uncertain (or interpolated) points. Crisp points on the antecedent space produce the highest degrees of fulfillment (LHS conditions are completely fulfilled in a rule) which gives the highest authority to a rule to contribute to the final control action. In a dynamic system where an independent variable is time, it becomes important how much time is spent to give a rule full authority when the system is traveling between crisp points A and B. This is depicted in Fig. 8.29 where a hypothetical antecedent variable *Temperature* has two membership functions called *Cool* and *Warm*.

As shown earlier, tolerant implication operators employ a high rate of change of area at low DOF values, which means they quickly move away from fuzzy regions. They also slow down at the vicinity of crisp regions where the corresponding control rules remain effective. Experience shows that fuzzy controllers using tolerant implication operators with such characteristics (especially Mamdani, Larsen, Zadeh and Godelian) exhibit favorable stability behavior [6].

When designing a fuzzy controller for feedback systems, implication operators whose rate of change of area slows down near crisp solutions and speeds up away from crisp solutions must be considered to be the best candidates as practical experience suggests.

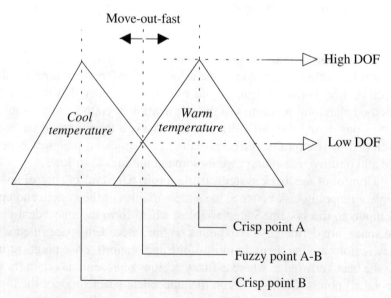

Figure 8.29 Fuzzy controller using the antecedent variable *Temperature* to make a control decision.

Note that the principle stated above is a reflection of the experience gathered over the operational records of many simple fuzzy controllers used in the industry. Besides simple fuzzy controllers, there is not enough practical knowledge about other types of problems to make a judgment on which implication operator works the best. In general, the first principle of fuzzy system design applies to the selection of an implication operator, which suggests that the selection is completely determined by a context-specific solution. In its determination however, tolerance/conservatism and rate-of-change characteristics can be considered in the presence of constraints imposed by the specific solution.

8.3.3 The Rule Activity Perspective

Another way to look at the rate-of-change criterion is to monitor rule activity for a range of input values similar to the example shown in Fig. 8.28. The rule activity curves in Fig. 8.30 (for three rules with the same consequent membership function shape) are in fact the same as the rate of change of area curves shown in Fig. 8.28.

Three rules are assumed to fire one after another in the course of input absorption in Fig. 8.30. A band of action is selected only to make a point. Degree of fulfillment of each rule changes linearly between 0 and

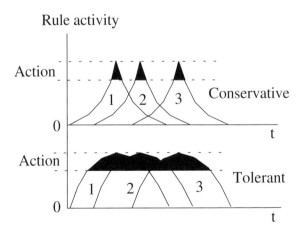

Figure 8.30 Conservative and tolerant implication operators producing different rule activity dynamics between 3 rules within the confined action-band.

1, like the case shown in Fig. 8.28. Between the two extreme implication operators, a conservative operator fires each rule in a more reserved manner than that of a tolerant operator (horizontal thickness of the shaded area can be used in Fig. 8.30 to visualize this conclusion). To make an analogy, the conservative operator acts like an alarm system by firing the rules when it is absolutely[2] necessary.

The designer must examine the properties of the problem at hand in an intuitive manner and decide whether the following principle applies to the problem at hand.

> When the cost of action is expensive, approximated solution is intolerable, or decision uncertainty is risky, then the selection of conservative implication operators over tolerant ones can inherently exert conservatism suitable for the implied objective.

Some optimization, pattern recognition, classification, and risk assessment problems can make use of this principle. Note that rule activity is not directly reflected at the output due to the aggregation process. Although the principle stated above holds in the rule activity domain, it is imperative to consider the aggregation process to accurately estimate the actual output activity.

8.3.4 Pendulum Analogy

The previous categorization of the implication operators depending on the proportionality between the area under the curve and DOF is quite informative when we need to classify the operators based on how they approach a crisp point. Again, our discussions are restricted to sequentially correlated events. As shown in Fig. 8.28, both the first and second category implication operators exhibit different rate-of-change (i.e., speed) characteristics. Some of them speed up whereas others slow down approaching a crisp (ideal) point. Therefore we present another categorization in Table 8.1: slow, fast, and steady approach implication operators. *Approach* here again refers to changing DOF from the most fuzzy condition (0) to the ideal case (1).

---

[2] Figure of speech, not a mathematical measure.

Chap. 8 Implication Process   435

Table 8.1 Implication Operators Categorized Based on Their Approach to Ideal Case

SLOW APPROACHERS	FAST APPROACHERS	STEADY APPROACHERS
Mamdani	Drastic	Larsen
Kleen-Dienes	Bounded	Standard
Godelian	Lukasiewicz	
Zadeh		
Gougen		

The dynamic behavior of the fast-approaching implication operators can be better understood by the pendulum analogy as shown in Fig. 8.31. A pendulum swings around its equilibrium point. According to the law of the conservation of energy, the fastest speed occurs while the mass is passing from its equilibrium point at all times until swing completely stops.

Because the fast-approaching implication operators speed up when DOF is maximum (i.e., at the crisp point of an antecedent membership function), the dynamic behavior resembles that of a pendulum. Note that a pendulum is inherently a stable system since it will eventually settle down

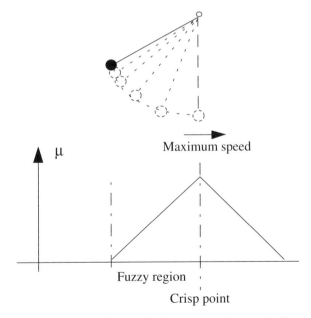

Figure 8.31 Fast-approaching implication operators behave similar to a pendulum motion.

at its equilibrium position. For example, if the crisp point in Fig. 8.31 represents the equilibrium point of an open-loop system, then a fuzzy controller using a fast-approaching implication operator will maintain the system at its equilibrium in a dynamic manner just like a pendulum motion.

> The fast approaching implication operators such as drastic, bounded, and Lukasiewicz can be used in a fuzzy inference design if the system dynamics are inherently stable and/or if spending increasingly longer time moving away from crisp points is tolerable.

System stability is an obvious requirement in application to control problems. Stability becomes vitally important if the crisp point of the antecedent membership function represents equilibrium, whereas fuzzy surfaces might correspond to states of nonequilibrium. That is because fast implication operators spend more time on fuzzy surfaces (assumed unstable regions) than on the vicinity of crisp points (assumed stable regions). In application to other problems, tolerance and cost are the two issues to consider. In data-driven design, if the stability of the system is not known for some crisp points, then the arguments presented here do not apply. However, this is rarely seen in practice and such a design is not recommended.

The dynamic behavior of the slow-approaching implication operators resembles that of an inverted pendulum. Just like balancing a stick on a vertical position, the control actions away from equilibrium are fast and they slow down as approaching to equilibrium. This is depicted in Fig. 8.32, where the equilibrium corresponds to the crisp point of a consequent membership function.

An inverted pendulum is inherently unstable; thus, control actions are capable of stabilizing an unstable system. Based on this analogy and assuming that the crisp point on the membership function corresponds to an equilibrium, slow-approaching implication operators are more suitable for control problems in which open-loop stability is not guaranteed.

> The slow-approaching implication operators such as Mamdani, Kleen-Dienes, Godelian, and Gougen can be used in a fuzzy inference design of dynamic processes with more confidence if the stability of open-loop dynamics is not guaranteed.

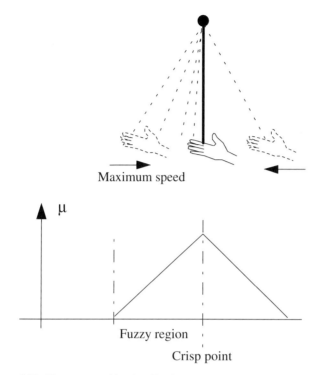

Figure 8.32 Slow-approaching implication operators behave like an inverted pendulum motion.

Note that there is no practical way to prove the stability of slow-approaching implication operators since overall system stability depends on several other factors.

8.4 AGGREGATION DESIGN

In Chapters 2 and 3, we defined aggregation to be a method for fusing different implication results into one final result. Designing of an aggregation process is trivial, because the couplings between implication operators and the accompanying aggregation operators are suggested in the literature, as shown in Table 8.2. The designer's task is to implement one of the pairs from this table.

Let's examine the aggregation operation to be able to discuss other design issues. An aggregation operation takes place among all of the con-

Table 8.2 Aggregation and Implication Operators

IMPLICATION OPERATOR	AGGREGATION OPERATOR
Mamdani	Union ($\vee$)
Larsen	Union ($\vee$)
Lukasiewicz	Intersection ($\wedge$)
Kleen-Dienes	Intersection ($\wedge$)
Standard Sequence	Intersection ($\wedge$)
Drastic Product	Union ($\vee$)
Zadeh	Intersection ($\wedge$)
Bounded Product	Union ($\vee$)
Gougen	Intersection ($\wedge$)
Godelian	Intersection ($\wedge$)

tributing rules of one consequent fuzzy variable. Considering the rule block definition, there is one aggregation operation for every rule block. An aggregation operator performs either a union (OR) or an intersection (AND) operation among the implication results, which are fuzzy sets. Consider a rule block of the form

$$X_i \bullet P_{i,x} \to Y_k \bullet P_{k,1}$$
$$X_j \bullet P_{j,y} \to Y_k \bullet P_{k,2}$$
$$\ldots$$
$$X_m \bullet P_{m,z} \to Y_k \bullet P_{k,N}$$
(8.15)

where the consequent Y_k is the only output variable used in all rules. Antecedents can be any input variable. For simplicity, we ordered the predicates from 1 to N, which may not be as neatly aligned in an actual application. The linguistic equivalence is illustrated below by a simple example with three fuzzy rules.

IF Temperature IS Low THEN Valve IS Closed
IF Humidity IS High THEN Valve IS Half_open
IF Pressure IS Medium THEN Valve IS Almost_Closed

The evaluation of these three fuzzy rules for a given set of inputs (T_1, H_1, P_1) along with the Mamdani implication operation are depicted in Fig. 8.33. The shaded areas represent individual implication results.

Now, if we look at the information gathered on the consequent domain it is apparent that it should be somehow fused (or composed) to obtain a final result. Union and intersection operations shown in Fig. 8.34 are the two practical options for aggregation.

Chap. 8 Implication Process 439

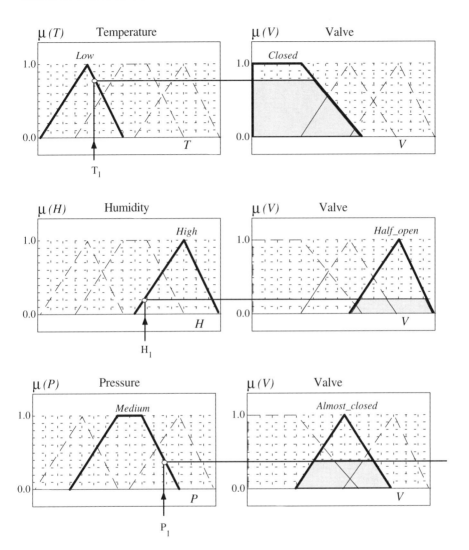

Figure 8.33 Evaluation of three rules in a rule block before aggregation.

The Mamdani implication operator is used with union aggregation. However, we have also shown the intersection in Fig. 8.34 for illustration purposes. The shaded area is a new fuzzy set (in either case) that represents the result of this inference procedure. The utility of such a result in practical life is an issue related to the decomposition-defuzzification topic discussed later in this chapter.

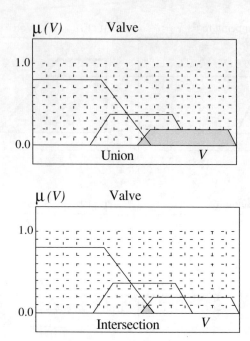

Figure 8.34 Union and intersection operations to obtain an aggregated result.

8.4.1 Multiple (Additive) Aggregation

In Eq. (8.15), the consequent predicates are aligned in an order (i.e., from 1 to N); however, there is no restriction in general on whether or not a predicate can be used more than once. Assume that the specific solution acquired from an expert is in a form in which some consequent predicates are used more than once. This is represented by the second and third lines in Eq. (8.16), using the predicate $P_{k,2}$ twice.

$$\begin{aligned}
X_i \bullet P_{i,x} &\to Y_k \bullet P_{k,1} \\
X_j \bullet P_{j,y} &\to Y_k \bullet P_{k,2} \\
X_n \bullet P_{n,w} &\to Y_k \bullet P_{k,2} \\
&\cdots \\
X_m \bullet P_{m,z} &\to Y_k \bullet P_{k,N}
\end{aligned} \quad (8.16)$$

Following the previous example, the linguistic equivalence can be stated as shown below. The predicate *Half_open* is used in two rules.

IF Temperature IS Low THEN Valve IS Closed
IF Humidity IS High THEN Valve IS Half_open
IF Flow IS Too_low THEN Valve IS Half_open
...
IF Pressure IS Medium THEN Valve IS Almost_Closed

The designer has two choices in such a case. The first avenue is to apply a standard aggregation method. If the rule composition such as the one above is made in an evidence collection manner—that is, when each evidence increases the strength of a decision as explained in Chapter 7—then the aggregation process must be modified to accommodate this effect [7]. Assume that the additional third rule produces an implication result as shown in Fig. 8.35.

One method to implement multiple aggregation is to add the contributions. Also called the *fuzzy additive method,* this operation is nothing but a bounded sum in which the outcome cannot exceed 1.0. In the example below, the sum of the heights $0.35 + 0.7 = 1.05$ yields 1.0, which is the full shape of the membership function as illustrated in Fig. 8.36.

The tip of the membership function in the middle is the multiple aggregation effect. If the number of evidence rules is high, then the bounded sum operation may not be adequate to produce the intended effect. In such a case, the multiple aggregation effect can be implemented during the decomposition/defuzzification stage using the center_of_area method. This is discussed in Section 8.5.1.

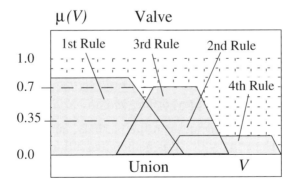

Figure 8.35 Multiple aggregation case.

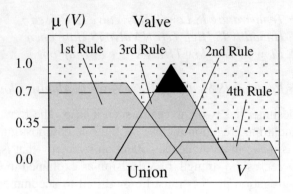

Figure 8.36 Fuzzy additive method for multiple aggregation.

8.4.2 Right-Hand-Side Operators

In a standard rule formation, logical operators after the THEN operator are not seen. However, in more expressive representations, the ELSE operator is sometimes used to denote aggregation. The basic fuzzy inference algorithm does not restrict the use of other logic operators such as AND and OR after THEN as long as their algorithmic functions are well defined. Let's first examine ELSE.

Although it has not been frequently used in the literature, the function of ELSE when used as an aggregation operator is not equivalent to its true linguistic meaning. ELSE means OTHERWISE in daily language as well as in traditional computer terminology implying unsatisfied conditions. The trouble with ELSE is that it is a crisp comparator unsuitable for fuzzy systems. For example, the use of ELSE below implies that the condition in the first rule has a priority such that if the first rule does not fire, then the statement following the ELSE operator will hold.

> *IF Weather IS Nice THEN I WILL Play_outside*
> *ELSE I WILL Stay_inside_and_watch_TV*

The problem is that rules can fire partially, and they seldom fire at full strength. Thus, fuzzy ELSE is nothing but a threshold control on the DOF value on the conditional part. If the fulfillment of the first condition is below the threshold, the second action will take place. This brings out a new problem. If the DOF value is below the threshold, will it be used for the second statement? One practical solution is to use 1−DOF value for the consequent of the second statement that models the logical proposition *otherwise* properly.

Similarly, there are cases of using AND and OR on the RHS of a rule without implying the aggregation operation. The operator AND is used to denote multiple actions as illustrated below.

IF Savings_account IS Low THEN Reduce_spending AND
Increase_savings

Obviously, the AND operator above does not imply intersection aggregation; rather, it produces two simultaneous actions having satisfied a single fuzzy condition. This structure can be modified to fit the basic fuzzy inference algorithm by composing two rules:

IF Savings_account IS Low THEN Reduce_spending
IF Savings_account IS Low THEN Increase_savings

Note that the consequents are different, therefore these rules belong to different rule blocks. Using *OR* logic on the RHS is more problematic since it implies a free choice of one of the actions.

IF Savings_account IS Low THEN Reduce_spending OR Increase_savings

Because the two consequents (*Spending, Saving*) are different, there is no union aggregation implied in this structure. There are simply two actions to be taken based on an unspecified selection criterion. In the framework of the basic fuzzy inference algorithm, this structure can be represented by taking the same avenue as in the previous case (i.e., composing two rules) and by including an output processing message with an "either" clause. Apparently, a direct hardware implementation of outputs—such in process fuzzy controllers—must deal with such a situation by specifying a selection criterion, or by resolving the rule structure during design.

8.5 DESIGNING A DEFUZZIFICATION/ DECOMPOSITION PROCESS

Aggregated results obtained from a fuzzy inference engine are the final results within the scope of fuzzy logic theory. A need for defuzzification arises from the fact that such results are not practically useful in real life applications. The defuzzification step is an approximation itself [8,9], based on the assumption that a scalar will represent a fuzzy set in an appropriate manner. When the aggregation process is viewed as the contribution of individual decisions, then defuzzification can be viewed as acquiring a popular vote or consensus. The defuzzification techniques established in the literature are developed with this view in mind.

Designing a defuzzification process is again a selection among a few viable options established in the literature as well as in industrial applications. We will first examine the properties of those most widely used, then discuss the selection criteria. These options are listed in Table 8.3.

Other methods such as far and near edge decomposition, and preponderance of evidence defuzzification are not as frequently implemented as the ones listed in Table 8.3. However, interested readers can find such methods in the literature [10].

8.5.1 Defuzzification Techniques

The most frequently used defuzzification method is the center of gravity (centroid) technique, which is analogous to finding the balance point by calculating the weighted mean of the fuzzy output (also known as composite moments).

$$x' = \frac{\sum_{i=1}^{N} x_i \, \mu_o(x)}{\sum_{i=1}^{N} \mu_o(x)} \qquad (8.17)$$

The membership function $\mu_o(x)$ represents the fuzzy set of the final output (either aggregated or single) of one fuzzy variable, and x is the location of each singleton on the universe of discourse. Because of two-dimensional output fuzzy sets, this method is also known as the center-of-area method. In Eq. (8.18), C is the center of gravity of each area A and Y is the balanc-

Table 8.3 Most Widely Used Defuzzification Design Options

DEFUZZIFICATION METHODS
Center of gravity (centroid), center of area
Maximum possibility
Center of maximum possibilities, mean of maximum possibilities
Center of mass of highest intersected region
Others

ing point. This formulation allows incorporating rule importance weights in the following manner.

$$Y = \frac{\sum_{j=1}^{N} w_j \overline{C_j} \overline{A_j}}{\sum_{j=1}^{N} w_j \overline{A_j}} \quad (8.18)$$

In Eq. (8.18), w is the rule importance weight. Note that the index j in these equations corresponds to one fuzzy rule in a rule block. Another common method finds the maximum possibility point on the universe of discourse as the answer to defuzzification.

$$x' = x_i \quad \mu_o(x_i) > \mu_o(x_j) \quad j = 1, \ldots, N \quad j \neq i \quad (8.19)$$

If the maximum point is nonsingular (plateau), then the defuzzified output is the average of maximums or the center of maximums. Also called the mean of maximums, this defuzzification technique is normally employed after union aggregation just like the centroid method.

The center of mass method finds the region that has the highest density of intersecting fuzzy sets. Thus, it is employed with intersection aggregation. The defuzzification computation using this method is performed in parallel with the aggregation process because the location of each individual solution (fuzzy set) needs to be known. The final region, which has the highest density of intersection, is computed by counting the frequency of inclusion, or simply by taking the highest possibility point from intersected fuzzy sets.

$$x' = x_i \quad \mu_o(x_i) \geq \max[\mu_1(x) \wedge \mu_2(x) \wedge \ldots, \mu_n(x)] \quad (8.20)$$

8.5.2 Geometric Interpretation

Defuzzification methods are often better understood by visualizing them using a hypothetical example. Suppose that a linguistic variable X has three predicates appear on the right-hand side of three rules. Also assume that the membership functions corresponding to the predicates were subject to an implication process where the three outputs before aggregation are shown in Fig. 8.37. The four defuzzification methods applied to the example are shown in Figs. 8.38 through 8.41.

Note that x' is in fact a varying entity $x'(t)$ with respect to the input absorption process. Thus, the behavior of the defuzzified output from one

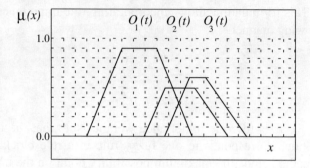

Figure 8.37 Three hypothetical implication results that belong to consequent X.

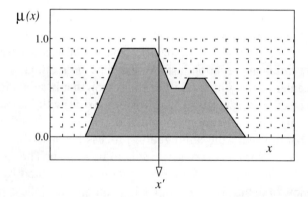

Figure 8.38 Union aggregation and center-of-gravity method for defuzzification.

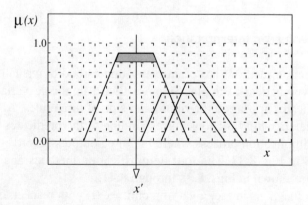

Figure 8.39 Center of maximum possibility method for defuzzification.

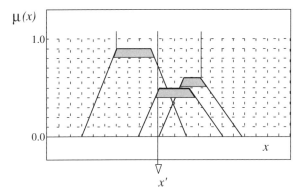

Figure 8.40 Center of maximum possibilities for defuzzification.

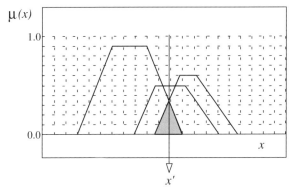

Figure 8.41 Center of mass with the highest density of intersection for defuzzification.

case to another is as important as the instantaneous output itself. There are other defuzzification techniques suggested in the literature; however, they are the product of theoretical advances with a scant history of application.

8.5.3 Design Criteria

The most important design criterion in selecting a defuzzification method is responsiveness versus sluggishness in response to the formation of implication results. Although it depends on the shape of the consequent membership functions, the speed of action from one initiation step to another is somewhat characteristic of each technique. This issue is illustrated in Fig. 8.42, where the first category of implication operators (Mamdani, Larsen, drastic, bounded) coupled with union aggregation produce results at three instances (from left to right).

Examining Fig. 8.42 reveals that the maximum possibility method is the fastest responding technique at the expense of disregarding competing results. Thus, it is suitable for competitive and prioritized rule formation strategies described in Chapter 7. The center-of-gravity method is slower than the maximum possibility method; however, it incorporates all the contributions, thus it is in harmony with cooperative rule formation strategy. The center of maximums method is the slowest one among the three because it is sensitive to the width of each plateau regardless of its height. It is an extremely cooperative method and can be the choice for the evidence collection strategy.

The second category implication operators (Standard, Zadeh, Lukasiewicz, Kleen-Dienes, Gougen, and Godelian) in conjunction with the intersection aggregation operator do not affect the discussions presented in the previous paragraph. That is, the maximum possibility

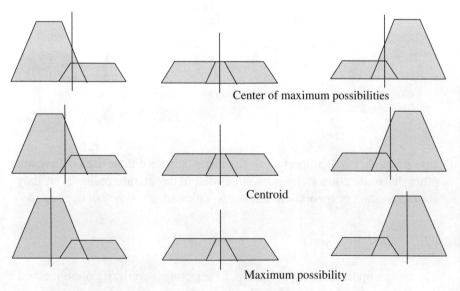

Figure 8.42 Defuzzification behavior using three methods with union aggregation and a first category implication operator (Mamdani).

Chap. 8 Implication Process 449

method is faster than that of the center-of-gravity method. Kleen-Dienes implication is illustrated in Fig. 8.43 for the same sequence of evaluations shown in Fig. 8.42. The intersection of two fuzzy sets in each graph is shown in dark shading. The center-of-gravity method is equivalent to the center-of-mass technique described previously for obvious reasons.

> The characteristics of a defuzzification/decomposition method are not a function of the implication or aggregation operator used in a fuzzy inference design.

This principle suggests that the choice of a defuzzification technique is a matter of how to interpret the geometrical shape of a fuzzy set that has already been determined by the choice of a pair of implication/aggregation operators; thus, the overall behavior can be considered to be a result of the superimposition of independent events.

Besides maintaining suitability between the rule formation strategy and a defuzzification method, another important design criterion is the rate of change of the output from one initiation step to another. This can

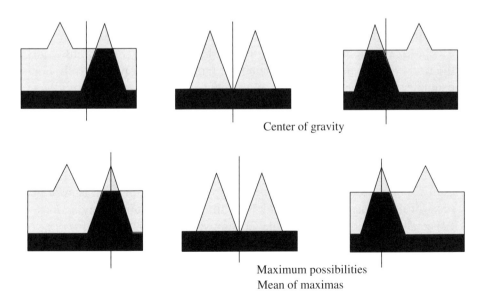

Center of gravity

Maximum possibilities
Mean of maximas

Figure 8.43 Defuzzification behavior using three methods with intersection aggregation and a second category implication operator (Kleen-Dienes).

be an important consideration for dynamic (or sequentially correlated) systems regardless of the problem type. Traditionally,[3] fuzzy control designs employ the center-of-gravity method for its smooth transition characteristics that leave no room for sudden jumps in control actions. When instability and actuator wear-out are of concern, this choice seems to be appropriate, as the experiential evidence suggests. Note that there can be no theoretical proof linking the stability of a system controlled by a fuzzy controller and the defuzzification method of choice simply because of the abundance of factors involved in determining stability. Open-loop dynamics, rule composition, membership function design, and choice of implication operator are some of them. However, given a simple linear model of open-loop dynamics, a stability analysis can be performed for that particular design that will only be valid for that design. The choice of a defuzzification method affects the consequent membership function design due to active range of action as explained in Chapter 5.

8.5.4 Other Approaches of Defuzzification

As alluded to earlier, the need for defuzzification emerges from the fact that aggregated implication results are fuzzy sets and are often not practical in real life. The difficulty, in fact, is due to the nature of the existing systems of information processing. Most of the existing systems do not accept fuzzy output as something useful. There are two distinct cases. In the first case, the actual action cannot be represented by partial truth. For example, the output from a risk assessment analysis is always fuzzy, yet the action to be taken may not be fuzzy (e.g., we either get on the plane or not). In the second case, the action can also be fuzzy; however, we are not accustomed to pursuing such an avenue because of traditional values often formed in a social context. For example, a defendant in a court must be found either guilty or not guilty,[4] and a verdict such as "somewhat guilty" is not feasible unless criminal laws are rewritten. Unfortunately, most of the decision-making situations in real life require fuzzy actions. Especially in fields like medicine, economy, law, and business, fuzzy actions—which consist of taking several partial actions—are unavoidable.

---

[3] In the perspective of the fuzzy system already built in the hardware domain.

[4] In some cases, such as "guilty of a second degree murder," the criminal system offers somewhat fuzzy results.

To better understand the depth of the problem, let's consider an aggregated implication result as shown in Fig. 8.44. The fuzzy set illustrated by a dark shaded area does not match any of the identified categories *Low, Medium, High, or Very_high*. This represents the majority of cases encountered during the lifetime of a fuzzy inference engine. The interpretation of such a result can be mathematically accomplished by devising a matching algorithm- -mainly based on the ratios of areas filled and void—that can tell us how *Low, Medium,* or *High* is the answer. Accordingly, the action for each category must be predetermined so that a distribution such as the one below can be implemented in a proportion similar to that of the aggregated result. There are several techniques in the literature, such as measuring the information entropy [11] of the output fuzzy set, determining the preponderance of evidence [10], and fuzzy regression models for optimal matching [12].

When it is unnecessary to decompose an output fuzzy set due to the objective of the design (e.g., in applications such as natural language processing), then alternative techniques, which directly utilize an output fuzzy set as shown in Fig. 8.44, may be considered. In such a case, one of the design philosophies is to develop an output processor that analyzes results in a comparative manner. The comparisons can be formulated between different initiation steps or between the results of different rule blocks. In some cases, the difference analysis may become more informative than interpreting a single result like the one in Fig. 8.44. For example, consider the two aggregated results shown in Fig. 8.45.

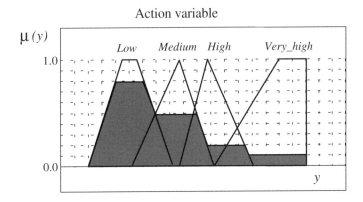

Figure 8.44 Aggregated result not matching any of the categories exactly.

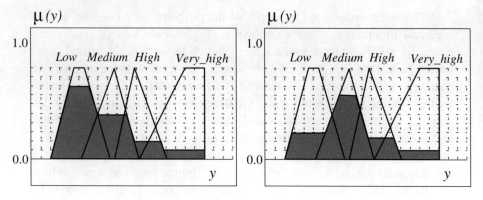

Figure 8.45 Aggregated results obtained through two different initiation steps.

It is obvious that the fuzzy set on the right is more *Medium* than the one on the left. The difference based on a trivial computation can be used to identify how much shift from *Low* to *Medium* has occurred between the two initiation steps of input absorption or between two sets of rule blocks. Just like working in the dimension of derivatives, many problems including control can be solved in this manner without using conventional defuzzification methods. If the difference is obtained through different time steps, then the shift plus the crisp value of the most prominent answer is the result applicable to dynamic systems. If the difference is obtained between two rule blocks for a given single input set, then the shift plus the crisp value of the most prominent answer is the result applicable to nondynamic problems such as classification. In such an approach, it is obvious that the total area of the original membership functions must be proportional to the total width of the universe of discourse.

8.6 INTERPRETING OUTPUT FUZZY SETS

Consequent membership function design issues, as they were presented in Chapters 5 and 6, shed a light on various effects produced by their shapes. Within the course of execution, a fuzzy inference engine may produce aggregated results that indicate the quality of the rules. A typical picture often encountered in the literature is illustrated in Fig. 8.46.

The hypothetical cases shown in Fig. 8.46 represent a single initiation step in which the formation of fuzzy implication-aggregation results

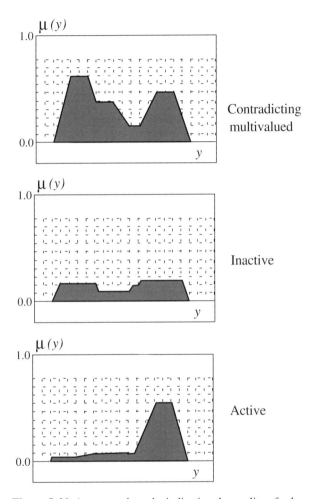

Figure 8.46 Aggregated results indicating the quality of rules.

was obtained. A contradiction is encountered when different rules strongly contribute to the extreme locations on the universe of discourse leaving low possibility regions in the middle. This is shown at the top of Fig. 8.46. Using one of the common defuzzification methods, the decomposed output will reside in the middle where the implication result is suppressed. This results in a contradiction. Unless the operational points of a fuzzy system are completely known, such a contradiction cannot be detected easily just by examining the rules. That is why this type of contradiction is referred to as *conflicts in disguise* in Section 7.6.3.

The second type of undesired performance is illustrated in the middle in Fig. 8.46, in which there is no significant contribution from any of the rules. This type of activity is indicative of two possible causes. First, the rules are not of full rank design (see Section 7.6.6) so that the inference engine is susceptible to void regions in the input space. Second, the input processor of the basic fuzzy inference algorithm has failed to detect an out-of-bounds input case. In both cases, the performance is undesired. A desired behavior of the aggregated implication result is shown at the bottom of Fig. 8.46, where there is a clear tendency toward a predefined output category.

Let's return to the case of contradiction. The detection of a contradiction suffers from another problematic situation in which the output is multivalued. For example, the aggregated implication result shown in Fig. 8.47 might be realistic when there might be two valid solutions.

For example, consider the following diagnostic fuzzy rules, in which the phase change phenomena from liquid phase to solid phase and from liquid phase to gas phase obey the same principle.

IF Phase_change HAS Occurred THEN Heat WAS Adequately_extracted
IF Phase_change HAS Occurred THEN Heat WAS Adequately_added

If Fig. 8.47 represents ΔQ (difference of energy), then the aggregated implication result is correct instead of being contradictory. Although it is easy to detect a multivalued outcome in this example, it may not be so easy for large rule-base systems with hundreds of rules com-

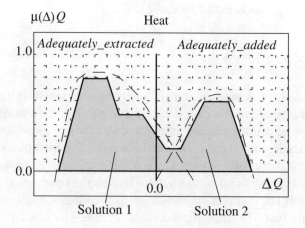

Figure 8.47 Multivalued solution.

posed by many experts. However, contradictions and multivalued outputs of this nature cannot be identified algorithmically unless they are identified during the knowledge acquisition process.

> Contradictions and multivalued outputs can only be appropriately detected in the design phase of a fuzzy inference engine during the knowledge acquisition process.

Therefore, any attempt to evaluate the performance of a fuzzy inference engine by examining the instantaneously formed aggregated implication results is not highly recommended.

8.7 SUMMARY

The last step of the basic fuzzy inference computation is the calculus of the RHS of a rule. This includes the implication process, aggregation, and defuzzification. This book includes 10 implication operators out of many found in the literature. They were selected based on their popularity as of today. Among them, Mamdani and Larsen are the most widely used operators, especially in the field of fuzzy control. Aggregation, which means the fusion of the elements of one decision variable, is often accomplished by union or intersection operation among the output fuzzy sets. Defuzzification, which is the process of obtaining a scalar representative of the aggregated fuzzy set, can be accomplished in a number of ways. Among them, the centroid, center of maximum, and mean of maximums methods are commonly used.

Choosing an implication operator out of many viable options is a tough task for the designer, not because there is a chance of selecting the wrong one, but because it is difficult to justify the choice. The first part of this chapter outlines the properties of each implication operator via a simple mapping example. The behavior analysis of the implication operators reveals that they exhibit two different types of behavior when the inference system is in transition (on fuzzy surfaces) from one crisp point to another in the input product space. Linguistic design criteria such as conservatism, tolerance, optimism, pessimism, precision, and cost can be associated with these two types of behavior of the implication operators.

Thus, the designer can evaluate the problem at hand to find the most suitable implication operator based on one of the linguistic design criteria, if such a criterion is applicable.

The topics including rate of change, rule activity, and pendulum analogy provide useful insight into the selection of an implication operator for sequentially correlated (dynamic) inference systems. Besides the associations to the linguistic design criteria, the rate-of-change property of an implication operator may give the designer some clues about the system's stability. The implication operators that fall into the category of "slow-approachers" such as Mamdani and Larsen behave similar to the movements to control an inverted pendulum (i.e., a capability that can handle inherently unstable open-loop dynamics). These discussions assume that the crisp points of mapping are known to yield stability, whereas the fuzzy regions may induce instability. Nevertheless, the stability is never guaranteed just by selecting an appropriate implication operator. Also note that the stability issue is applicable to all sorts of dynamic problems ranging from ecological studies examining the population of certain fish to financial models estimating a possible stock market crash.

The last topic in this chapter is the choice of a defuzzification technique among the options found in the literature. Very similar to the considerations related to the implication operators, a defuzzification technique can be sluggish or abrupt in application to sequentially correlated systems. For example, the centroid technique that does not allow sudden jumps during transition is sluggish, and therefore it is preferred in control systems with actuator constraints. Other arguments presented in this section of the book can be considered by the designer to apply the most suitable defuzzification technique for the problem at hand.

REFERENCES

[1] Mizumoto, M., "Fuzzy Controls Under Various Reasoning Methods." *Information Sciences* 45 (1988): 129–141.

[2] Kruse, R., J. Gebhardt, and F. Klawonn. *Foundations of Fuzzy Systems*, Chichester: John Wiley & Sons, 1994.

[3] Lee, C.C. "Fuzzy Logic in Control Systems: Fuzzy Logic Controller— Parts I and II." *IEEE Trans. on Systems, Man and Cybernetics* 20 (2), March/April 1990: 404–418,

[4] Ruan, D., and E.E. Kerre. "Fuzzy Implication Operators and Generalized Fuzzy Method of Cases." *Fuzzy Sets and Systems* 54 (1993): 23–37.
[5] Bandler, W., and L. J. Kohout. "Fuzzy Power Sets and Fuzzy Implication Operators." *Fuzzy Sets and Systems* 4 (13) (1970): 141–164.
[6] Tanaka, K., and M. Sugeno. "Stability Analysis and Design of Fuzzy Control Systems." *Fuzzy Sets and Systems* 45 (1992): 135–156.
[7] Chen, S. J., and C. L. Hwang. *Fuzzy Multiple Attribute Decision Making.* Berlin: Springer, 1992.
[8] Yager, R. R., and D. P. Filev. "On the Issue of Defuzzification and Selection Based on a Fuzzy Set." *Fuzzy Sets and Systems* 55 (1993): 255–271.
[9] Runkler, T. A., and M. Glesner. "A Set of Axioms for Defuzzification Strategies, Towards a Theory of Rational Defuzzification Operators," *Proceedings 2nd IEEE International Conference on Fuzzy Systems,* San Francisco, IEEE Press, New York, 1993: 1116–1161.
[10] Cox. E. *The Fuzzy Systems Handbook.* New York: AP Professional, 1994.
[11] Kosko, B. *Neural Networks and Fuzzy Systems.* Englewood Cliffs, NJ: Prentice-Hall, 1992.
[12] Ross, J. Timothy. *Fuzzy Logic with Engineering Applications.* New York: McGraw-Hill Inc., 1995.

SELECTED BIBLIOGRAPHY

Weber, S. "A General Concept of Fuzzy Connectives, Negation and Implication Based on t-Norms and t-Conorms." *Fuzzy Sets and Systems* 11 (1983): 115–134.

Yager, R. R. "Some Procedures for Selecting Fuzzy Set Theoretic Operators." *Int. Journ. of General Systems* 8 (1982): 115–124.

Pfeiffer, B. M. and R. Isermann. "Criteria for Successful Applications of Fuzzy Control." *First European Congress on Fuzzy and Intelligent Technologies,* (Aachen, 1993): 1403–1409.

Appendix

The Basic Fuzzy Inference Algorithm

OVERALL ALGORITHM

The details of the basic fuzzy algorithm and the manner in which it works are described in this Appendix. As explained in a more compact form in Chapter 3, the overall algorithm consists of five concomitant steps. The working mechanism of each block can be programmed in different ways. We present here one possible algorithmic flow that can be used as a reference for development (Fig. A.1).

Due to the limited space and scope of this book, some topics such as adaptive solutions, auxiliary parameters, navigating rules, and prioritized thresholds are not included. Nevertheless, such effects can be easily incorporated by the designer because they consist of trivial computations. The basic fuzzy inference algorithm, as presented in this Appendix, can be programmed in computer languages such as C, C++, Pascal, FORTRAN, and in other machine-specific computer languages such as Assembler or HC11. The inner workings of each block, when the fuzzy inference engine contains hundreds of rules, can be organized to yield an optimized memory usage and fast processing speed. One of the major considerations in fuzzy system development has been code optimization, which is a topic of computer programming and is also beyond the scope of this book.

Appendix The Basic Fuzzy Inference Algorithm

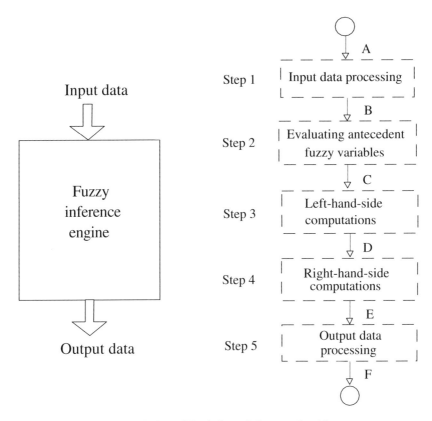

Figure A.1 Overall basic fuzzy inference algorithm.

STEP 1: INPUT DATA PROCESSING

Figure A.2 elaborates Step 1 of Fig. A.1. Most of the operations in Fig. A.2 are trivial. The events outlined in Fig. A.2 are described in more detail as follows:

> Level A1 *Receiving an input data set is one of the two physical boundaries between the external world and the fuzzy system. Everything else after this point assumes a digital computing domain even though fuzzy systems, in general, are not restricted to this domain only. A very important point is that the data set received from the external world represents one initiation step (time, iteration, or sequence).*

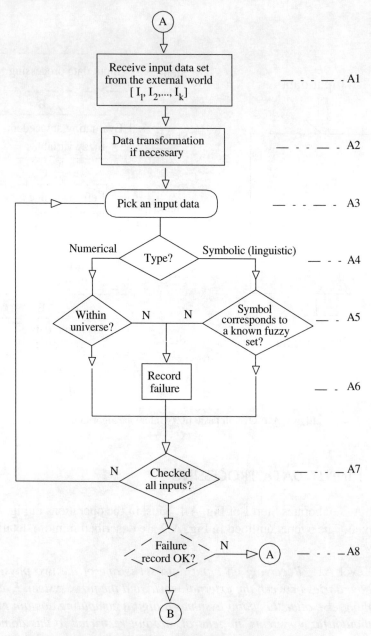

Figure A.2 Input data processing in the basic fuzzy algorithm.

Level A2 *Data transformation, if necessary, is accomplished in this step. Depending on the design, two- or higher-dimensional numerical data is converted into possibility units. A detailed description of the data transformation process is given in Chapter 3.*

Level A3 *A loop starts here by selecting one input data element per cycle. This loop will continue until all the elements of the input data set are processed.*

Level A4 *A trivial checking operation takes place at this level to identify the type of data (numerical or symbolic).*

Level A5 *For numerical data, the universe of discourse of the corresponding antecedent variable is used to check if data is between the upper and lower boundaries of the universe. The boundaries of the universe of discourse for each variable are in the memory of the fuzzy system. If data is nonsingular, that is, if there is more than one data point, the same operation is employed. For symbolic data, a recognition step is taken in which the symbol, which is often a linguistic statement, is translated into a numerical form. This is only possible if the translation of the symbol was embedded in the memory of the fuzzy system during the design. In other words, if* High *is the linguistic input data of the antecedent variable* temperature, *it must be known by the fuzzy system what* High *means in terms of temperature values. This is analogous to speaking in daily language to someone who must understand the words. Thus, an inference engine, in general, has a language memory. If the symbol is recognized, its numerical translation will automatically be within the universe of discourse by design.*

Level A6 *In the case of detecting inappropriate data, a failure record is kept at this step. Failure record keeping, which represents a general input processing strategy, allows the option of continuing fuzzy computations in case of a partially unrecognized input data set.*

Level A7 *Loop ending condition checks to see if all the input data are analyzed.*

Level A8 *The last step before declaring the validity of the input data set is somewhat more involved than the trivial operations explained above. When all the input data elements are found appropriate, the input data set is acceptable. However, when a fraction of the input data set is inappropriate, an acceptability criterion is ap-*

plied. For applications in which a missing data situation is tolerable, the acceptability criteria may be made flexible to allow the propagation of input data. In such cases, data transformation of the unacceptable input will produce zero possibility, which is a valid value for inference computations. When the input data set is rejected, the algorithmic flow returns to collecting a new data set. This decision-making process based on failure record, which is drawn with dashed lines in Fig. A.2, is subject to design.

STEP 2: EVALUATING ANTECEDENT FUZZY VARIABLES

Figure A.3 elaborates Step 2 of Fig. A.1. The events outlined in Fig. A.3 are described below:

Level B1   *An input data set, which is validated through the previous step, is obtained from the external world for the initiation time t. An input data element is picked.*

Level B2   *The properties of the corresponding antecedent fuzzy variable are retrieved from the memory of the fuzzy system. These properties include the number of membership functions, the shape of each membership function (coordinates), their layout on the universe of discourse, and the threshold properties.*

Level B3   *A simple test is applied to identify whether the picked input element contains a single data point or a distribution of points (fuzzy set).*

Level B4   *Operations in this level involve fuzzy computing and they are subject to design. The evaluation of antecedent variables for a single data point or for a distribution of data points is performed in this level. There are two types of design issues: (1) design of membership functions, and (2) method of evaluating them given input data. Although the first one represents a large variety of design options, the latter is quite standard. The design is discussed in Chapters 5 and 6. The method of evaluation is described in Chapter 3.*

Level B5   *To avoid undesired residual truth (very small possibility values), a threshold filtering is applied at this level. This process is subject to design and is explained in Chapters 3 and 5.*

Appendix The Basic Fuzzy Inference Algorithm

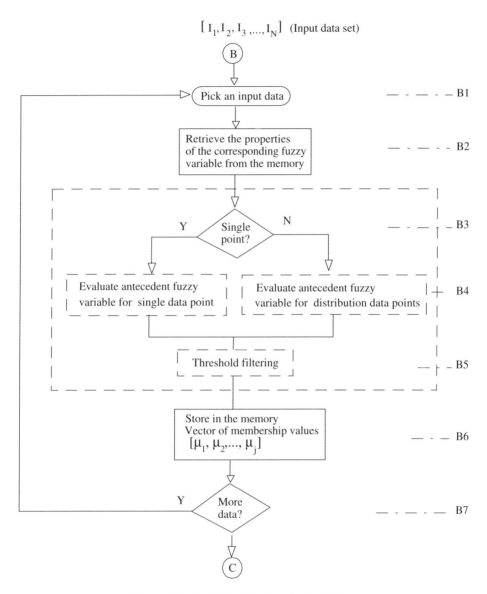

Figure A.3 Evaluating all antecedent variables.

Level B6   *A membership value vector obtained in the last step is stored in the memory of the fuzzy system for future use.*

Level B7   *Loop ending condition.*

STEP 3: LHS COMPUTATIONS

To conduct the LSH computations, the design information in the memory must be organized in a certain way. Here, we define four matrices in the following sections, namely the membership value matrix, design index matrix, logic operator matrix, and auxiliary parameter matrix. Although there is no matrix algebra involved, the matrix representation facilitates the memory organization and problem setup.

Antecedent Variable List

The antecedent variable list includes all fuzzy variables that are on the left-hand side of each rule with respect to the THEN operator. This list, which may be defined as a vector of character fields in the memory, is only important for labeling purposes. It serves as a road map to which all other index definitions are referenced. To maintain a consistent terminology, we will define X to be a character (linguistic) variable in the definition of the antecedent variable vector

$$\overline{X}_A = \begin{bmatrix} X_1 \\ X_2 \\ ... \\ X_N \end{bmatrix} \quad (A.1)$$

The order in Eq. (A.1) is arbitrary in the memory of the fuzzy system and is not related to the way rules are composed. The index N indicates the number of antecedent fuzzy variables in the memory.

Membership Value Matrix

The computation of a membership value vector for each antecedent fuzzy variable was illustrated in the previous section. If there are N antecedent fuzzy variables then there are N vectors, each with different vector size depending on the number of membership functions defined. The membership values are obtained through the input absorption process for a given initiation step. The numerical values will be different at the next initiation step depending on the input data set from the external world. Each row of the membership value matrix denoted by $\overline{\mu}$ corresponds to a different fuzzy antecedent variable. Each column corresponds to different membership functions. Note that this matrix may have empty locations based on the number of membership functions defined for each variable.

$$\overline{\mu} = \begin{bmatrix} \mu_{1,1} & \mu_{1,2} & \cdots & \mu_{1,j} \\ \mu_{2,1} & \mu_{2,2} & \cdots & \mu_{2,j} \\ \cdots & & & \\ \mu_{i,1} & \mu_{i,2} & \cdots & \mu_{i,j} \end{bmatrix} \quad (A.2)$$

Antecedent Predicate Matrix

Another important illustration to facilitate the computational aspects of the basic fuzzy inference algorithm is the antecedent predicate matrix. A predicate, which can be a fuzzy qualifier or quantifier, is denoted by P and takes linguistic values. Unlike the membership value matrix, the predicate matrix is fixed by design and it also serves as an address to memory where the data of its corresponding membership functions are stored.

$$\overline{P} = \begin{bmatrix} P_{1,1} & P_{1,2} & \cdots & P_{1,j} \\ P_{2,1} & P_{2,2} & \cdots & P_{2,m} \\ \cdots & & & \\ P_{N,1} & P_{N,2} & \cdots & P_{N,n} \end{bmatrix} \quad (A.3)$$

The predicate matrix of Eq. (A.3) has the same dimensions as the membership value matrix of Eq. (A.2) because each predicate corresponds to one membership function at the same matrix location.

LHS Design Index Matrix

Although the representation given above describes what is fixed by design and what is dynamically updated, we will go one step further and define the LHS design matrix, which only contains the index references to the membership value matrix.

$$\overline{I}_{LHS} = \begin{bmatrix} I_{1,1} & I_{1,2} & \cdots & I_{1,l} \\ I_{2,1} & I_{2,2} & \cdots & I_{2,m} \\ \cdots & & & \\ I_{Z,1} & I_{Z,2} & \cdots & I_{Z,n} \end{bmatrix} \quad (A.4)$$

This matrix exactly corresponds to the rule layout. The elements of $\overline{I}_{LHS}$ can have a negative sign in front indicating a negatively constructed fuzzy statement (i.e., using IS NOT). Every row is a new rule. Columns correspond to fuzzy statements. Because the length of a rule can be anything, the column indices l, m, and n are not the same. Note that all the elements in Eq. (A.4) are the statements before the THEN operator.

Logic Operator Matrix

This matrix contains composition operator indices such as AND and OR. Note that each operator index contains two kinds of information: (1) type of operator, and (2) order of computation.

$$\overline{\Theta}_{LHS} = \begin{bmatrix} \Theta_{1,1} & \Theta_{1,2} & \cdots & \Theta_{1,l-1} \\ \Theta_{2,1} & \Theta_{2,2} & \cdots & \Theta_{2,m-1} \\ \cdots & & & \\ \Theta_{Z,1} & \Theta_{Z,2} & \cdots & \Theta_{Z,n-1} \end{bmatrix} \quad (A.5)$$

The order of computation is related to the grouping of different operators. Grouping of different operators, which is normally indicated by parentheses in computer language terminology, requires a particular sequence of computations. The sequence is determined by the innermost statements (highest degree of nesting). One of the most practical ways of determining the sequence is to count the left and right parentheses, then to assign an order number to the composition operators. As an example, consider the fuzzy rule shown below.

IF [*Temperature IS High* OR *Flow IS Low*] AND *Valve IS Open* THEN ....
      S_1 S_2 S_3

This rule is formed such that the first computation must be the OR operation between S_1 and S_2, and the second computation must be the AND operation between the result of the first computation and S_3. The hierarchical order of solutions is illustrated in Fig. A.4 for an arbitrary rule with three nested parentheses.

The logic operator index Θ, which is a symbolic variable, can be made to carry both kinds of information for simplicity. For example, the two logic operators in the rule shown in the example above would be

$$\Theta_1 = \text{OR-001} \quad \Theta_2 = \text{AND-002}$$

This type of coding is practical from the computer programming point of view and it has nothing to do with the fuzzy inference engine design.

Appendix The Basic Fuzzy Inference Algorithm

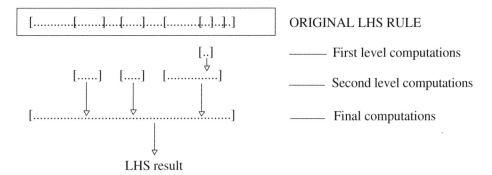

Figure A.4 Grouping different logic operators requires a hierarchical solution scheme.

LHS Auxiliary Parameter Matrix

When fuzzy statements are subject to modifications by linguistic hedges or by adaptive algorithms, the type of modification must also be stored in the memory of the fuzzy system. Because auxiliary parameters are assigned to each fuzzy statement, they follow the statement indices. The LHS auxiliary parameter matrix is given below.

$$\bar{\lambda}_{LHS} = \begin{bmatrix} \lambda_{1,1} & \lambda_{1,2} & \cdots & \lambda_{1,w} \\ \lambda_{2,1} & \lambda_{2,2} & \cdots & \lambda_{2,x} \\ \cdots & & & \\ \lambda_{Z,1} & \lambda_{Z,2} & \cdots & \lambda_{Z,h} \end{bmatrix} \quad (A.6)$$

Every row is a new rule, whereas columns correspond to the statements only (not to operators or parentheses). Thus, the indices w, x, and h are not same as the ones in $\bar{I}_{LHS}$.

LHS Computations

Consequently, the LHS portion of the entire rule base system can be symbolically represented by four entities: (1) membership value matrix; (2) design index matrix; (3) logic operator matrix; and (4) auxiliary parameter matrix. The indices in these matrices correspond to the antecedent variable list (vector) initially constructed by the designer.

$$\bar{\mu} = \begin{bmatrix} \mu_{1,1} & \mu_{1,2} & \cdots & \mu_{1,j} \\ \mu_{2,1} & \mu_{2,2} & \cdots & \mu_{2,j} \\ \cdots & & & \\ \mu_{i,1} & \mu_{i,2} & \cdots & \mu_{i,j} \end{bmatrix}$$
(1)
Membership Function Matrix

$$\bar{I}_{LHS} = \begin{bmatrix} I_{1,1} & I_{1,2} & \cdots & I_{1,l} \\ I_{2,1} & I_{2,2} & \cdots & I_{2,m} \\ \cdots & & & \\ I_{Z,1} & I_{Z,2} & \cdots & I_{Z,n} \end{bmatrix}$$
(2)
LHS Design Index Matrix

$$\bar{\lambda}_{LHS} = \begin{bmatrix} \lambda_{1,1} & \lambda_{1,2} & \cdots & \lambda_{1,w} \\ \lambda_{2,1} & \lambda_{2,2} & \cdots & \lambda_{2,x} \\ \cdots & & & \\ \lambda_{Z,1} & \lambda_{Z,2} & \cdots & \lambda_{Z,h} \end{bmatrix}$$
(3)
LHS Auxiliary Parameter Matrix

$$\bar{\Theta}_{LHS} = \begin{bmatrix} \Theta_{1,1} & \Theta_{1,2} & \cdots & \Theta_{1,l-1} \\ \Theta_{2,1} & \Theta_{2,2} & \cdots & \Theta_{2,m-1} \\ \cdots & & & \\ \Theta_{Z,1} & \Theta_{Z,2} & \cdots & \Theta_{Z,n-1} \end{bmatrix}$$
(4)
LHS Operator Matrix

The LHS computation of each rule is independent from others. This yields a vector of results of the entire rule base with one element per each rule. Once the overall LHS computations are characterized using these four matrices by an algorithmic function $\psi(\bar{\mu}, \bar{I}, \bar{\lambda}, \bar{\Theta})_{LHS}$, the resultant vector is expressed as

$$\psi(\bar{\mu}, \bar{I}, \bar{\lambda}, \bar{\Theta})_{LHS} = \begin{bmatrix} r_1 \\ r_2 \\ \cdots \\ r_Z \end{bmatrix} = \bar{r} \qquad (A.7)$$

where the total number of rules is Z. Step 3 of the basic fuzzy algorithm between the terminal points C and D (Fig. A.1) computes the $\bar{r}$ vector which contains DOF values. The elements of this vector are numerical, and they will change from one initiation step to another as the fuzzy system absorbs new input data sets from the external world and produces the numerical elements of the membership value matrix $\bar{\mu}$.

Using the definitions given above and Eq. (A.7), the block diagram representation of the LHS computations of the basic fuzzy inference algorithm is illustrated in Fig. A.5. Operations between the terminal points C and D are explained as follows:

Level C1 *The first step is to retrieve the properties fixed by design from the memory of the fuzzy system. These include the LHS operator matrix, LHS fuzzy probability matrix, and the LHS design index matrix. Among these matrices, the LHS operator matrix is the road map to follow.*

Appendix The Basic Fuzzy Inference Algorithm

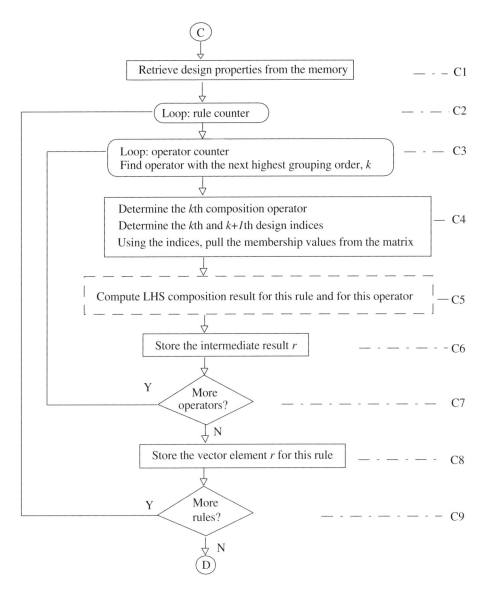

Figure A.5 LHS computations in the basic fuzzy inference algorithm.

Level C2 *Outer loop starts from the first line (first rule) of the LHS operator matrix.*

Level C3 *Inner loop picks one operator at a time until all the operators are exhausted in the selected rule. Picking an operator is* done based on the highest grouping order or the most nested structures in case of mixed operator types. This information may be attached to the operator data in the memory to facilitate computations. Assuming such a structure, the OR operator will be picked first between the AND-1 and OR-3 in a hypothetical rule. Here, 1 and 3 indicate the grouping order or the number of parentheses. The location (k) of the picked operator in the logic operator matrix will be used in levels C4 through C7.

Level C4 At this step, all necessary information for the LHS computations is collected for the selected rule and for the selected logic operator. This includes determining the type of the operator by encoding the data (i.e., extracting AND from AND001) which belongs to the kth operator index in the memory. Having found the location k, the indices $(i, j)_k$ and $(i, j)_{k+1}$ are extracted from the LHS design index matrix in the memory. The indices $(i, j)_k$ and $(i, j)_{k+1}$ are used to locate the membership values $\mu(i, j)_k$ and $\mu(i, j)_{k+1}$ in the LHS membership value matrix.

Level C5 This is the only block between the terminal points C and D that involves fuzzy computations. Composition between two statements is computed using $\mu(i, j)_k$, $\mu(i, j)_{k+1}$, λ_k, λ_{k+1}, and Θ_k at this step. Between the two most commonly used operators AND and OR, the min and max operations are applied, respectively. If the fuzzy statements are not to be externally modified, λ_k and λ_{k+1} are equal to unity. In the case of nested structures, the computation takes place between the $\mu(i, j)_k$, Θ_k, and r', where r' is the previous result, or between r', Θ_k, and $\mu(i, j)_{k+1}$ depending on which side (k or $k+1$) was computed previously. Note that only a single result is computed at this step.

Level C6 The intermediate result r' is stored in the fuzzy memory at this step.

Level C7 The loop termination condition is applied. When all the operators in the selected rule are processed, the operator loop terminates.

Level C8 The resultant r value is stored in the appropriate vector location corresponding to the selected rule.

Level C9 Loop termination condition is applied. When all the rules are processed, the rule loop will terminate.

Appendix The Basic Fuzzy Inference Algorithm

EXAMPLE A.1

A hypothetical inference engine will be examined in this example to illustrate the entire LHS composition computations. Suppose that we are analyzing two fuzzy rules as shown below.

IF (Temperature IS High OR Flow IS Medium) AND Valve IS Open THEN
$$\text{Operation IS Fine}$$
IF Valve IS Closed THEN Operation IS Questionable

Because we are dealing with the LHS operations, the THEN operator and the statements after the THEN operator are not considered. The fuzzy variables in the memory (or the antecedent variable list) are:

temperature
flow
valve

The LHS predicate matrix in the memory is:

$$\begin{bmatrix} P_{1,1} & P_{1,2} & P_{1,3} \\ P_{2,1} & P_{2,2} & P_{2,3} \\ P_{3,1} & P_{3,2} & P_{3,3} \end{bmatrix} \equiv \begin{bmatrix} low & high & \\ low & medium & high \\ open & half & closed \end{bmatrix}$$

Note that this matrix contains elements beyond what is defined in the rule base in order to make the point that the LHS predicate matrix can be designed to include the definitions of the substitute or reserved elements. Every row above corresponds to one fuzzy variable in the memory. Every column indicates the membership function name. Therefore, the first variable *temperature* has two membership functions by design that are *low* and *high*, the second variable *flow* has three membership functions by design that are *low*, *medium*, and *high*, and the third variable *valve* has three membership functions by design that are *open*, *half*, and *closed*. Their properties are also stored in the memory to be used during the input absorption process.

The LHS design index matrix in the memory is:

$$\begin{bmatrix} I_{1,1} & I_{1,2} & I_{1,3} \\ I_{2,1} & & \end{bmatrix} = \begin{bmatrix} 1,2 & 2,2 & 3,1 \\ 3,3 & & \end{bmatrix}$$

This matrix exactly coincides with the rule base. Each row above corresponds to one fuzzy rule whereas each column corresponds to one fuzzy statement. The indices describe the rules. The first indice $I_{1,1}$ is 1,2 indicating the row and column indices in the membership value matrix. It also means that the first fuzzy statement in rule 1 is *Temperature IS High*.

The logic operator index matrix in the memory is:

$$\begin{bmatrix} \Theta_{1,1} & \Theta_{1,2} \\ \Theta_{2,1} & \end{bmatrix} = \begin{bmatrix} OR-2 & AND-1 \\ 0 & \end{bmatrix}$$

which represents all possible variations. The first operator in the first rule is OR, which has a grouping order of 2. The second operator in the first rule is AND, and its grouping order is 1. The second rule has no operators.

All the information above is fixed via design and resides in the memory permanently. Now we will implement the C–D block diagram for one initiation step.

Let's suppose the input data for the antecedent variables at the initiation step T are:

$$temperature = 27° \text{ F}$$
$$flow = 10 \text{ m/sec}$$
$$valve = 30\% \text{ open}$$

Also suppose that the membership values are computed during the input absorption process to yield the following membership value matrix:

$$\begin{bmatrix} \mu_{1,1} & \mu_{1,2} & \mu_{1,3} \\ \mu_{2,1} & \mu_{2,2} & \mu_{2,3} \\ \mu_{3,1} & \mu_{3,2} & \mu_{3,3} \end{bmatrix} \equiv \begin{bmatrix} 0.12 & 0.75 & 0.0 \\ 0.21 & 0.66 & 0.0 \\ 0.77 & 0.14 & 0.06 \end{bmatrix}$$

This matrix has the same outline as that of the predicate matrix. Note that the last element in the first row is 0.0 because the number of membership functions of the first antecedent variable is 2 and there is no third membership function in the rules. This element will be zero at all times. Other zeros, whenever they appear, will be the result of the input absorption process.

Level C2: The algorithm starts from the first rule (first line of the operator index matrix).

Level C3: The highest grouping order k in the first rule of the operator matrix is 2 at the location 1,1.

$$\begin{bmatrix} \Theta_{1,1} & \Theta_{1,2} \\ \Theta_{2,1} & \end{bmatrix} = \begin{bmatrix} OR-2 & AND-1 \\ 0 & \end{bmatrix}$$

Level C4: The type of the operator at this location is OR. The indices at the location 1,1 of the design index matrix are 1,2. The indices at the adjacent location are 2,2.

$$\begin{bmatrix} I_{1,1} & I_{1,2} & I_{1,3} \\ I_{2,1} & & \end{bmatrix} = \begin{bmatrix} 1,2 & 2,2 & 3,1 \\ 3,3 & & \end{bmatrix}$$

The 1,2 element in the membership value matrix is 0.75, the adjacent 2,2 element in the same matrix is 0.66.

Level C5: The composition computation is 0.75 OR 0.66 = 0.75.
Level C6: The result 0.75 is stored.
Level C7: Yes there is one more operator in the first rule. Go back to the loop.
Level C3: The next highest grouping order k in the first rule of the operator matrix is 1 at the location 1,2.

$$\begin{bmatrix} \Theta_{1,1} & \Theta_{1,2} \\ \Theta_{2,1} & \end{bmatrix} = \begin{bmatrix} OR-2 & AND-1 \\ 0 & \end{bmatrix}$$

Level C4: The type of the operator at this location is AND. The indices at the location 1,2 of the design index matrix are 2,2. However, this index was previously used to produce r′ (0.75). The indices at the adjacent location are 3,1.

Appendix The Basic Fuzzy Inference Algorithm   473

$$\begin{bmatrix} I_{1,1} & I_{1,2} & I_{1,3} \\ I_{2,1} & & \end{bmatrix} = \begin{bmatrix} 1,2 & 2,2 & 3,1 \\ 3,3 & & \end{bmatrix}$$

The 3,1 element in the membership value matrix is 0.77, the previously computed result r′ is 0.75.
Level C5: The composition computation is 0.75 AND 0.77 = 0.75.
Level C6: The result 0.75 is stored.
Level C7: No, there are no more operators.
Level C8: The first element in the result vector is 0.75 and it is stored in the memory.
Level C9: There is one more rule to process. Go back to the rule loop.
Level C3: The highest grouping order k in the second rule of the operator matrix is 0 at the location 2,1.

$$\begin{bmatrix} \Theta_{1,1} & \Theta_{1,2} \\ \Theta_{2,1} & \end{bmatrix} = \begin{bmatrix} OR-2 & AND-1 \\ 0 & \end{bmatrix}$$

Level C4: There is no operator at this location. The indices at the location 2,1 of the design index matrix are 3,3. There is no adjacent location.

$$\begin{bmatrix} I_{1,1} & I_{1,2} & I_{1,3} \\ I_{2,1} & & \end{bmatrix} = \begin{bmatrix} 1,2 & 2,2 & 3,1 \\ 3,3 & & \end{bmatrix}$$

The 3,3 element in the membership value matrix is 0.06.
Level C5: The DOF computation is 0.06.
Level C6: The result 0.06 is stored.
Level C7: No, there are no more operators.
Level C8: The second element in the result vector is 0.06 and it is stored in the memory.
Level C9: There are no more rules to process.
As a result, the DOF result vector for the initiation step τ is

$$\bar{r} = \begin{bmatrix} 0.75 \\ 0.06 \end{bmatrix}$$

which will be used on the RHS operations described next. Note that the numerical values in this vector will change as the fuzzy system absorbs new input data from the external world.

STEP 4: RHS COMPUTATIONS

Consequent Variable Vector

All consequent fuzzy variables are kept in the memory of the fuzzy system along with their properties, such as membership function shapes and threshold levels. The order of listing these variables in the memory by means of a vector constitutes a road map which all other index definitions are referenced to.

$$\overline{X}_C = \begin{bmatrix} X_1 \\ X_2 \\ ... \\ X_N \end{bmatrix} \quad (A.8)$$

The order in Eq. (A.8) is arbitrary and is not related to the way rules are composed. Note the index N, which indicates the number of consequent fuzzy variables in the memory. N does not have to be equal to the number of rules.

EXAMPLE A.2

Suppose that we have a fuzzy inference engine described by the following five simple fuzzy rules:

IF Temperature IS Low	THEN Valve IS open
IF Flow IS High	THEN Valve IS Half
IF Temperature IS Medium	THEN Operation IS Bad
IF Temperature IS High	THEN Valve IS Closed
IF Flow IS Low	THEN Operation IS OK

Remember that we are analyzing the right side of the THEN operator. The consequent variable vector in the memory is

$$\overline{X}_C = \begin{bmatrix} Valve \\ Operation \end{bmatrix}$$

where the order is arbitrarily selected. However, once selected the rest of the implication process will be based on this order.

Consequent Predicate Matrix and Membership Functions

The consequent predicate matrix is fixed by design in the same way as the consequent variable vector and it resides in the memory of the fuzzy system. This definition is useful both during the RHS computations and during the output processing stage. The consequent predicate vector $\overline{P}_C$ is expressed as

$$\overline{P}_{RHS} = \begin{bmatrix} P_{1,1} & P_{1,2} & ... & P_{1,x} \\ P_{2,1} & P_{2,2} & ... & P_{2,y} \\ ... & & & \\ P_{N,1} & P_{N,2} & ... & P_{N,w} \end{bmatrix} \triangleright \begin{bmatrix} \mu_{1,1} & \mu_{1,2} & ... & \mu_{1,x} \\ \mu_{2,1} & ... & & \mu_{2,y} \\ & & & \\ \mu_{N,1} & ... & & \mu_{N,w} \end{bmatrix} \quad (A.9)$$

Appendix The Basic Fuzzy Inference Algorithm

where N is the number of consequent variables and the indices x, y, w are the number of predicates (membership functions) of each variable. As illustrated above, the consequent predicate matrix corresponds to the consequent membership functions residing in the memory.

EXAMPLE A.3

Using the same rules in Example A.1, the consequent predicate matrix in the memory is

$$\bar{P}_{RHS} = \begin{bmatrix} Open & Half & Closed \\ Bad & OK & ---- \end{bmatrix}$$

where each row corresponds to one consequent fuzzy variable. Thus, the predicates in the first line belong to the variable *Valve*, whereas the second line corresponds to *Operation*.

RHS Design Index Matrix

This matrix reflects the rule composition structure and is based on the two previous definitions. The elements of this matrix are the row numbers of the LHS result vector $\bar{r}$ given by Eq. (A.7).

$$\bar{A} = \begin{bmatrix} A_{1,1} & A_{1,2} & \cdots & A_{1,k} \\ A_{2,1} & A_{2,2} & & A_{2,l} \\ \cdots & & & \\ A_{N,1} & A_{N,2} & & A_{N,m} \end{bmatrix} \begin{matrix} k \leq Z \\ l \leq Z \\ m \leq Z \end{matrix} \qquad (A.10)$$

In Eq. (A.10), each row corresponds to one consequent fuzzy variable in the variable vector, and each column corresponds to one of the consequent membership functions. Because the length of the LHS vector is limited by the number of rules Z, the width of the matrix cannot be larger than Z. This matrix is the road map for implication computations just like the logic operator matrix is a road map for the LHS computations.

EXAMPLE A.4

Using the same fuzzy rules in Example A.1, we will form the RHS design index matrix by examining the RHS statements. Suppose that the LHS result vector [r_1 r_2 r_3 r_4 r_5] has already been computed at this point as shown below

......	r_1	THEN Valve IS Open
......	r_2	THEN Valve IS Half
......	r_3	THEN Operation IS Bad
......	r_4	THEN Valve IS Closed
......	r_5	THEN Operation IS OK

where each *r* is a DOF value. Recalling the consequent variable vector and the predicate matrix

$$\bar{X}_C = \begin{bmatrix} Valve \\ Operation \end{bmatrix} \quad \bar{P}_{RHS} = \begin{bmatrix} Open & Half & Closed \\ Bad & OK & \_\_\_\_ \end{bmatrix}$$

the RHS design index matrix is formed

$$\bar{A} = \begin{bmatrix} A_{1,1} & A_{1,2} & A_{1,3} \\ A_{2,1} & A_{2,2} & 0 \end{bmatrix} = \begin{bmatrix} 1 & 2 & 4 \\ 3 & 5 & 0 \end{bmatrix}$$

indicating that the first consequent predicate $A_{1,1}$ (membership function) will be subject to the implication process along with r_1.

Aggregation Vector

This connector determines the type of aggregation for each consequent fuzzy variable.

$$\bar{\Im} = \begin{bmatrix} \Im_1 \\ \Im_2 \\ \dots \\ \Im_N \end{bmatrix} \quad (A.11)$$

The size of this vector is equal to the number of consequent variables *N*. All the elements are designed to be the same (union operation) in most of the applications.

RHS Auxiliary Parameter Vector

Because the RHS part of the rule base is assumed to be reduced to one statement per rule, the RHS auxiliary parameters are in a vector form.

$$\bar{\lambda}_{RHS} = \begin{bmatrix} \lambda_1 \\ \lambda_2 \\ \dots \\ \lambda_Z \end{bmatrix} \quad (A.12)$$

The dimension of this vector is equal to the number of rules *Z*. Note that this generalized representation does not impose any restrictions on how to compute the auxiliary parameter effects in a fuzzy inference engine.

Appendix The Basic Fuzzy Inference Algorithm 477

Weight Vector

In certain applications, some rules may be considered to be more important than others. The basic fuzzy inference algorithm, in its most generalized form, allows weight assignments during the design (or during the implementation in adaptive fuzzy systems). The weight vector with each element corresponding to one rule is expressed as

$$\overline{W} = \begin{bmatrix} w_1 \\ w_2 \\ \dots \\ w_Z \end{bmatrix} \quad (A.13)$$

Note that weights are not restricted to probabilities or possibilities. A more detailed discussion on weights is given in Chapter 7.

RHS Computations

First, let's summarize what is residing in the memory of the fuzzy system to perform the implication computations. The RHS of the entire rule base is symbolically represented by seven entities: (1) consequent variable vector; (2) RHS consequent matrix; (3) RHS design index matrix; (4) aggregation vector; (5) RHS auxiliary parameter vector; (6) weight vector; and (7) LHS result vector. The first six entities are fixed by design, and the seventh entity, the LHS result vector, is dynamically updated (at each initiation time τ) by the input absorption process. Besides these seven entities, the properties of consequent membership functions are also stored in the memory.

$$\overline{X}_C = \begin{bmatrix} X_1 \\ X_2 \\ \dots \\ X_N \end{bmatrix} \quad \overline{P}_{RHS} = \begin{bmatrix} P_{1,1} & P_{1,2} & \dots & P_{1,x} \\ P_{2,1} & P_{2,2} & \dots & P_{2,y} \\ \dots & & & \\ P_{N,1} & P_{N,2} & \dots & P_{N,w} \end{bmatrix} \quad \overline{A} = \begin{bmatrix} A_{1,1} & A_{1,2} & \dots & A_{1,k} \\ A_{2,1} & A_{2,2} & \dots & A_{2,l} \\ \dots & & & \\ A_{N,1} & A_{N,2} & \dots & A_{N,m} \end{bmatrix} \begin{matrix} k \le Z \\ l \le Z \\ m \le Z \end{matrix}$$

(1)   (2)   (3)

$$\overline{\mathfrak{I}} = \begin{bmatrix} \mathfrak{I}_1 \\ \mathfrak{I}_2 \\ \dots \\ \mathfrak{I}_N \end{bmatrix} \quad \overline{\lambda}_{RHS} = \begin{bmatrix} \lambda_1 \\ \lambda_2 \\ \dots \\ \lambda_Z \end{bmatrix} \quad \overline{W} = \begin{bmatrix} w_1 \\ w_2 \\ \dots \\ w_Z \end{bmatrix} \quad \overline{r}(\tau) = \begin{bmatrix} r_1(\tau) \\ r_2(\tau) \\ \dots \\ r_Z(\tau) \end{bmatrix}$$

(4)   (5)   (6)   (7)

The important information embedded in these definitions from the algorithmic solution point of view are the indices that indicate the layout of design data, and the distinction between the static data (determined by design) and dynamic data (driven by external inputs). The RHS design index matrix is the road map in the basic fuzzy inference algorithm. If we denote the algorithmic implication function by Φ, then the result of the implication computations is given in a vector form as illustrated below

$$\Phi(\overline{P}_{LHS}, \overline{A}, \overline{\lambda}_{LHS}, \overline{\Im}, \overline{W}, \overline{r}(\tau)) = \overline{O}(\tau) = \begin{bmatrix} O_1(\tau) \\ O_2(\tau) \\ ... \\ O_N(\tau) \end{bmatrix} \qquad (A.14)$$

where τ is the initiation step (not necessarily clock time), and N is the number of consequent fuzzy variables. Note that the elements of the output vector are new fuzzy sets produced by the implication relation. Step 4 of the basic fuzzy inference algorithm shown in Fig. A.1 computes the output vector of Eq. (A.14).

The block diagram representation of the implication computations is shown in Fig. A.6. The description of each block is given as follows:

Level D1   *Design data including the RHS design index matrix, predicate matrix, LHS auxiliary parameter vector, weight vector, threshold information for each consequent variable, consequent membership functions, and aggregation operator vector are retrieved from the memory of the fuzzy system.*

Level D2   *The variable loop starts here. These are the consequent fuzzy variables as defined by the consequent variable vector. Each consequent variable constitutes one output of the inference engine. Note the difference between the LHS algorithm and the RHS algorithm. The outer loop in the LHS algorithm counts the rules whereas the outer loop in the RHS algorithm counts the consequent variables.*

Level D3   *Each consequent fuzzy variable has its corresponding aggregation operator defined in the memory. At this step, the type of the aggregation operator is determined for this variable.*

Level D4   *The column counter loop operates on the RHS design index matrix, in which every column corresponds to one predicate (and one membership function) in the RHS predicate matrix.*

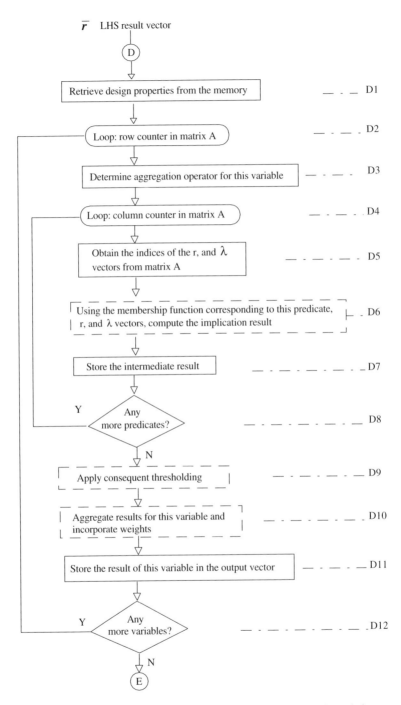

Figure A.6 The implication computation procedure in the basic fuzzy inference algorithm.

Level D5 *The elements of the A matrix contain symbolic data indicating the index number in reference to the LHS result vector. This index number also indicates the corresponding elements in the auxiliary parameter vector. The location of each element in the A matrix (column counter) indicates which predicate (which membership function) will be used in the implication computation along with the LHS result, weight value, and auxiliary parameter. For example, $A(1,3) = 2$ means that the implication computation will involve the second element in the LHS result vector, the second element in auxiliary parameter, and the membership function that corresponds to the $P(1,3)$ element in the RHS predicate matrix.*

Level D6 *Using all the information gathered above, the implication computation takes place at this step for each predicate. This step involves fuzzy computations and is subject to design. The implication computations are explained in more detail in Chapters 3 and 8.*

Level D7 *The result from the previous step is a new fuzzy set and is stored in the memory to be used later in the aggregation process.*

Level D8 *The Loop termination condition is employed at this step.*

Level D9 *This step involves fuzzy computations and is subject to design. A threshold is applied to all individual implication results to avoid residual truth. Note that the thresholding of consequent fuzzy variables is different from that of the antecedent fuzzy variables because they produce different effects.*

Level D10 *This step also involves fuzzy computations and is subject to design. All individual results (fuzzy sets) belonging to this consequent variable are aggregated in this step using the aggregation operator determined previously. Weight factors are also incorporated by obtaining numerical values from the weight vector. Which weight element to use from the weight vector is also determined by the RHS design index matrix. The result is a new fuzzy set to be used next in the output processing step.*

Level D11 *The result from this consequent variable is stored in the memory.*

Level D12 *Loop terminates when all the consequent fuzzy variables are processed.*

Appendix The Basic Fuzzy Inference Algorithm

EXAMPLE A.5

Referring back to the Example A.1, we will analyze the RHS computations by following the algorithmic steps shown in Fig. A.6. The rules were stated as follows.

IF (Temperature IS High OR Flow IS Medium) AND Valve IS Open THEN
Operation IS Fine

IF Valve IS Closed THEN Operation IS Questionable

The hypothetical input case had resulted in a composition result vector:

$$\bar{r} = \begin{bmatrix} 0.75 \\ 0.06 \end{bmatrix}$$

Let's complete the design definitions next. The consequent vector has one element that is the variable *Operation*.

$$\overline{X}_C = [Operation]$$

The consequent predicate matrix is:

$$\overline{P}_{RHS} = [Fine \quad Questionable]$$

The RHS design index matrix is:

$$\overline{A} = [1 \quad 2]$$

We assume that the aggregation vector contains union operation, the auxiliary parameter vector is unity, and the weight vector contains elements all equal to 1.0. All these definitions are retrieved from the memory at step *D1*. Now we will walk through the algorithm.

Level D2: The loop counter is 1, the variable is *Operation*.
Level D3: The ELSE operator is union.
Level D4: The column counter in matrix A is 1.
Level D5: The index in A matrix is 1, the others are not applicable.
Level D6: Taking the first r value (0.75) and the first membership function *Fine*, the implication result is computed by clipping the membership function at the level 0.75. This is the Mamdani implication operator, and the clipping yields the following shape.

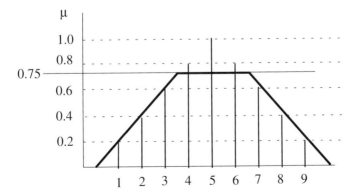

Thus, the original membership function

$$\mu_{Fine} = 0.2/1 \cup 0.4/2 \cup 0.6/3 \cup 0.8/4 \cup 1.0/5 \cup 0.8/6 \cup 0.6/7 \cup 0.4/8 \cup 0.2/9$$

has been converted into the clipped form:

$$\mu_{Fine} = 0.2/1 \cup 0.4/2 \cup 0.6/3 \cup 0.75/4 \cup 0.75/5 \cup 0.75/6 \cup 0.6/7 \cup 0.4/8 \cup 0.2/9$$

Note the fourth, fifth, and sixth singletons above, which are all bounded by 0.75.
Level D7: This new fuzzy set is stored in the memory.
Level D8: Are there any more predicates? Yes. Go back to *Level D4*.
Level D4: The column counter is increased by 1, which becomes 2.
Level D5: The second index in A matrix is 2.
Level D6: Taking the second r value (0.06) and the second membership function *Questionable*, the implication result is computed by clipping the membership function at the level 0.06. The clipping yields the following shape.

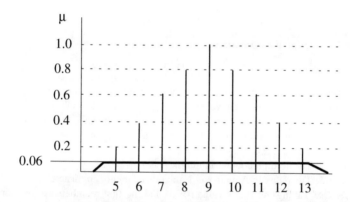

Thus, the original membership function

$$\mu_{Ques} = 0.2/5 \cup 0.4/6 \cup 0.6/7 \cup 0.8/8 \cup 1.0/9 \cup 0.8/10 \cup 0.6/11 \cup 0.4/12 \cup 0.2/13$$

has been converted into the clipped form:

$$\mu_{Ques} = 0.06/5 \cup 0.06/6 \cup 0.06/7 \cup 0.06/8 \cup 0.06/9 \cup 0.06/10 \cup 0.06/11 \cup 0.06/12 \cup 0.06/13$$

Note that all the singletons are 0.06 due to low-level clipping. Also note that this membership function occupies different space on the universe of discourse.
Level D7: The clipped membership function is stored in the memory.
Level D8: There are no more predicates.
Level D9: Consequent thresholding, if applied in this case by assuming 0.2 level alpha-cut, would result in eliminating the last clipped membership function because all

Appendix The Basic Fuzzy Inference Algorithm   483

the singletons (0.06) are smaller than 0.2. However, we will skip this elimination to illustrate aggregation.

Level D10: Aggregation of the clipped membership functions is shown below.

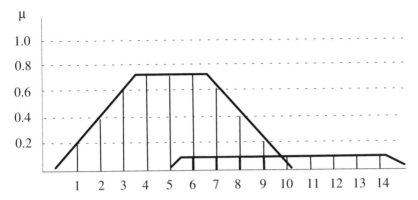

Because the aggregation operator is union, the final fuzzy set is computed as follows:

$$\mu_{Agg} = (0.2/1) \cup \qquad \mu_{Agg} = (0.2/1) \cup$$
$$(0.4/2) \cup \qquad (0.4/2) \cup$$
$$(0.6/3) \cup \qquad (0.6/3) \cup$$
$$(0.75/4) \cup \qquad (0.75/4) \cup$$
$$(0.75/5) \cup \qquad (0.75/5) \cup$$
$$(0.75/6 \vee 0.06/6) \cup \qquad (0.75/6) \cup$$
$$(0.6/7 \vee 0.06/7) \cup \qquad (0.6/7) \cup$$
$$(0.4/8 \vee 0.06/8) \cup \qquad (0.4/8) \cup$$
$$(0.2/9 \vee 0.06/9) \cup \qquad (0.2/9) \cup$$
$$(0.06/10) \cup \qquad (0.06/10) \cup$$
$$(0.06/11) \cup \qquad (0.06/11) \cup$$
$$(0.06/12) \cup \qquad (0.06/12) \cup$$
$$(0.06/13) \cup \qquad (0.06/13) \cup$$
$$(0.06/14) \qquad (0.06/14)$$

Note the effect of the union operation where both clipped fuzzy sets overlap. The fuzzy set on the right is the result obtained from this process. To illustrate weight effects, let's assume that the weight vector is as shown below.

$$\overline{W} = \begin{bmatrix} 0.9 \\ 1.0 \end{bmatrix} \qquad \overline{A} = [1 \; 2]$$

The A matrix tells us which row element of weight vector affects which membership function. In this case, weight 0.9 corresponds to *Fine* and 1.0 corresponds to *Questionable*. Accordingly, the height of the clipped membership function *Fine* will be adjusted to its 90 percent value, that is: $0.75 \times 0.9 = 0.675$. The aggregated result therefore has the following appearance.

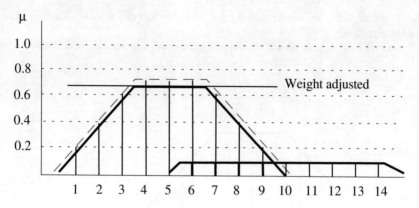

which is computed as follows.

$$\mu_{Agg} = \{(0.2 \times 0.9)/1\} \cup \qquad \mu_{Agg} = (0.18/1) \cup$$
$$\{(0.4 \times 0.9)/2\} \cup \qquad (0.36/2) \cup$$
$$\{(0.6 \times 0.9)/3\} \cup \qquad (0.54/3) \cup$$
$$\{(0.75 \times 0.9)/4\} \cup \qquad (0.675/4) \cup$$
$$\{(0.75 \times 0.9)/5\} \cup \qquad (0.675/5) \cup$$
$$\{(0.75 \times 0.9)/6 \vee 0.06/6\} \cup \qquad (0.675/6) \cup$$
$$\{(0.6 \times 0.9)/7 \vee 0.06/7\} \cup \qquad (0.54/7) \cup$$
$$\{(0.4 \times 0.9)/8 \vee 0.06/8\} \cup \qquad (0.36/8) \cup$$
$$\{(0.2 \times 0.9)/9 \vee 0.06/9\} \cup \qquad (0.18/9) \cup$$
$$(0.06/10) \cup \qquad (0.06/10) \cup$$
$$(0.06/11) \cup \qquad (0.06/11) \cup$$
$$(0.06/12) \cup \qquad (0.06/12) \cup$$
$$(0.06/13) \cup \qquad (0.06/13) \cup$$
$$(0.06/14) \qquad (0.06/14)$$

STEP 5: OUTPUT DATA PROCESSING

The output processing is the final step in the basic fuzzy inference algorithm. It includes the interpretation of the aggregated implication result. The defuzzification method is one of the standard practices to obtain a scalar from the implication result (a fuzzy set). This process is outlined in Fig. A.7.

Level E1 *If weights are chosen to be computed during defuzzification, then the weight vector is retrieved from the memory at this stage. Note that aggregation matrix A is also needed along with the*

Appendix The Basic Fuzzy Inference Algorithm   485

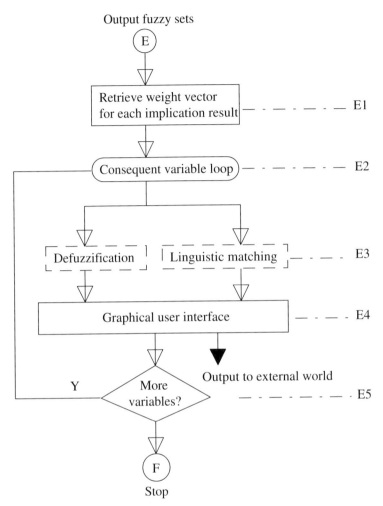

Figure A.7 Output processing in the basic fuzzy inference algorithm.

weight vector to determine which implication result gets which weight value.

Level E2   *Consequent loop counter.*

Level E3   *Defuzzification is implemented at this stage. If a linguistic output is also expected from this inference engine, then an appropriate matching method is employed. This level of computations* is subject to design and corresponding methods are explained in Chapter 8. The two most widely used methods—the center-of-gravity defuzzification and maximum possibility linguistic matching—are illustrated in Example A.6.

Level E4 *Results obtained through previous steps are processed in a GUI to be presented in the external world.*

Level E5 *Loop termination condition.*

EXAMPLE A.6

Let's continue the case examined in the previous example. Since there is only one consequent variable in this example, the output processor receives one output fuzzy set as shown below.

$$\mu_{Agg} = (0.18/1) \cup$$
$$(0.36/2) \cup$$
$$(0.54/3) \cup$$
$$(0.675/4) \cup$$
$$(0.675/5) \cup$$
$$(0.675/6) \cup$$
$$(0.54/7) \cup$$
$$(0.36/8) \cup$$
$$(0.18/9) \cup$$
$$(0.06/10) \cup$$
$$(0.06/11) \cup$$
$$(0.06/12) \cup$$
$$(0.06/13) \cup$$
$$(0.06/14)$$

Level E1: Weights were already incorporated in the formation of the output fuzzy sets shown above.

Level E2: The consequent loop counter is 1.

Level E3: Center-of-gravity defuzzification is performed as follows:

$y_1 = 0.18 \times 1.0 + 0.36 \times 2.0 + 0.54 \times 3.0 + 0.675 \times 4.0 + 0.675 \times 5.0 + 0.675 \times 6.0 +$
$\quad 0.54 \times 7.0 + 0.36 \times 8.0 + 0.18 \times 9.0 + 0.06 \times 10.0 + 0.06 \times 11.0 + 0.06 \times 12.0 +$
$\quad 0.06 \times 13.0 + 0.06 \times 14.0 = 24.525$

$y_2 = 0.2 + 0.4 + 0.6 + 0.675 + 0.675 + 0.675 + 0.6 + 0.4 + 0.2 + 0.06 + 0.06 + 0.06 + 0.06$
$\quad + 0.06 = 4.485$

$$y = \frac{y_1}{y_2} = \frac{24.525}{4.485} = 5.468$$

Appendix The Basic Fuzzy Inference Algorithm 487

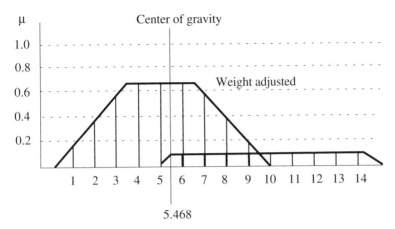

Note that the active range of actions is in between the two centers (5.0–10.0) considering the shape of the two consequent membership functions in this example. The linguistic output is the predicate *Fine* with the maximum possibility 0.675.

Level E4: The graphical user interface presents the outputs obtained in this initiation step to the external world:

INPUTS
Temperature = 27° F
Flow = 10 m/sec
Valve = 30% Open

OUTPUT
Operation = 5.468 (numerical)
Operation = *Fine* (linguistic), Possibility = 0.675

Index

A
Accuracy, 30, 133
Active range of consequents, 267
Adaptive fuzzy systems, 73
Adaptive strategy for climate control, 380
Additive aggregation, 440
Aggregation, 118, 404
  effect of weights, 124
Aggregation design, 437
Aggregation operators, 36, 139, 438
Alpha cuts, 45, 262
Ambiguity, 3, 30
AND, 59, 105
Antecedent, 19, 136
Approximation, 29
Aptronix, 196
ASK, 81
Assembler, 15
Autoassociative problems, 133, 189
Automatic alarm system
  in a chemical facility, 370
Auxiliary variable design
  for comparison, 261
  for conflicts, 258

B
Background requirements, 139

Basic fuzzy inference algorithm, 13, 66, 83, 458
Basic logic operators, 324
  AND, 324
  consistency, 328
  mean, 329
  OR, 326
  product AND, 325
  Yager, 326
Bayesian, 80
Boolean logic, 5
Bucciarelli, L. L., 196

C
C, 15
C++, 15
CADIAG, 81
Canonical form, 19, 115
Cartesian product, 48
Cascade rule block architecture, 355
Category, 154
Center
  of area, 127
  of gravity, 70, 127
  of largest area, 70
  of mass, 70
  of weights, 70
Chain rule syllogism, 111, 356

Chaos, 3, 30
Characteristic function, 23
Classical set theory, 2
Classification, 7, 185
Clipping, 113
Cluster, 154
Clustering, 7, 191
  c-means clustering, 13, 191, 205
Commensurateness, 307
Complex problems, 133, 192
Composing fuzzy rules, 232
Composite moments, 72
Composite rule, 19, 116
Composition, 19
Compositional inference, 52, 343
Compositional rule of inference, 62
Conceptual design, 132
Conditional reasoning, 64
Conditional statements, 4
Conjunctive inference, 61
Consequent, 19
Consequent fuzzy variable, 137
Conservatism, 133
Consistency principle, 28
Contraposition, 64
Control, 7, 180
  feedback, 182
  of feedback systems, 431
  optimal, 183
  of a waste treatment process, 364
Conventions, 148
Convex shapes, 282
Cost, 133
Cox, E., 196
Crisp versus fuzzy sets, 22
Cross section data, 89
CubiCalc, 196

D

Data compression, 7
Data driven design, 204, 206
Data processing of mixed inputs, 89

Dead zones, 262, 264
Decision boundary, 19, 204, 206
Decision making in social sciences, 347
Decomposition, 127
Deductive reasoning, 64
Defuzzification, 70, 127, 139, 404
  design criteria, 447
  geometric interpretation, 445
Defuzzification techniques, 444
  center of area, 444
  center of gravity, 444
  center of mass, 444
  center of maximum possibilities, 444
  composite moments, 444
  maximum possibility, 444
  mean of maximum possibilities, 444
Degree of fulfillment (DOF), 19, 103, 105, 108
Degrees of commensurateness, 19
Dempster Shaffer's body of evidence theory, 248, 342
DENDRAL, 78
Derivative control, 382
Design, 12
  elements, 135
  options, 139
  principle, 134
  process, 139
Designable elements, 17
Designing a consistency operator, 333
Designing a decomposition process, 443
Designing a defuzzification process, 443
Designing a mean operator, 336
Designing the product AND and Yager AND operators, 332
Development shells, 16

Dilution, 29
Distribution data, 98
DOF comparison method, 385
Dominance
  among the components of a rule, 398
  in dynamic systems due to membership function design, 396
  due to membership function design, 395
  among rules, 394
Drainkov, M., 196
Dubois, D., 196

E

Edmunds, G., 349
ELSE, 105, 119, 442
Embedded programming, 15
EMYCIN, 78
Entailment principle, 60
Equal distance partitioning, 210
ES/KERNEL, 81
Estimation, 7, 177
Evaluating antecedent fuzzy variables, 92
Evaluation of a membership function, 96
Event tree, 370
Excluded middle law, 41, 386
Exclusive-Or, 207
EXPERT, 81
Expert systems, 77
Extension principles, 45, 62

F

Far and near edge decomposition, 444
Fault tree, 370
Feature selection, 185
FIDE, 196
Fingerprint classification, 360
Forecasting, 7, 177

Formation
  of control rules from data, 350
  of individual rules, 344
  of rule blocks, 354
FORTRAN, 15
Forward problems, 133, 176
Foundation of fuzzy systems, 4
Frequency histogram, 93
Frequency of occurrence, 241
Full rank design, 390
Fuzzification techniques, 41
Fuzziness, 1
FUZZLE, 196
Fuzzy additive method, 441
Fuzzy controller design, 223
Fuzzy data, 208
Fuzzy financial risk assessment, 343
Fuzzy implication, 64
Fuzzy Kalman filter, 10
Fuzzy knowledge builder, 196
Fuzzy measures, 28
Fuzzy PI controller, 161
Fuzzy proposition, 57
Fuzzy regression models, 451
Fuzzy relations, 46
  max-min transitive, 47
  reflexive, 47
  symmetric, 47
Fuzzy rule, 105, 110
Fuzzy set, 35
  complement, 35
  convex, 34
  level, 45
  normal, 34
  operations, 34
  power, 35
  scalar product, 35
  support, 34
Fuzzy set theory, 2
Fuzzy sets, 37
  algebraic mean, 37

Fuzzy sets (*cont.*)
  algebraic product, 37
  intersection of, 37
  properties of, 40
  union of, 37
Fuzzy statements, 104
Fuzzy system design, 11
Fuzzy system design and elements, 134
Fuzzy systems and algorithms, 66
Fuzzy systems at work, 7
Fuzzy Systems Engineering, Inc., 196
Fuzzy values, 57
Fuzzy variable design, 201
Fuzzy voting, 144
fuzzyTECH, 196

G

Gaussian, 203
Generalized *modus ponens,* 63
Generalized *modus tollens,* 63
Geometrical approach, 117, 405
Gibson, H. M., 403
Granularity, 203
Greater than, 57

H

Hall, L. O., 196
HC11, 15
Hedges, 19, 136
  concentration, 253–54
  dilution, 253–54
  intensification, 253, 255
  negation, 253, 256
  relaxation, 253, 256
  restriction, 253, 257
Heteroassociative problems, 133, 184
Homogeneous rules, 345, 347
Hybrid systems, 9
HyperLogic, 196

I

Identifying antecedents and consequents, 147
Implication, 19
  computation, 112
  operators, 66, 139, 408
    area under the curve analysis, 426
    arithmetic product, 411
    behavioral properties, 425
    Boolean operator, 413
    bounded product operator, 420, 431
    drastic product operator, 416, 431
    first category operators, 409, 426, 431
    Godelian operator, 423, 431
    Gougen operator, 421, 431
    Kleen-Dienes operator, 413, 431
    Larsen operator, 409, 431
    Lukasiewicz operator, 411, 431
    Mamdani operator, 408, 431
    pendulum analogy, 434
    rate of change analysis, 430
    and rule activity perspective, 433
    second category operators, 409, 428, 431
    standard sequence operator, 415, 431
    Zadeh operator, 418, 431
  process, 403
  relation, 407
  with thresholding, 120
Importance weights, 139
Imprecision, 3, 30
Inaccuracy, 3, 30
Inductive reasoning, 64
Inference engine, 19, 78

Index 493

Inference using fuzzy variables, 60
INFORM, 196
Information compression, 189
Information entropy, 451
Input data processing, 87
Input data set, 88
Input fuzzifier, 137
Integrated Systems, Inc., 196
Intensification, 29
Interpreting output fuzzy sets, 452
Inverse problems, 133, 180

J
Jamshidi, M., 196

K
Kandel, A., 81
Knowledge acquisition, 14, 133, 140, 142
    essays, 146, 150
    extracting rules from data, 154–55
    interpreting data, 152
    interpreting formulas, 158
    questioning approach, 144
Kosko, B., 196
Kruse, R. J., 196

L
Left-hand-side computations, 103
Left-right maxima, 70
Less than, 57
LHS, 19
Linguistic design, 204
Linguistic design criteria, 133, 166
    accuracy, 169
    adaptability, 172
    application of, 174
    computability, 172
    conservatism, 166
    cost, 169
    learning from experience, 173
    optimism, 167
    pessimism, 167
    precision, 169
    responsiveness, 168
    robustness, 170
    sensitivity, 170
    sluggishness, 168
    tolerance, 166
    understandability, 171
Linguistic fuzzy variable design, 249
Linguistic hedge effect, 105, 109
Linguistic hedges, 253
Linguistic input libraries, 137
Linguistic variable, 19, 56, 202
LISP, 78
Logic, 5
    boolean, 5
    deontic, 5
    epistemic, 5
    modal, 5
    tense, 5
Logic operator design issues, 330
Logic operators, 56, 105, 138

M
Mamdani operator, 112
Management of subjective probabilities, 246
Mapping, 212
    different cases of mapping, 212
Marginal fuzzy restrictions, 48
Marvin, F. S., 2
Math Works, Inc., 196
MATLAB, 196
Max-* composition, 56
Maximum possibility, 70
Max-min composition, 55
McNeill, F. M., 196
Mean of maximums, 70
Membership function, 4, 19, 23
    design, 234
        frequency of occurence, 241
        minimum entropy, 244

Membership function (*cont.*)
  probabilistic, 239
  height, 136–37, 283
    bottleneck effect, 288
    consequent fuzzy variables, 289
    paralysis, 290
    truth categorization, 284
    truth characterization, 285
  line style, 136–37, 295, 298
    antecedent variables, 310
    consequent membership functions, 304
    consequent variable, 313
    design by fulfillment table, 300
    designing curves, 299
    development of a fulfillment table from data, 301
    overlap, 308
    overlap index, 310
    standards, 303
    triangle versus trapezoidal, 302
  location, 136–37
  overlap, 136, 137
  shape analysis, 276
  shape effects, 391
Membership functions, 220
  combining membership functions, 227
  interactive, 220
Membership value, 19
Modeling, 7, 184
MODiCO, Inc., 196
Modus ponens, 83, 87, 110
Momentum investment method, 367
Monotonic reasoning, 341
Multiple aggregation, 387, 440
Multivalued solution, 454
MYCIN, 78

N
Navigation, 347
Necessity, 28
Nested structures, 346
Neural networks, 10, 178, 376
Neuro-fuzzy systems, 18
Nonlinear filter, 91
Normalization, 138

O
On-line performance monitoring in a large-scale plant, 373
OPS5, 78
Optimism, 133
Optimization, 7
OR, 59, 105
OTHERWISE, 442
Output processing, 110, 126, 128, 138

P
Paradoxical cases, 382
  amount of evidence, 383
  conflicts in disguise, 386
  conversion from multiple rules to OR combination, 387
  empty input space, 389
  mutual feedback, 388
  outliers, 384
Parallel rule block architecture, 355
Paralysis, 269, 387
Partial truth, 1
Partitioning, 204, 209
Pattern recognition, 7, 186
Personal construct theory, 142
Pessimism, 133
PID control, 183, 382
Pi-shaped functions, 281
Plausibility, 28
Point membership functions, 214–15
Possibilistic interpolation, 342
Possibility, 28

Possibility theory, 27
Practical design considerations, 258
Practical fuzzy measures, 29
Prade, H., 196
Precision, 30, 133
Predicate, 19
Predicate calculus, 5
Prediction, 7, 177
Premise, 19
Preponderance of evidence, 444, 451
Probability, 28
Projection, 61
Projection principle, 48
PROLOG, 15, 81
Proportional inference, 341
PROSPECT, 78
PROTIS, 81
Prototypes, 320

R
Randomness, 3, 30
Relational inference, 46, 338
    for diagnostics, 340
Responsiveness, 133
RHS, 19
Right-hand-side computations, 109
Right-hand-side operators, 442
Robustness, 88, 133
Ross, T. J., 196
RT/Fuzzy, 196
Rule block, 343
Rule composition, 138
Rule composition strategies, 138, 357
    adaptive rules, 375
    competitive rules, 357
    cooperative rules, 363
    hierarchical rules, 372
    prioritized rules, 369
    weighted rules, 366
Rule discovery algorithm, 77

Rule formation per inference type, 338
Rule importance weights, 125

S
SEEK, 142
Selecting implication operators, 405
Sensitivity, 133
Sequentially correlated data, 89
Sequentially correlated events, 430
Set-theoretic and average operators, 35
Singleton, 23
Sluggishness, 133
SPERIL, 81
SPHINX, 81
S-shaped functions, 281
Statistical favoritism, 393
Stock picking, 367
SYNTEX, 81
System definition, 176
Systems ontology, 174

T
Tabular representation of fuzzy inference engine, 390
Tank liquid level controller, 350
TEIRESIAS, 142
Terano, T., 196
Terminology and conventions, 19
THEN, 59, 105, 442
Threshold, 136, 138
    design, 262
    filtering, 94, 101
    operation using alpha cuts, 103
Thresholding and aggregation, 123
TIL shell, 196
Togai InfraLogic, Inc., 196
Tolerance, 133
Truth categorization, 31, 239
Truth characterization, 239
Truth representation, 31

Truth value, 19
Tsoukalas, L. H., 196

U

Unconditional statements, 4
Unconstrained predicates, 251
Undecidedness, 3, 30
Universe of discourse, 19, 136
Useful tools supporting design, 195

V

Vagueness, 3, 30
VLSI chips, 15

Y

Yager, R. R., 196
Yen, J. R., 196

Z

Zadeh, L. A., 2, 196
Zimmerman, H. J., 196